BIOLOGY

PATTERNS IN THE ENVIRONMENT

BIOLOGY

PATTERNS IN THE ENVIRONMENT

Evelyn Morholt
Paul F. Brandwein
Lawrence S. Ward

HARCOURT BRACE JOVANOVICH, INC.
New York Chicago San Francisco Atlanta Dallas

About the Authors

EVELYN MORHOLT, Chairman, Department of Biology and General Science, Fort Hamilton High School, Brooklyn, New York.

PAUL F. BRANDWEIN, Adjunct Professor, Conservation and Education, University of Pittsburgh; President, Center for the Study of Instruction; formerly Director, The Pinchot Institute for Conservation Studies, Milford, Pennsylvania.

LAWRENCE S. WARD, Assistant Director of Mathematics and Science, Nairobi College, East Palo Alto, California; formerly teacher, East Orange High School, East Orange, New Jersey, and Peterson High School, Santa Clara, California.

About the Cover

The photographs were taken by Tom McCarthy. They show a beach on St. Thomas, U.S. Virgin Islands.

What is the red area in the picture on the front cover? What is your reasoning? You can check your answer by looking at the back cover.

The picture on the front cover was taken with infrared film. This special film responds to heat energy given off by things in the environment. The areas where the most energy is given off are red.

What pattern in the environment would explain the red color of the trees? What other patterns can you see in this environment?

Acknowledgments

We would like to thank members of the biology department and students at Fort Hamilton High School, Brooklyn, New York, for their cooperation and assistance.

PRINTED IN THE UNITED STATES OF AMERICA

ISBN 0-15-362440-X

Contents

The Capture of Energy 107

Unit 4

The Earth As Man's Home 143

Unit 5

Man Remaking the Environment: Control of Disease 199

Unit 6

1

Likenesses

INTRODUCTION

Telling Ahead of Time

Can you be sure of anything? Our world is always changing. The moon was out of reach when you were born. But now, men walk on the moon! ■

New changes follow old ones. Radios used to be very large. But today radios can be very small. Small enough to fit in your pocket. This one is the size of a watch. ●

Television sets also used to be very large. And these early TV sets had only black-and-white pictures. Now **scientists** (SY·en·tists) have found a way to make tiny TV sets. Look at the picture. ▲ This TV is the size of a book. And now there is color TV.

One day you will be able to talk to the TV set. You will say "Change to Channel 2," and the channel will change. Or say "Good night," and the set will shut off.

Here is one more change. There are now phones that can take messages. If you aren't there to answer your telephone, it takes a message for you. And there are also phones that show pictures of the people who are speaking. You can see who you are talking to.

New ideas follow old ideas. New changes follow old changes.

Can you be sure of anything in your lifetime? Yes.

●

▲

■

●

You can be sure of *change*. Is there a way you may see change coming? Perhaps there is.

You do know some things ahead of time. For instance, you know the day changes. The sun rises in the morning and it is light. ■ The sun sets and it is night. ● Daylight is followed by darkness—at least in the United States.

As sure as you are alive, daylight follows darkness. After darkness, daylight comes again. This you can tell ahead of time. You can **predict** (pri·DIKT) that the sun will rise. You can predict that it will set.

We call this a **pattern** (PAT·urn). A pattern is something that happens again and again. The sun rises and sets, day in and day out. This is a pattern.

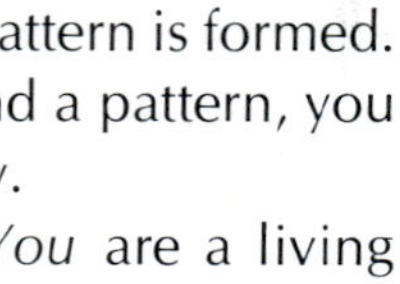

Pattern

The change of seasons is a pattern. Summer follows spring. Winter follows fall. Change follows change.

When changes repeat themselves, a pattern is formed. You can look for patterns. When you find a pattern, you can predict some things that will follow.

Are there patterns in living things? *You* are a living thing. Of course, you are. You are alive. Do you see any patterns in yourself?

Are there patterns in you? Let us see.

1. PATTERNS

You will not go to school all your life. Nobody does. You learn to be on your own as you grow up. One thing you learn is how to learn by yourselves. This is the main thing to learn in school.

Much of this book is written with that goal in mind. It is written so that you can learn on your own.

This book is about **biology** (by·OL·uh·jee). Biology is the study of living things. You will study biology on your own. You will take part in some **independent** (in·dih·PEN·dent) **study.** This means you will work on your own. You already do many things on your own. ▲ Now you will begin to study on your own.

After all, to learn to be *on your own* is to learn to be *independent*. Try the next few pages *on your own.*

▲

▲

AN INDEPENDENT STUDY

A Pattern in Your Body

Has a doctor ever held your wrist between his fingers? He was probably taking your pulse. When your heart beats, it forces blood through your body. As the blood passes certain points of the body, you can feel a beat. You can feel a beat at your wrist. This is your pulse. The doctor counts your pulse beat to find out how many times your heart beats in one minute.

What is your pulse beat? Do all people have the same pulse beat? Is there a pattern in pulse beats? Let us see.

First, take your own pulse. It is important that you all do it the same way. You can see how it is done in the picture. ■ Hold your wrist the way the picture shows you. Do you feel a steady beat? You will soon start counting the beats. You will need a clock or watch with a second hand.

In the next few pages you can learn on your own. The pages are like these two pages. You will find questions on the left-hand page. Place a piece of paper over the page on the right. This will be your answer sheet. When you have the answer, write it on your answer sheet.

How can you find out if your answer is correct? You can check your answer with the answer in the box on the right-hand page. Check each answer before you go on to the next question.

Let's go through the lesson. You will see how this kind of independent study works. (Please write your answer on your answer sheet—*not* in the book.)

1

First, hold the place on your wrist where you feel the pulse beat. Hold your wrist and count the beats for one minute. Watch the second hand on the clock as you count.
My pulse beat for 60 seconds (or 1 minute) is _____.

(Write your answer on your answer sheet. Lift your answer sheet and check box 1 on the right-hand page.)

1

This is the answer to number 1.

You now have your pulse beat for 60 seconds (1 minute). But you will need to take it a few more times.

(Now turn the page to question 2.)

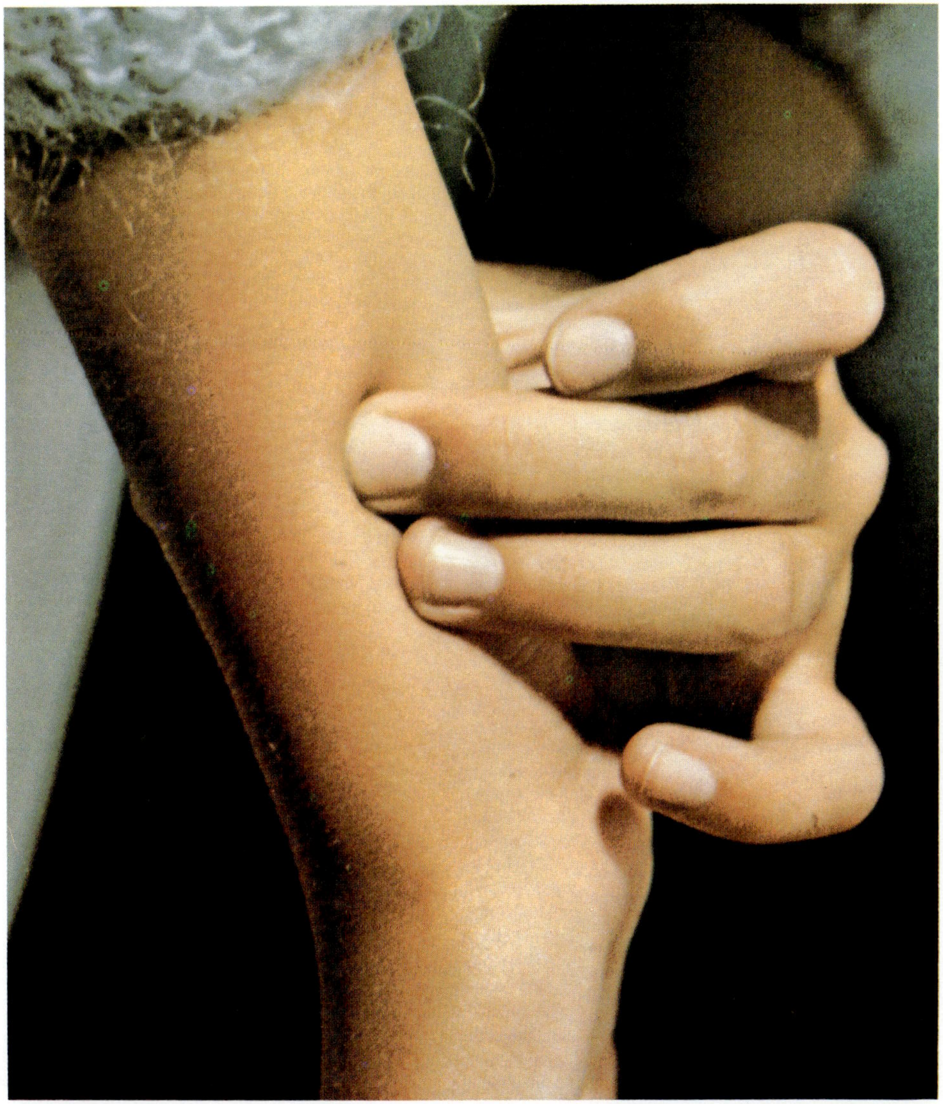

2

Now take your pulse beat four *different times*. You may find that the count changes each time.

1st time ____________ 3rd time ____________

2nd time ____________ 4th time ____________

(Please write the number of your pulse beats on your answer sheet. Look across the page to box 2.)

3

Why should you take your pulse four times? Isn't once enough? No, once is never enough. By doing a count more than once, you make fewer mistakes. You have less chance of making errors (ERR·urz). Scientists do things more than once. This is one way scientists try to make fewer mistakes.

The reason you take your pulse more than once is to make fewer e-----.

(Write your answer on your answer sheet. Check across the page to box 3 to see if you are right.)

4

You take more than one count to make more sure of being right. You need to check yourself.

Here is an example. One girl took her weight. It was 100 pounds. A boy also took his weight. He took it *three* times. The first time it was 120 pounds. Then it was 121 pounds, and the third time it was 121 pounds.

Who had a better chance of being right, the *boy* or the *girl*?

(Write your answer on your answer sheet. Now check across the page to box 4 to see if you are right.)

2

Now you have taken four pulse counts.

You were asked to take four counts for a special reason.

(Go to question 3 to see why.)

3

errors

You want to make fewer mistakes.

(Now go to question 4.)

4

the boy

The boy had a better chance of being right. He weighs around 120 pounds. Not 80 pounds. Not 150 pounds.

(Turn the page to question 5.)

5

Why did the boy have a better chance of being right? He checked himself. He took his weight *three* times. If he got about the same weight more than once, he probably wasn't wrong.

Remember, when you take *more* than *one count* you are checking yourself. There is less chance for *error*. Ten counts are better than three counts. The more counts you take, the less chance there is for ____________.

(Write your answer down on your answer sheet. Check across the page to box 5. Are you right?)

6

In a school in Nebraska, there were 284 boys and girls in the 10th grade. They all took their pulse beats. Their pulse beats were *different*. The lowest pulse beat found was 60 beats a minute. The highest was 80 a minute.

You can see that everyone does not have the same pulse beat. All boys and girls have pulse beats. But the pulse beat for each person can be d--------.

(Write your answer on your answer sheet. Check across the page to box 6 to see if you are right.)

7

Many people have pulse beats that are alike. People *are* alike in some way. Boys are alike in many ways, but they are also d--------.

(Write your answer down on your answer sheet. Check across the page to box 7 to see if you are right.)

5

mistakes
or
errors

You do this to be more sure.

(Go on to question 6.)

6

different

We are not all exactly alike.

(Go on to question 7.)

7

different

You know this.

(Turn the page to question 8.)

8

The 10th grade boys and girls in Nebraska wanted to find out if there was a *pattern* in pulse beats. They wanted to know if people have a *pattern* in their pulse beats.

Each boy and girl took his or her pulse beat 5 times. They did this to make sure they were counting right. They found there was a pattern in their ____________ beats.

(Write your answer down on your answer sheet. Check across the page to box 8 to see if you are right.)

9

We do not know the people in your class. But we can make a good guess about you. This is our guess. Most of you have a pulse beat from 60 to 80 beats a minute.

How can we guess this? Because we know something about the pat---- of pulse beats in people.

(Write your answer down on your answer sheet. Now check it by looking across the page to box 9.)

10

We know something about the pulse beats of people. Scientists have studied the pulse beats of many, many people. They have found that for *most people* the pulse beat is between 60 and 80 beats a minute.

What is the pattern of pulse beats per minute for *most* people? The pattern of pulse beats is between 60 and ____ beats a minute.

(Write your answer down. Check across the page to box 10 to see if you are right.)

8

pulse

(Go on to question 9.)

9

pattern

A pattern, you know, repeats itself.

(Go on to question 10.)

10

80

This is the pattern for most people.

(Turn the page to question 11.)

11

What is **normal** (NOR·mul) for most people? The word *normal* means something that is expected or usual. Normal for people is something expected or usual in people.

Which is *normal* in people—a pulse beat of *60 a minute* or of *80 a minute?* or *both?*

(Write your answer down on your answer sheet. Look at box 11 to see if you are right.)

12

Why? What is *normal* for one person may not be normal for another. What is normal for you might not be normal for someone else. It is normal for *one* person to have a pulse beat of 60 a minute. Another person may have 70. Another might have a pulse beat of 80. Each is normal for that person.

Is your pulse beat normal for you? What do you think? *Yes* or *no?*

(Write your answer on your answer sheet. Look across the page to box 12.)

13

Scientists and doctors have counted the pulse beats of thousands of boys and girls of your age.

They have found the pulse beat of most boys and girls of your age to be about 70. Some pulse counts are *normally* low at 60. Some are normally high at 80. Only a doctor can tell if a person's pulse beat is n-----.

(Write your answer down on your answer sheet. Check across the page to box 13 to see if your answer is right.)

11

Both are probably normal.

(Go on to question 12 to see why.)

12

Yes, probably.

However, only a doctor can tell what is normal.

(Go on to question 13.)

13

normal

(Turn the page to question 14.)

14

Can your pulse beat change? Or does it stay the same? See for yourself.

First, take your pulse beat again. Write the count on your answer sheet.

Now, stand in one place and *hop* on one foot. Hop about 20 times. Take your pulse beat once again. Write the count on your answer sheet. Then take your pulse beat again. Write the count for this time.

Did your pulse beat go *up* or *down* after you hopped?

(Write your answer on your answer sheet. Check across the page to box 14 to see if you are right.)

15

Why did your pulse beat go up? When you *move faster, your heart beats faster.* The faster heart beat makes the pulse faster. Your pulse may be 60 beats a minute *before* you exercise. It may go to 90 beats a minute *after* you exercise.

Is 90 beats *normal* after you exercise? *Yes* or *no?*

(Write your answer on your answer sheet. Check across the page to box 15 to see if you are right.)

16

Why is 90 beats a minute normal after hopping? Because it is normal for the pulse beat to go *up* after ex------.

(Write your answer on your answer sheet. Check across the page to box 16 to see if you are right.)

14

up

Your pulse beat went up after you hopped.

(Go on to question 15.)

15

yes

90 beats is normal after exercise.

(Go on to question 16.)

16

exercise

When you exercise, your pulse beat goes up.

(Turn the page to question 17.)

17

What happens to your pulse beat when you rest or sleep?

Would your pulse beat be *higher* or *lower* than when you exercise?

(Write your answer on your answer sheet. Check across the page to box 17 to see if you are right.)

18

Is it *normal* for the pulse beat to be lower when you rest or sleep?

Yes or *no?*

(Write your answer on the answer sheet. Check across the page to box 18 to see if you are right.)

19

The heart beat goes *up* during exercise. And it goes *down* during rest.

This happens in most people. This is a ________ in most people.

(Write your answer on your answer sheet. Check across the page to box 19 to see if you are right.)

17

lower

Your pulse beat goes down when you sleep.

(Go on to question 18.)

18

yes

It is normal for your pulse to be lower when you sleep.

(Go on to question 19.)

19

pattern

The pattern for the heart beat is for it to go up during exercise. And down during rest. This is why the pulse changes.

2. LIKENESSES—ANIMALS

You and your classmates have been studying some patterns. We can use another word for these *patterns.* We call these patterns **likenesses** (LIKE·niss·ez).

All people have a pulse beat. A pulse beat is one of the likenesses of people. People also have noses, eyes, and legs. Noses, eyes, and legs are likenesses of people.

True, people are different in some ways. But look around you. You and your teacher have likenesses. All people have likenesses. They have likenesses besides noses, eyes, and legs. Name at least two more likenesses in people.

Now we will begin our study of patterns, or likenesses, in *other living things.*

Likenesses in Other Living Things

Let's look at some other living things. Are there likenesses in the animals in these pictures? Yes, of course. All the animals have heads. They all have four legs and tails as well. You can find the nose, eyes, and ears in each animal. All these animals have this pattern. Look again. Do you also notice that all these animals have fur?

We can say these animals have many likenesses. All the animals have parts of the body that are alike in some ways. A part of the body is called a structure (STRUCK·chur). Eyes, ears, and legs are structures. All these animals have structures that are alike in some ways.

Likenesses Inside the Body

You can see structures such as eyes, ears, and legs. And you can see likenesses in these structures. Can you, however, tell if there are likenesses *inside* these animals?

You can't see the inside body parts, or structures. For instance, you can't see the hearts of these animals. But you can *predict* some things about animals. Have you felt the heart beat of a dog? or a cat? or a pigeon?

They have heart beats. And they have pulse beats.

Have you seen a lion on TV? What would happen to its *pulse beat* if it began to *run?*

Suppose all the animals in a forest began to run as fast as they could. What would happen to their pulse beats? to their heart beats?

Suppose the animals came to rest. Would the rate of their heart or pulse beats change? Would the rate go up or down?

Remember the study of your pulse beat? You felt it change when you hopped. What can you predict about the heart beat of these animals?

Let us reason this way. You know your pulse rate changed. Your pulse rate changed when *you* moved fast. The pulse rate of these animals should also change when *they* move fast.

After all, we are like these animals in some of our body parts. We know these animals have hearts. Their hearts *probably* act in much the same way as ours.

What is the use of knowing how living things are alike? If you know how animals are alike, you may *predict* how they will behave. Is this true? Let us see.

Likenesses in Animals

Take the lion in the picture.● You have watched lions on TV. You may even have been in Africa, where lions live. Or maybe you have seen a lion from Africa in the zoo. You may know something about lions.

You know how they *move*. A zebra can smell a lion. Then the zebra moves fast—away from the lion. Lions move. So do zebras, deer, and horses. So do eagles, frogs, crabs, and worms. Animals can move about. They have this likeness. They have a pattern of action—they move. **Movement** is a likeness in animals.

Why are zebras afraid of lions? Lions attack zebras for food. Lions need food. So do zebras, deer, and foxes. So do tigers, toads, fish, and robins. It seems as if all animals need to get food.▲ Now we can say that **food-getting** is another likeness in animals.

●

▲

Lions, like other animals, are not born full-grown. They are cubs before they are lions. ■ Lions are sometimes called "big cats." The mother lion, the female, looks like a big cat, doesn't she? Her baby looks like a kitten. But baby lions are called cubs.

Lions have cubs, and dogs have pups. Cows have calves. Deer have babies that are called fawns.

The baby cubs grow up into adult lions. Calves grow up into bulls or cows. A fawn grows up to be a deer. ■ And you? You grow up to be a man or a woman. Here you can see another likeness among animals—**growth.** Animals grow after they are born.

To have their young, male and female animals mate. This is the way they **reproduce** (ree·pruh·DOOSE), or make more of their own kind.

Another pattern, or likeness, among animals is **reproduction** (ree·pruh·DUCK·shun). Animals reproduce, or make more of their own kind. ●

What main likenesses have you found among animals? You have found *movement, food-getting, growth,* and *reproduction*. True?

Are there any other likenesses? Yes. Study *yourself* again to find another likeness. Try doing each thing we ask.

Get up from your chair, please.

Now sit down, please.

Turn your head to the left.

Now lift your right arm.

Did you *do* something each time? Did you **respond** (ri·SPOND) to what we asked? Respond means to do something or to answer to something or someone.

● After class, call your friend's name. Will he not answer to his name? He *responds* to your call by answering.

Do animals respond? A zebra will *respond* to the smell of a lion by running away. The cubs of a lioness will *respond* by running toward her—especially when they are hungry. Have you ever seen how some kinds of insects respond to light? how they fly to light? Look at the picture. ▲

▲ Animals seem to *respond* to a *change* around them.

◆

★

Do you know any animals that do not?

You can add **response** (ri·SPONSE) to the list of likenesses in animals.

The animals you have been studying are *alike* in these patterns:

They move *(movement).*◆

They get food to eat *(food-getting).*★

They grow *(growth).*

They reproduce *(reproduction).*

They respond *(response).*

All animals act in these ways. These are likenesses, or patterns, in living things. We call these activities (ak·TIV·ih·teez) of animals **life activities.** Activities, here, means how animals act and live.

Can you think of any animal that doesn't grow or move? Or one that doesn't get food and reproduce? All animals have these likenesses.

Now, if this is so, you can tell some likenesses in animals you have never seen. Try to predict once more.

■

A Try at Prediction

1. Suppose men finally land on Mars. Suppose in their first talk to earth they say there are strange animals. They will tell more about the animals next week.

What do you think they may say about the life activities of the animals?

What *likenesses* can you predict the animals on Mars might show?

2. The animal in the picture is the King Lizard. ■ It belongs to a group of animals called *dinosaurs* (DIE·nah·sores). As you can see, some dinosaurs were giant animals. There were different kinds of dinosaurs. They all lived long ago.

Dinosaurs

Now they have died out. However, you know about the life activities of other animals. What could you say about the life activities of the dinosaurs?

You can predict many things. Now try to give some reasons, or **explanations** (ek·splah·NAY·shunz).

A Try at Explanation

Some animals reproduce much the same way. For example, dogs, cats, tigers, and horses reproduce much the same way. They reproduce the way human beings reproduce. Their young are born alive. The fawn in the picture is only a few minutes old. ●

Eagles, robins, ducks, and owls reproduce the same way. Their young are hatched from eggs. ▲

Suppose, just suppose, the animals on Mars are bird-like. Would they reproduce the way the *tiger* does or the way the *eagle* does? Why do you say so?

●

▲

3. LIKENESSES — PLANTS

Plants, too, are living things. Does this mean there are some likenesses in plants? Are there patterns among plants?

Grow a few plants and see for yourself.

Some Likenesses in Plants

Soak five radish (RAD·ish) seeds overnight in water. Now place the radish seeds between a wet paper towel and the side of a glass. The picture shows you how to do it. ♦ Do you see that each radish seed is between the paper towel and the glass? You should also put some water in the bottom of the glass. Why does the paper towel stay wet? What happens to the water at the bottom of the glass?

Keep the seeds in the glass for two or three days. Be sure that you always have water in the bottom of the glass.

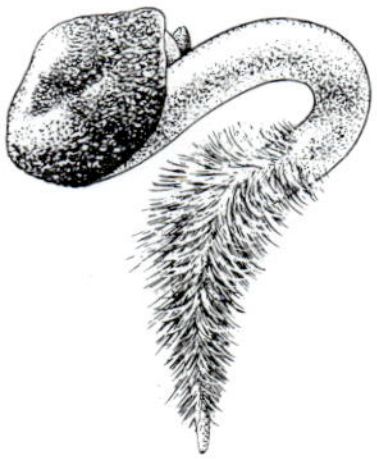

Seedling

Are your radish seeds growing the way those in the picture are growing? ★ This glass belongs to a boy in a school in Ohio. He saw leaves and roots growing from the seeds. From each seed came a young radish plant. A new plant is called a **seedling** (SEED·ling).

A seedling is a young plant. Seeds grow into seedlings. You can do the same thing with bean seeds. ◊ or corn grains. ⊙ You may find what you see in the pictures. Do the seeds or grains grow into seedlings?

Bean seeds and corn grains are like radish seeds in one way. They grow. In fact, as you look at plants, you find a likeness among them. They *grow*.

Seedlings have leaves and roots. They all have leaves and roots. Are the leaves all the same? Do you see differences in the leaves of the radish, the bean, and the corn seedling? In what ways are the leaves the same? How are they different?

♦

★

◊

⊙

Where do we get seeds? Where do you think seeds come from? The next time you eat a tomato or an apple, look for seeds.

Where are the seeds in the picture? ■ Of course. The seeds are inside the *fruits* of the different plants.

A fruit has one or more seeds. Plant the seeds. They grow. They grow into young seedlings.

The seedling grows. Then the plant has flowers. The fruit comes from the flower. The seeds are in the fruit.

These are likenesses in flowering plants.

Peas in a Pod

This *pod* is the fruit of a pea plant. ● The pod is open so you can see inside it. What do you find inside? The peas in the pod are *seeds*. The pod grew from the flower of a pea plant.

If you can get the flower of a pea plant, look inside. In the middle of the flower you will see a part that looks like the pod. ▲ Take off the petals. Now you can see this pod-like part even better. ♦

It is very small now. It will grow into a pod. The pod is the fruit. Seeds will grow in the pod. The seeds are inside the fruit. Most flowering plants reproduce about this same way.

Now, what life activities can you say that plants have?

Plants grow.

They reproduce.

Plants are alike in these life activities. Animals are, too. What other *likenesses,* or patterns, can you see between plants and animals?

Do plants *respond?* Do plants *move?* Do plants as well as animals *get food?* What do you think?

■

●

▲

♦

★

◊

How a Plant Moves

Put a green plant in the window of your classroom. Be sure its leaves are turned the way those in the picture are turned. ★ Be sure to move the plant so its leaves are turned *away* from the window.

Look at your plant each day. We can predict that in a few days the leaves will have turned back to the light. ◊ Why do you think we can predict this? Most *green* plants act this way. Their leaves turn (or move) to the light. We can say that plants *move*.

Animals can move from place to place. Most plants do not move from place to place. But they do move in their own way. Just as the leaves turn to the light. We can say leaves *respond* to light.

What do you see in this picture? ⊙ Which way do the roots of the radish seedlings grow? Toward the wet paper towel? Or toward the dry one? They grow toward the wet paper towel. Can we say that the roots *respond?* The roots grow toward water.

Plants do *move*. They also *respond*. But they do these life activities in different ways from animals.

Do plants need food? You would probably say that they do. They need *food* to *grow* and *reproduce*. How could they *move* without food? How could they respond?

You need food to grow, and so do plants. See for yourself by looking at a special kind of plant.

⊙

Growing Mold Plants

■

●

▲

Try this. Add a few drops of water to a piece of *bread*. Put it in a plastic (PLAS·tick) bag. ■

Someone else in your class can use a piece of boiled *potato*. Put it in a plastic bag. ●

Still another classmate might put a piece of *cheese* in another plastic bag. ▲

Look at the bags after a few days. What has happened?

You can see some gray, white, black, or green spots on the food in each bag. These spots are made mostly of tiny *plants*. The plants are called *mold* plants.

These pictures show what three girls in Kansas found when they grew mold plants. ◆

Mold plants use bread, potato, and other foods to grow on. They use these foods as their own food, just as we do. Molds are plants, but they are not *green* plants.

Mold plants need food just as we do. They use the same foods we do.

Green plants get their food in another way. *Green plants make their own food.* We shall study the way green plants make food at another time.

How Plants and Animals Are Alike

It seems that plants are like animals in these ways:

Animals *move*. So do plants.

Animals *respond*. So do plants.

Animals *need food*. Plants need food.

Animals *grow*. Plants do, too.

Animals *reproduce*. So do plants.

Plants and animals are alike in these life activities. Of course, plants do these things in different ways from animals. But they have the same patterns—the same likenesses.

This should not surprise you. Plants and animals are living things.

◆

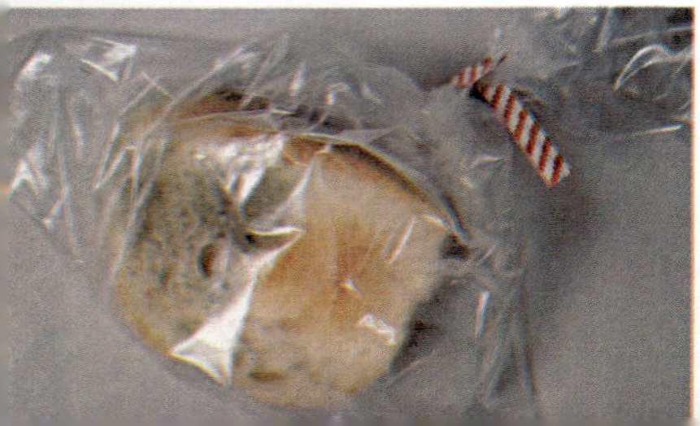

◆

◆

★

◆

A Try at Prediction

You may never have seen a plant that is called rust. It is very much like a mold plant. What could you predict about the life activities of a rust plant?

A Try at Explanation

1. A car *moves*. It *responds* to a push on the gas pedal. Is a car alive? Why not?

2. What life activities are shown in the apple trees? ★ in the mold plants? ◆ in the corn plants? ⊙

Does an animal like a pigeon carry on these same life activities? Do all animals and plants carry on the same life activities?

4. PUTTING LIKENESSES TOGETHER

Have you ever played baseball? or basketball? volleyball? softball? stickball? soccer? Most of you probably have played at least one of these games. Does it seem strange to bring up games in a book about living things? Let's go on a little further.

An orange is round. This you know. But have you ever seen an Osage (OH·sage) orange tree? Very few people have seen this tree and its fruit.

Suppose we tell you that the fruit of the Osage orange tree is shaped like a baseball. It is green in color. Also, the fruit of the Osage orange tree is a little larger than a baseball.

Would this give you an idea of how an Osage orange looks? Probably. *Even though you may never have seen one*. Turn the page. See how close your idea of an Osage orange comes to the picture.

⊙

⊙

⊙

■

●

Here is an Osage orange. ■ Did you think it might look like this?

You might explain your thinking this way:

"I know what a baseball is. ● I could guess what an Osage orange would look like. I could guess it would look like a large, green baseball."

But what you did is something that only man can do well. There are scientists who study man and the way he thinks. They study how and why we act the way we do. They study how people get ideas. They would say that you had formed a **concept** (KON·sept). They would say you had formed a concept of the *roundness* of a baseball. And that you had formed a concept of what *green* is. Concepts are formed in the brain.

How Are Concepts Formed?

How do you think you might have formed the concept of *roundness?* and the concept of the color *green?*

Concepts begin to be formed when you have an *experience* (ik·SPEAR·ee·uhns). You do something or hear or see something—you have experiences.

You felt the round shape of a baseball in your hand. Or the round shape of a basketball. You felt the likenesses in the two balls. You remembered these experiences. And you formed the concept of *roundness*.

In the same way you learned the color *green*. You saw and learned that leaves of plants are green. You learned through experience about the color *green*.

How did you learn what a dog is? You learned *dog* the same way you learned *green*. Through experience you formed the concept *dog*. Now you can guess what a *brown dog* or a *black dog* might look like.

You could even tell about a dog the size of a horse. ▲ This dog would be a large one! But it would have fur and it would bark. It would if there were such a dog!

Concept: Putting Together Likenesses

You form a concept when you put together likenesses. Once you have a concept—any concept—you can use it

in many ways, as you will see. Once a concept is formed, it saves you a lot of time.

Do you know why you can imagine a dog the size of a *horse*? You have a concept of a *horse*. That is, you know what a horse is. How did you, on your own, form a concept of *horse*?

What experiences with horses have you had? Have you seen many horses? Have you seen pictures of horses?

We used the word *pattern* first. Now we can call patterns something else. We can call them *concepts*.

Remember taking your pulse beat? You formed a *concept* of the *pulse beat*. Once you had formed the concept, you were able to think with it. For example, you could predict what the pulse rate of another person might be. You could then say that most people have a pulse beat between 60 and 80 beats a minute. You would *not* say 40 beats a minute, or 100 beats a minute.

Your Own Ideas of Living Things

You are beginning to form your own concept of *life*. You are forming this concept out of your experiences with living things. Of course, you began to form this concept a long time ago. You have had experiences with many living things from your birth. From the many likenesses in living things, you are forming a concept—a concept of life.

Would you say living things are alike in many ways? Both *plants* and *animals* are alike in what they can do.

These are their likenesses:

(They can move.)	*movement*
(They can respond.)	*response*
(They need food.)	*food-getting* ■
(They can grow.)	*growth* ●
(They can reproduce.)	*reproduction* ■●

A concept brings together likenesses.

A concept brings together the likenesses in your mind. You have, in a way, a "picture" of the likenesses in your mind.

The concept *living things* gives you such a "picture." The "picture" brings together the likenesses of all living things. What are they?

The likenesses are *movement, food-getting, growth*—and two more. Can you name the other two patterns?

■

●

▲

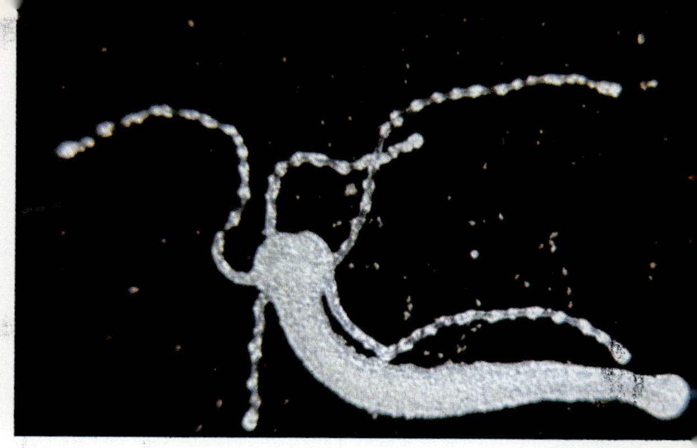

◆

Concepts Help You Think

When you study biology, you study living things. Biology deals with many concepts that help you understand living things and life. The science of biology is what this book is about. Biology is the science of life.

When you have finished this work you have begun, you will understand many concepts. These concepts will help you understand the living things around you. You will be able to predict about many things you never have seen.

You have already done what seems to be a simple thing. You can predict the pulse beat of most people. You are able to predict when the heart beat will be slower or faster. You are able to predict the patterns that make living things alive.

What you have learned is not a simple thing. *You have begun to learn how to form a concept.*

An Exercise in Forming a Concept

1. Look at the picture of a brown **hydra** (HY·dra). ▲ The hydra is a small animal that lives in fresh water. It lives in lakes and ponds. The pin next to it gives you an idea of how small a hydra is.

Now look at a picture of a *white* hydra. ◆ It is much the same as the brown hydra. About what size is it?

Don't turn the page yet. On the other side of this page is a picture of a green hydra. *Draw a green* hydra. After you have drawn it, see how close your drawing comes to the picture.

Now turn the page.

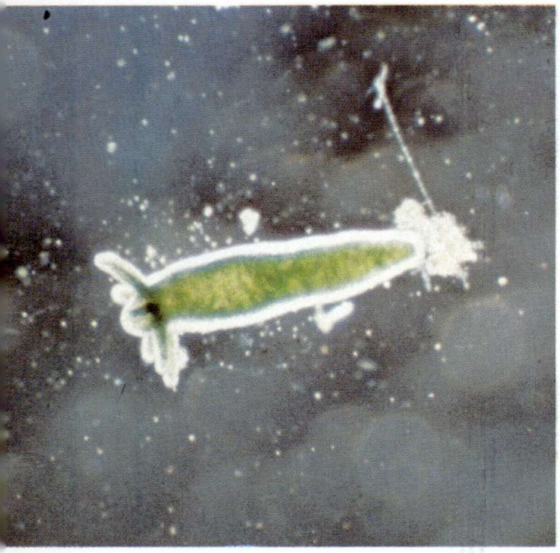

Were you right? You formed the concept of a hydra. It makes no difference if the hydra is green or white or brown. You can draw a hydra. You can now use this concept you have. You can use it to predict how other hydras look.

You have also formed a concept of a living thing. You can use this concept now to predict the way hydras act.

A concept, you can begin to see, is formed in the brain. A concept brings together the likenesses in the things you study.

Try another.

2. The Todas (TOE·duz) are a group of people who live in India. You probably will never meet a Toda. They keep to themselves. Yet you already know about them because you have a *concept of people—of human beings*. Describe a Toda boy or girl.

3. This grasshopper is a little over an inch long. It weighs less than an ounce. The alligator is about 8 feet long. And it weighs more than 200 pounds. Grasshoppers and alligators are animals. From your concept of *animal,* tell how the little grasshopper and the big alligator are alike.

5. ANOTHER CONCEPT

A Look at Your Breathing

You know you breathe. You take air into your lungs. You let air out of your lungs.

Try to hold your breath. You cannot hold it very long. Most people cannot hold it very long. You breathe without thinking. It's a good thing you do not have to stop to think about breathing.

When you take air into your lungs, you **inhale** (in·HALE). ♦ When you breathe out air, you **exhale** (eks·HALE). ♦

The air you exhale looks like the air around you. But it is not the same.

You can see this for yourself. Do another independent study.

♦

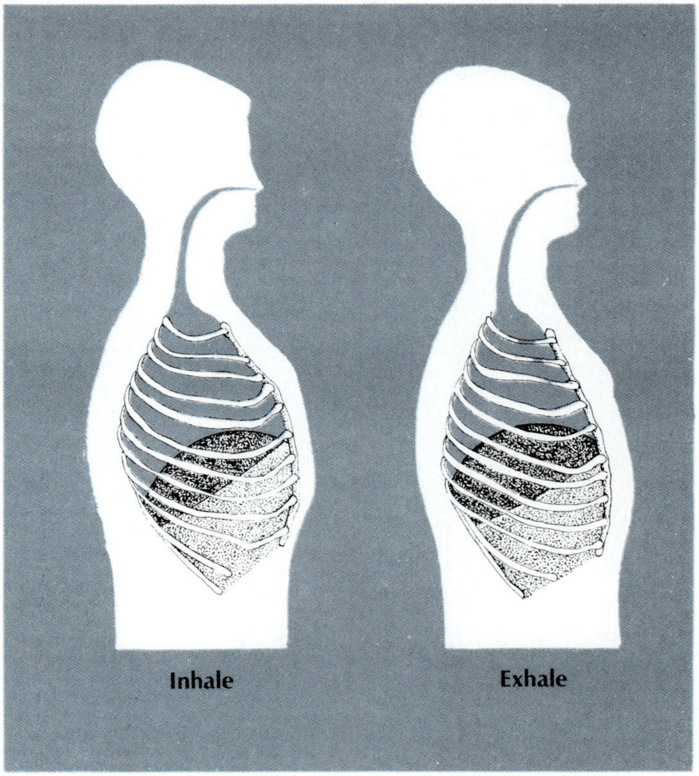

AN INDEPENDENT STUDY

Breathing

This study is for you. Put a piece of paper over the right-hand page. This will be your answer sheet. Then read the questions about breathing. Put your answer on your answer sheet. Next, check your answer in the boxes on the right-hand page. Then go on to the next question. Be sure to check each answer before you go on to the next question.

1

First, take two jars. Pour *water* into the two jars so they are half full. Put a straw in jar *B*. Breathe out through the straw into jar *B*. Breathe out slowly. Breathe out, or *exhale,* 15 times into jar *B*.

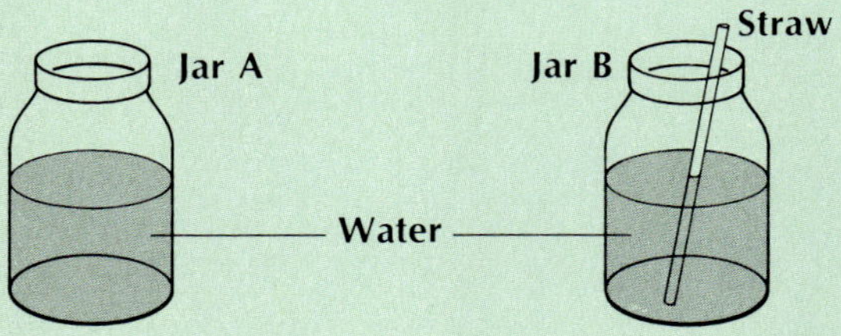

Now look at both jars. Is jar *B* the same as jar *A?*

(Write your answer on your answer sheet. Look across the page to box 1. Is your answer right?)

2

There are gases in the air around you. The gases in the air have no color. You cannot *see* the gases. The air you breathe in looks the same as the air you exhale. You cannot see the differences.

The air you *breathe out,* or <u>ex----</u>, looks like the the air around you.

(Write your answer on your answer sheet. Check across the page to box 2 to see if you are right.)

1

The jars **look** the same.

(You can't be sure.)

(Go on to question 2.)

2

exhale

The air you exhale looks the same as the air you *inhale*.

(Turn the page to question 3.)

3

But there is a difference. You can see the difference if you use a special chemical (KEM·ih·kul). This chemical is *brom* (BROME) *blue*.

Add a little brom blue to the water in jar *A*. Add the same amount of brom blue to jar *B*. Look at the two jars now. Is there any difference? *Yes* or *no?*

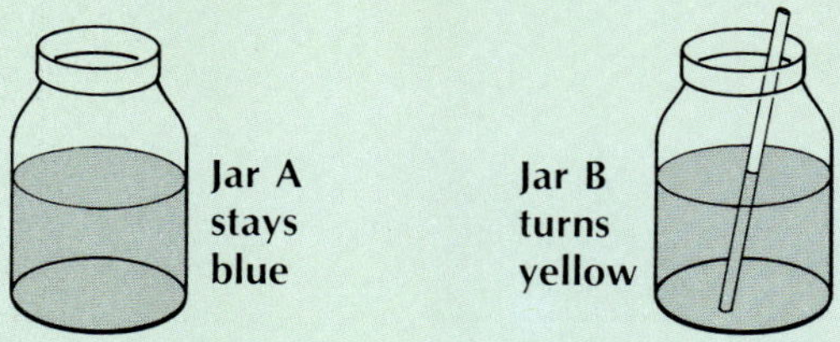

(Write your answer down and then check it across the page.)

4

Why did jar *B* turn yellow? Because you put something "extra" into jar *B*.

What is the "extra" in jar *B?*

(Write your answer on your answer sheet. Then check across the page to box 4 to see if you are right.)

5

You exhaled into jar *B*. The air that you exhaled made the difference.

This exhaled air has the gas *carbon dioxide* (KAR·bun die·OK·side). Carbon dioxide is a gas that you cannot see or smell. Brom blue turns yellow when *carbon dioxide* is added.

Brom blue turns yellow in jar *B* because there is more c––––– d–––––– in jar *B*.

(Write your answer on your answer sheet. Then check across the page to box 5 to see if you are right.)

3

yes

Brom blue turns *yellow* in jar *B*.

(Go on to question 4.)

4

the air you breathed out, or exhaled

That is the "extra" in jar *B*.

(Go on to question 5.)

5

carbon dioxide

(Turn the page to question 6.)

6

There is little or no change in jar *A*. Why did the brom blue in jar *A* not change color?

Because there is little or no carbon dioxide in jar *A*. Is this *right* or *wrong*?

(Write your answer on your answer sheet. Check across the page. Are you right?)

7

Air in the room mixes with brom blue in jar *A*. This air does not have much carbon dioxide in it. This is the air you inhale.

The air you inhale goes to your lungs. The air you exhale, or breathe out, comes from your l----.

(Write your answer on your answer sheet. Check across the page to box 7 to see if you are right.)

8

The air you inhale does not have much carbon dioxide. Does this mean your lungs give off *more* carbon dioxide than they take in?

Yes or *no?*

(Write your answer down and check it across the page.)

6

right

There is not much carbon dioxide in jar *A*.

(Go on to question 7.)

7

lungs

(Go on to question 8.)

8

yes

Your lungs do give off more carbon dioxide than they take in.

(Turn the page to question 9.)

9

You found that exhaled air has more carbon dioxide than the air you inhale.

How do you know this? You found this when you breathed into brom blue. You can see this again. This time breathe through a straw into a jar of brom blue.

What happens to the brom blue?

(Write your answer and check it across the page.)

10

Did the same thing happen for the other boys and girls in your class?

Does brom blue turn yellow when they breathe into it? *Yes* or *no?*

(Write your answer and check across the page to box 10 to see if you are right.)

11

The brom blue turned yellow for everyone. You could have predicted this. You have learned this concept: *There are likenesses in living things.* You all have lungs that work the same way. Human beings take air into their lungs. There is a difference in the air human beings *inhale* and *exhale*. The air we *exhale* has more __________ ____________ in it than the air we inhale.

(Write your answer on your answer sheet. Check across the page to box 11 to see if you are right.)

9

It turns yellow.

(Go on to question 10.)

10

yes

This is a likeness in all of us.

(Go on to question 11.)

11

carbon dioxide

(Turn the page to question 12.)

12

All people exhale more carbon dioxide than they inhale. Would you say this is what should happen? Would you say this is normal? *Yes* or *no?*

(Write your answer down. Check across the page to box 12 to see if you are right.)

13

Why is it normal? It is normal if it is a way that all human beings act.

You could predict that a new-born baby will exhale *more* carbon dioxide than it breathes in, or i------.

(Write your answer down. Check across the page to box 13 to see if you are right.)

14

It is normal if it is the way all human beings act. It is true, then, of an old man, a student, a young child, or an astronaut (AS·truh·not).

They all exhale air with more ____________ ________________ in it.

(Write your answer on your answer sheet. Check across the page to box 14 to see if you are right.)

15

The air you breathe in has more of the gas *oxygen* (OK·sih·jin). In your lungs the *oxygen goes out of* the air. And *more carbon dioxide goes into* the air. The air you breathe out has more carbon dioxide.

This change takes place in your l----.

(Write your answer on your answer sheet. Check across the page to box 15 to see if you are right.)

12

yes

People do exhale more carbon dioxide than they inhale. This is normal.

(Go on to question 13.)

13

inhales

(Go on to question 14.)

14

carbon dioxide

(Go on to question 15.)

15

lungs

(Turn the page to question 16.)

16

You now have this concept about breathing, haven't you? There is a likeness in people: *People exhale air with more carbon dioxide in it.*

Perhaps this is true for other animals—say a mouse or a bear. Both animals have lungs like ours. You might predict that a mouse would exhale more ____________ ________________, just as we do.

(Write your answer down on your answer sheet. Check across the page to box 16 to see if you are right.)

17

Suppose you could put a mouse in a jar. Look at the picture. Here is a mouse in a jar. In the picture you can see an open tube, marked tube *O*. The tube lets air into the jar so that the mouse can breathe.

What if the mouse stays in the jar? Would the air in the jar have *more* or *less* carbon dioxide? What do you think?

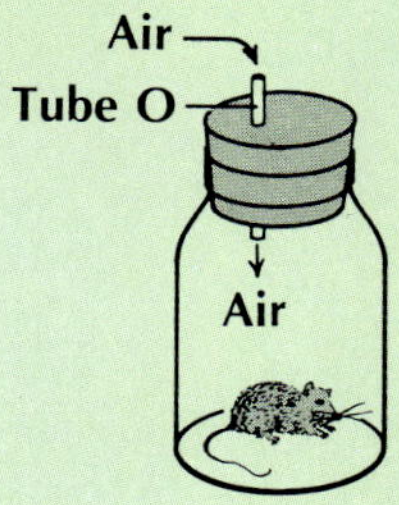

(Write your answer on your answer sheet. Check across the page to box 17. Are you right?)

16

carbon dioxide

(Go on to question 17.)

17

more

There would be more carbon dioxide in the jar.

(Go to question 18 on the next page to see why.)

18

It's easy to find out. Suppose you could take out the air from the jar with the mouse. Look at the picture below. Now there are two tubes in the jar with the mouse—tube *O* and tube *X*. We will call the way out of the jar tube *X*. We put this out-tube *X* into some brom blue.

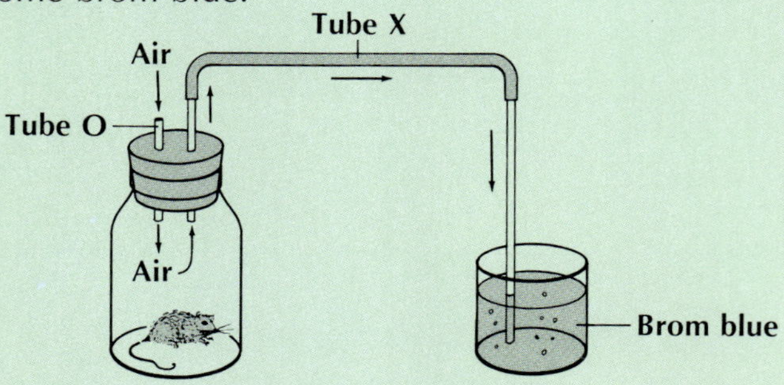

But wait. You need a pump. You need to keep air moving *into* the jar for the mouse to breathe. You also want to get some of the air out into the brom blue. You want to see if the brom blue turns yellow. You need a *pump* to pull the air through.

Now look at the picture below. We added a pump. Air is coming to the mouse from the outside, through tube *O*. Exhaled air from the mouse is pulled through tube *X*.

Now exhaled air is going into a bottle of brom blue.

The pump pulls air through the bottle of brom blue. What happens to the brom blue?

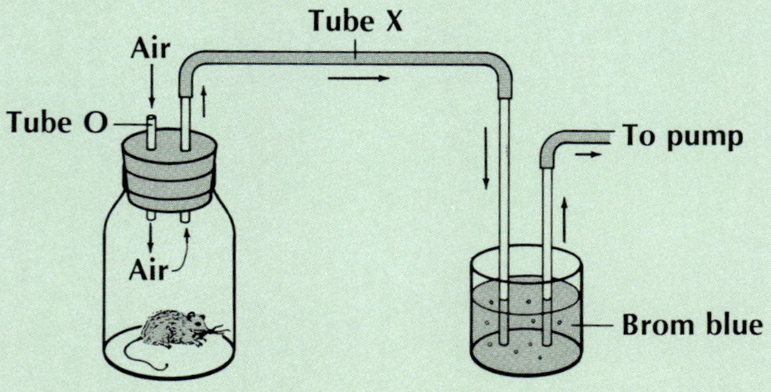

(Write your answer down and check it in box 18 across the page.)

18

It turns yellow.

The brom blue turns yellow when the exhaled air from the mouse is pumped through it.

(Turn the page to question 19.)

19

Why does the brom blue turn yellow? There is more carbon dioxide in the air the mouse exhales. *You could predict this.* You have a concept about breathing in human beings. You could predict a likeness in the mouse.

Is this concept true for other living things? Is it true for a bear? If you could measure the carbon dioxide in the air the bear breathes out, what do you think you would find?

The bear would give off ____________ carbon dioxide than he breathes in.

(Write your answer down. Check to see if you are right.)

Once More: The Meaning of a Concept

Somehow air is changed in the body. This is really nothing new to you.

You already know that your body takes in food. The food is changed into more of you. We call this *growth*. The body changes *food*.

Air is changed in the body. All living things change the food and air they take in.

Food and air are kinds of *matter*. We can say that *living things change the matter they take in*. The body seems to change the *matter* it takes in.

Do you think this is true for *all* matter that living things take in? You shall see. But it is true for air, is it not?

An Exercise in Prediction

You have in mind this concept: All living things *change* the air they take in.

Now try to predict what will happen. What will happen to the brom blue in each jar in the pictures? Try one of these yourself to check what you predict. See for yourself.

19

more

Just like us.
Just like the mouse.

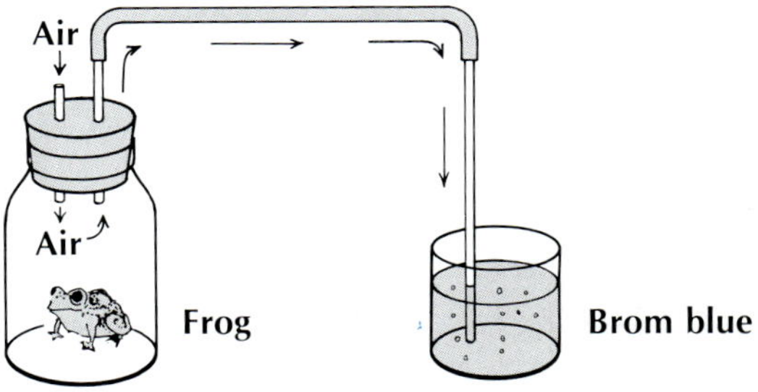

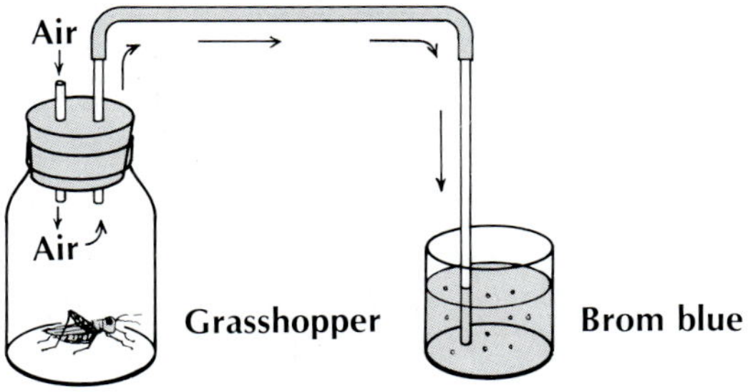

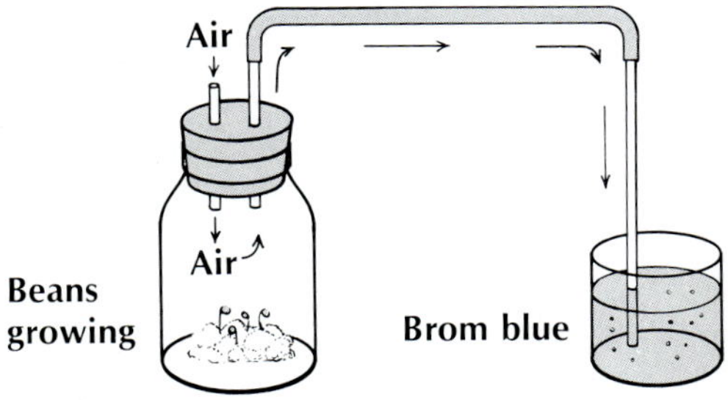

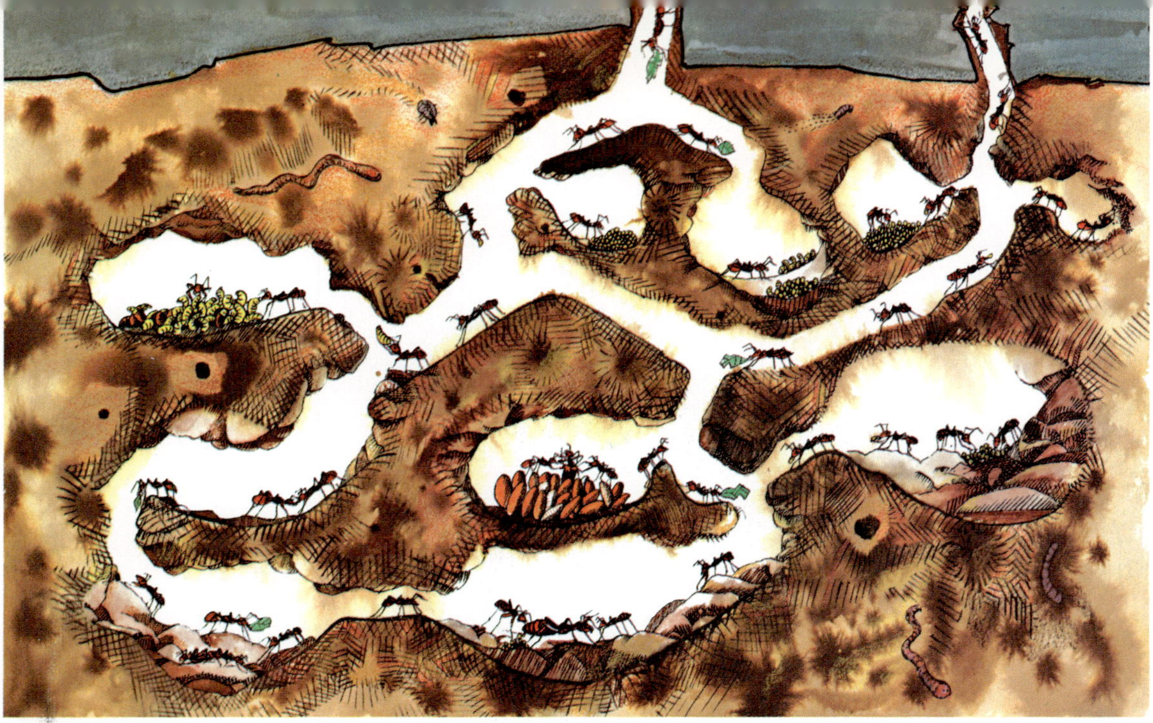

2

Centers of Life

INTRODUCTION

Why the Hurry?

Suppose you were smaller. Suppose you were small enough to get into an ant nest. You could then watch the work of the ants.

What do you think you would notice most? You would notice the ants hurrying around in the nest. ■

Ants seem always to be hurrying. Sometimes they hurry back and forth in long lines. Inside the nest, movement is going on all the time. Why do ants hurry so much? If you watch the ants long enough you will know why. The reason they hurry is a simple one: *to get enough food*. Without food, the ants would soon die.

This is a main highway into a city in the morning. ● Again there is a lot of hurrying. Cars move in long lines to the city. And away from the city. There is movement along the highway all the time.

There are lots of reasons why people go to the city or town. But an important reason is *to get food*. That is, people go to earn money to buy food, mostly.

Ants and people do not live the same way. And we are not saying that they behave the same way. But they both *move* to get food. And they both need food.

You and ants need food. Do sharks and elephants and trees and horses need food? Of course! Do all these living things use food in the same way? What do you think?

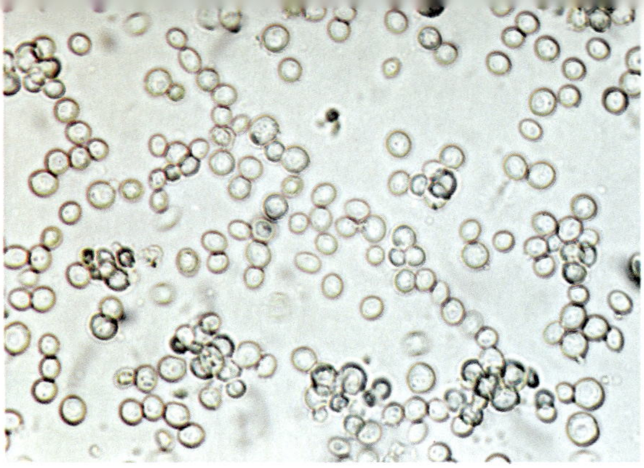

■

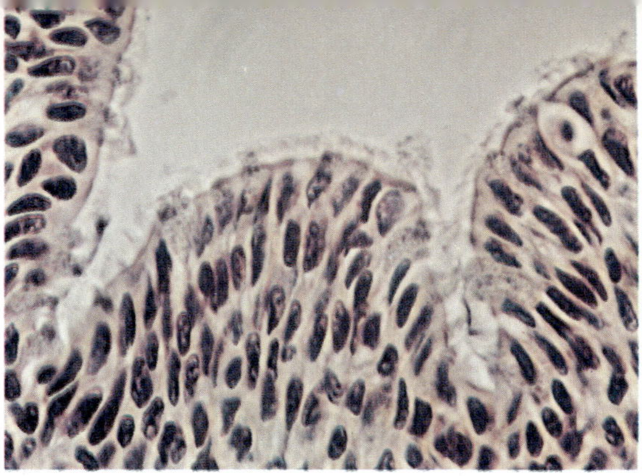

●

1. A LUMP OF YEAST AND YOU

▲

♦

★

What does *yeast* (YEEST) have to do with you? A lot. Yeast and you are both made up of **cells.** Do you know what cells are? Cells are tiny parts of living matter. All living things are made up of cells.

Yeast is really made up of many tiny one-celled plants. ■ Are you one-celled? No. You are made up of many, many tiny animal cells. ●

Yeast cells and your cells do many of the same jobs. Yeast cells do work to keep alive. Your cells and yeast cells use food. There is this likeness in yeast and you. So we can compare yeast cells and your cells.

First to the Yeast

The yeast you buy at the store looks like this. ▲

Roll two lumps of sugar in some yeast. Next, drop both lumps of sugar with the yeast on them into a glass of warm water. Stir the water with a spoon until you can no longer see the sugar.

Then, pour this mix into a bottle as shown in the picture. ♦ Put a one-hole stopper with a tube in the bottle of yeast and sugar. Set up a small jar with brom blue. (You used brom blue before. See page 36.) Put the tube from the yeast and sugar water into the brom blue.

Your setup should look like the one in the picture. Make sure the tube goes *into* the brom blue.

A girl and boy in New York grew yeast cells in sugar water. This is what they found. ★ Do you find the same thing happens? Each time this is done the same thing will happen.

But look at these pictures. When sugar water is used alone, there is no change. ◇ When yeast cells are used alone, there is no change. ⊙ The brom blue does not change when *only* yeast cells are in water. There is no change when sugar only is in the water.

The yeast cells in sugar water must give off "something." This "something" moves across the tube *X* into the brom blue. ◆ This "something" must turn the brom blue to *yellow.* ★

What is it? It is *carbon dioxide.* The same gas you give off in the air you exhale.

Why do you think brom blue changes color when the yeast cells *and* sugar are together? The yeast cells use sugar for food. As they use the sugar, yeast cells give off carbon dioxide.

Why doesn't brom blue change with only yeast cells in water? or with only sugar water? Without sugar, the yeast cells have no food. They do not give off carbon dioxide. Sugar alone does not give off carbon dioxide.

◇

⊙

And Now to You

You have seen a change take place in brom blue. A change with the yeast cells and sugar. Now, let's talk about *you.*

Take a straw and breathe into a jar of brom blue. What happens to the brom blue?

This picture shows a boy in Oklahoma. ⬟ He is breathing in. When he breathes in, or *inhales,* air goes into his lungs. This air comes through bottle *1.*

In this picture, the same boy is breathing out. ⬢ When he breathes out, or *exhales,* air goes into the brom blue in bottle *2.*

But look carefully at bottle *1*. Air from the room goes into this bottle. The air the boy breathes in passes through the brom blue in bottle *1*. Predict what will happen to the brom blue in bottle *1*.

Predict what you think will happen to the brom blue in bottle *2*. The boy breathes out into this brom blue.

Were you right? Turn the page. But don't turn the page until you predict what happened.

⬟

Bottle 1   Bottle 2

⬢

Bottle 1   Bottle 2

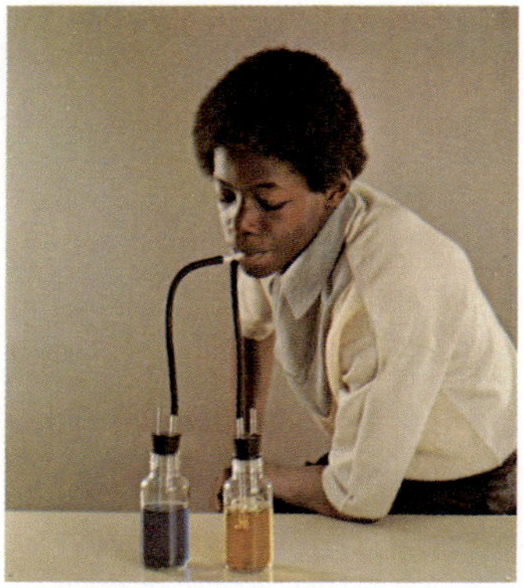

Bottle 1 Bottle 2

The brom blue turns yellow only in bottle *2*. This is the bottle that the boy breathes into. ■ There is no change in bottle *1*. The air from the room goes into the brom blue in bottle *1*. This air does not change the brom blue.

The air he *breathes out* does change the brom blue. This air has *more* carbon dioxide than the air from the room. The gas, carbon dioxide, turns the brom blue to yellow.

What is alike about what
yeast cells do?
you do?

What do you think? From what you have done and seen so far, you can make a good guess. You can make a guess based on your *experiences*. Why is this kind of guess called a "knowing," or "educated," guess?

Here is a chart that shows what was tested and what happened. ● This chart shows where the brom blue turned yellow.

WHAT WAS TESTED	**COLOR CHANGE OF BROM BLUE**
yeast alone	no
sugar alone	no
sugar *and* yeast	*yes* (to yellow)
you inhaling	no
you exhaling	*yes* (to yellow)

●

2. CENTERS OF ENERGY: CELLS

Track stars drink sweet orange juice for quick **energy** (EN·er·gee). Orange juice has sugar. The sugar is in the food, orange juice.

Heat is one form of energy. Heat is produced by your body. Heat is produced by your cells. The food you eat gives your body heat. Your body has a *temperature* (TEM·purr·ih·chur) of 98.6 degrees Fahrenheit (FAR·en·height). (You probably know that temperature tells how hot or cold something is.) Unless you are sick, the temperature of your body is about the same. Some of the food is used to keep this temperature the same.

How does your body keep this temperature? How does your body get heat? To get heat, you need food. Food gives energy. Some of this energy is heat. Some of this energy from food is used for growth. Clearly, food is needed for your body.

What is the food for yeast cells? Sugar. Sugar gives the yeast cells energy. Some of this energy is heat. Some of the energy from sugar is for growth. It is the same for yeast cells and for your cells. *Food gives all living things the energy for living.*

▲

How does sugar give energy? Let's see.

You and the yeast cells use sugar. You and the yeast cells give off heat. Heat is one kind of energy. Heat is only one form of energy given off when sugar is used. What else is given off? What else besides carbon dioxide and energy?

When sugar or some other food is used by your cells, water forms. ▲ When food is used by your cells, water is given off. And so is carbon dioxide and energy. It is the same for yeast cells.

We write what happens when food is used in this way:

$$\text{FOOD USED} \xrightarrow[\text{gives}]{} \text{ENERGY} + \underset{CO_2}{\text{CARBON DIOXIDE}} + \underset{H_2O}{\text{WATER}}$$

You know now that when food is used, carbon dioxide and water are formed. When food is used, there is *energy* for the cells. There is energy for all life activities. You, as well as the yeast cells, need energy. That energy comes from food.

Getting Energy from Food

Energy is needed for walking, keeping warm, thinking, growing. Energy is also needed for everything that goes on in cells.

There is more to the story. You see that water and carbon dioxide are given off. What is the gas that you use from the air? You inhale air that has **oxygen** in it. Oxygen is needed in this story of how cells get energy from food.

We can write what happens in this short way:

FOOD + OXYGEN (O_2) —gives→ ENERGY + CARBON DIOXIDE (CO_2) + WATER (H_2O)

Suppose we put **sugar** in as the food. This is what most living cells use as food. Now you may write:

SUGAR + OXYGEN (O_2) —gives→ ENERGY + CARBON DIOXIDE (CO_2) + WATER (H_2O)

Scientists have learned part of what happens to food in your body. They have learned part of the story of how living cells use the food you eat.

One way yeast cells get energy to grow is by using sugar like this:

SUGAR + OXYGEN (O_2) —gives→ ENERGY + CARBON DIOXIDE (CO_2) + WATER (H_2O)

You, like the yeast cells, get energy by using sugar this way. This is not strange. You and the yeast cells are both alive. It is not strange that each of you gets energy from food in much the same way.

Sugar has a lot of energy in it. Cells can store up this energy. After all, cells need a storehouse of energy. Cells need a supply of energy. It cannot be used up all at once. How do cells keep a store of energy?

Energy for Cells

You know your body is warm. So you know the cells in your body give off heat.

Try this. Put your hands together. Rub them together quickly. Heat! Yes. This is one way to get heat.

But this is not the way your cells get heat. And this is not the way your cells get any other kind of energy.

What about a car? How does a car get energy? The engine of a car gets energy when gasoline is broken down. Carbon dioxide and water are also given off. Here is how we show it:

GASOLINE —gives→ ENERGY + CARBON DIOXIDE (CO_2) + WATER (H_2O)

A Store of Energy: ATP

Your cells get energy by breaking down something too. Your cells, however, cannot get energy from gasoline. What do your cells get energy from? From food—but especially from sugar.

Your cells can then store up this energy in a special kind of matter. This matter is called **ATP** (AYE·TEE·PEE). Scientists call this matter ATP for short.

Cells need energy to run on. They need a store of energy. This store of energy is in ATP. Energy is locked up in ATP. It is in every cell.

Cells in your body break ATP down. Yeast cells break down ATP. All living cells break down ATP. When ATP breaks down, cells get energy. Cells get energy to use. ■

■

You can think of cells as centers where energy is changed. In the cells, ATP is built up. ATP is stored. ATP is also broken down. Your body gets heat and other kinds of energy when ATP breaks down.

Your body needs energy for everything you do. You need energy to work, to run, to think. ● *ATP is broken down to get this energy.* ATP is broken down in cells to get this energy. You can write it this way:

ATP ———— breaks down in your cells ————→ ENERGY

When gasoline is broken down, the car goes. When ATP is broken down, you go.

Keeping Up the Store of Energy: ATP

Now you know what breaks down in the cells. ATP breaks down. If ATP *only* breaks down, you would have a real problem.

●

You need ATP all the time. Then what must happen to the ATP? ATP has to be built up again.

Let's look at it this way:

ATP breaks down in the cells.

When ATP breaks down, it gives up energy.

ATP has to be built up again.

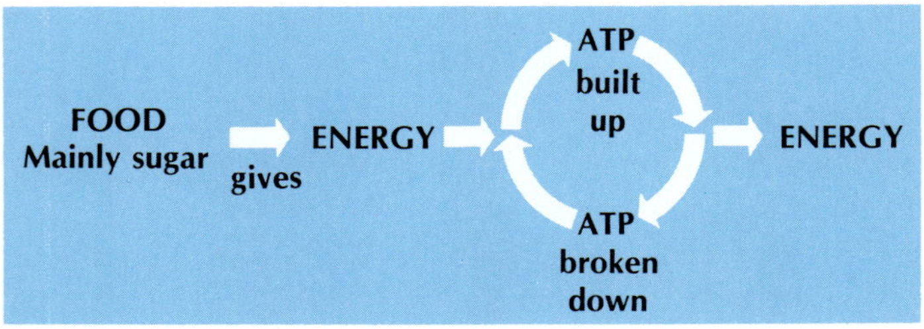

What happens after a cell has used up its ATP? When a car runs out of gas, it cannot move. When ATP is used up, the cell needs more food.

But living things can get more food. As you already know, food has stored energy. Sugar is what your body uses most easily for energy. Sugar is used most easily to get energy to make ATP.

When sugar is changed in your cells, energy is given off. This energy is stored up. Where? It is stored in ATP. Energy is used to build up ATP. You see, your cells run on the energy from ATP. Your cells make more ATP. Over and over again.

ATP is always there in your cells. It is there to give your cells a steady supply of energy—energy for all the work of cells.

You see then an important use for food. Food gives energy. This energy is stored up in ATP. And you now know that ATP is stored in cells.

To learn even more about cells, you must use a **microscope** (MY·krah-skope). A microscope is used to make objects look many times larger than they are. You can see cells under a microscope. And you must look at cells to get the whole story. This is what you are to do next.

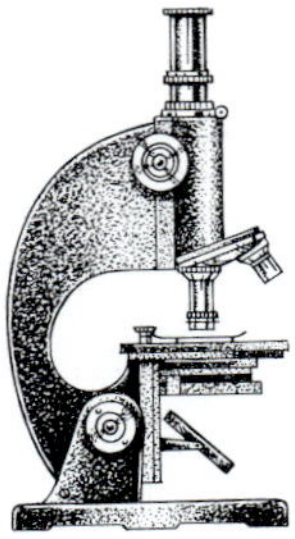

Microscope

3. WITHIN THE BODY—A CLOSER LOOK

Do you have a microscope? Let's look at something together.

Get a toothpick. Use the wide end of the toothpick. Scrape the inside of your cheek—*very, very,* carefully.

Now, put a drop of water on a clean microscope slide. Mix the matter on your toothpick in this drop of water. ♦

Add one drop of iodine (EYE·uh·dine) to the drop of water.

Take a coverslip for your glass slide. Place it carefully on the drop. ★ Now, look through the microscope. Try to get a clear view. What do you see? You are looking at your own *cheek cells.*

One of these pictures was made by a student. ◊ And the other was made by a scientist. ⊙ They are pictures of cheek cells. Which picture looks like your cheek cells?

You may say they look pretty much alike. They do, don't they?

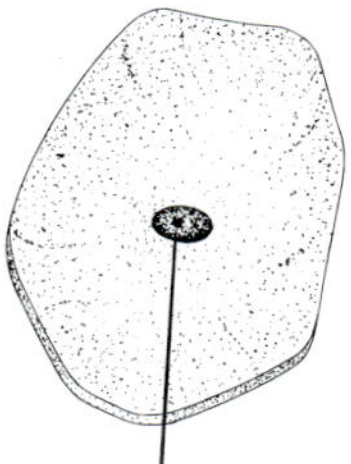

Look very closely at each cell. ⊙ Do you see what the iodine does? It changes the color of the cell. The round part in the middle of each cell gets very dark. This round part is the **nucleus** (NEW·klee·us). All the cells in your cheek have a nucleus. Almost all cells of living things have a nucleus.

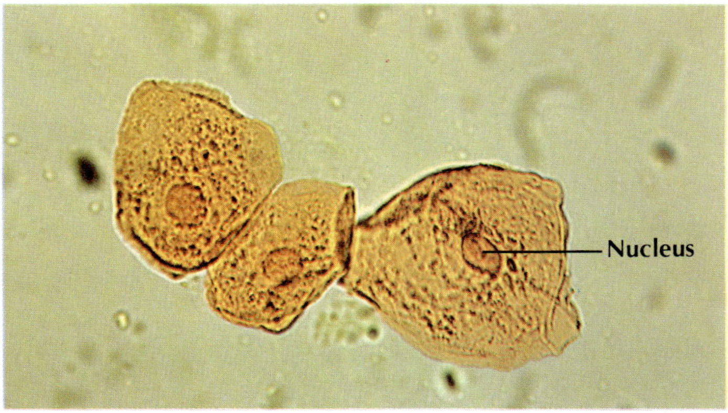

⊙

♦

★

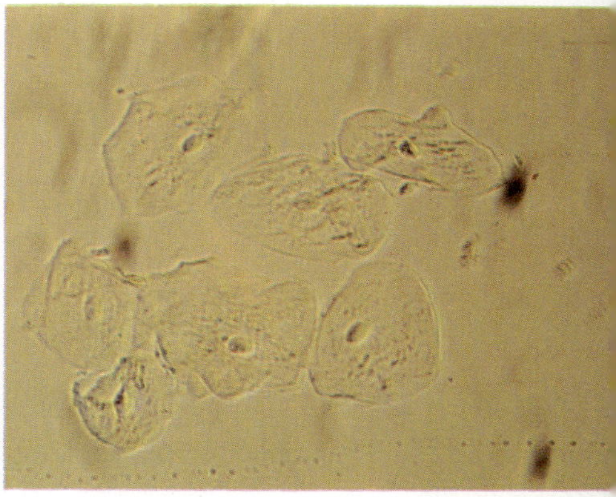
◊

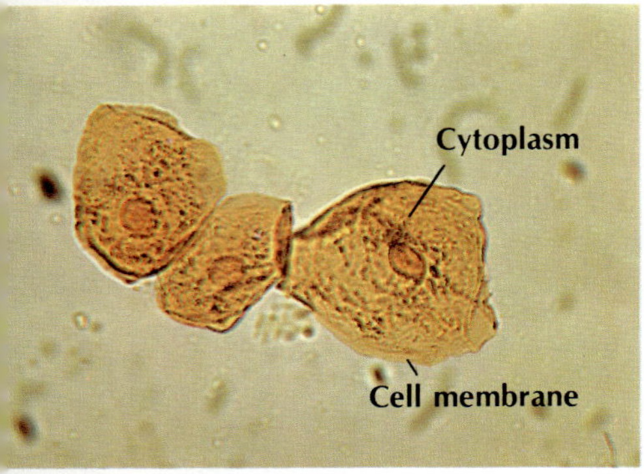

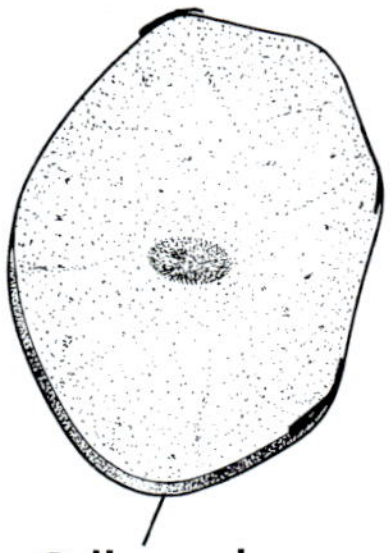

Each cell also has a covering. You can see that the inside of the cell does not run out into the water. Each cell has a shape. Each cell is held in shape by its covering. We call this covering a **membrane** (MEM·brain), a **cell membrane.** ■ This membrane is so thin you cannot see it with the microscope you have.

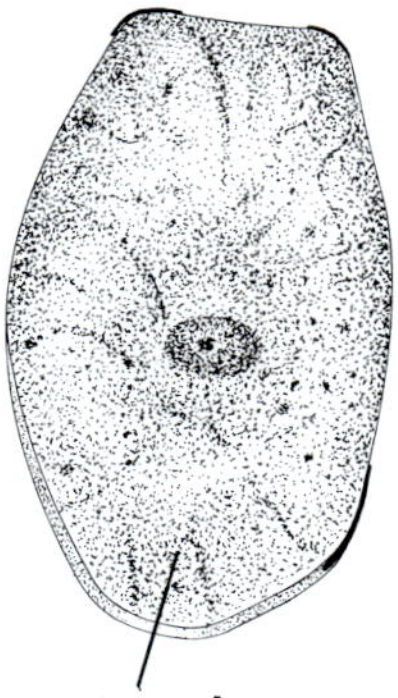

Look closely at the cell. Do you see some tiny grains between the cell membrane and the nucleus? The iodine has colored the grains. They are part of the **cytoplasm** (SY·tuh·plaz·um). This cytoplasm is the living matter in the cell. The nucleus is alive, too. So is the cell membrane. But the *nucleus* and the *cell membrane* are not part of the cytoplasm. Scientists do not think of them as part of the cytoplasm.

You would need a better microscope than yours to see the cytoplasm clearly. This microscope makes things look much larger. Then the grains do look different. You would see many small parts inside the cytoplasm. Some would look a lot like tubes. Some would look like rods. All these small parts have a name. Each does some special work in cells.

You need to know this. Most often, a cell has three parts: the *nucleus, cytoplasm,* and the *cell membrane*.

The cells of most plants and animals have these three parts. Most of your cells have them. So do the cells of a worm, a rosebush, and an elephant.

Cells — Cells — Cells

What if you were to study the body of any living thing? What if you could study it with a microscope? If you could look at every part of it, what would you find? What do you think?

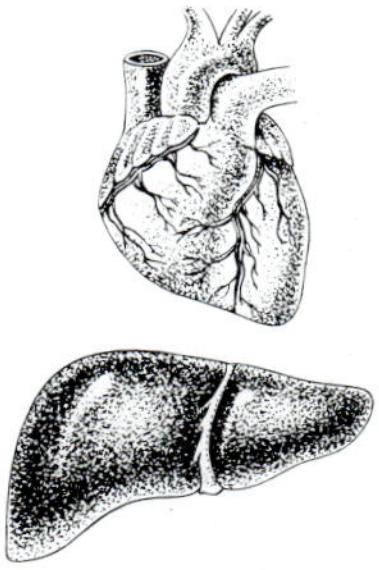
Organs

Scientists have studied plants and animals in this way. They have taken them apart. They have studied their **organs** (OR·gunz). An organ is a part of your body that is fitted to do a special job. The heart is an organ. So is the liver. There are many kinds of organs. An organ like the heart or liver has a special use in the body. Both plants and animals have organs.

Scientists have studied organs like those in your body. They have studied the eyes, the lungs, the liver, and the brain. These are all organs.

What do organs do? Each organ of the body has a job of its own. You can say every organ has a job to do, or a **function** (FUNGK·shun). The word "function" means the use that a part of the body has. Each organ has a use, or function, in your body.

You know, for example, that the engine of a car has a function. This means the engine has a use. How could the car run without the engine?

Look at a part of an organ under the microscope. What do you see? *Cells. Organs are made of cells.*

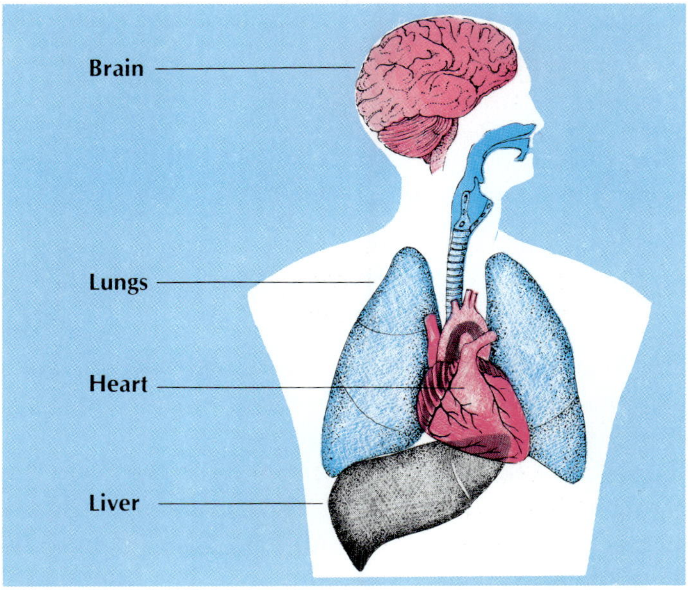

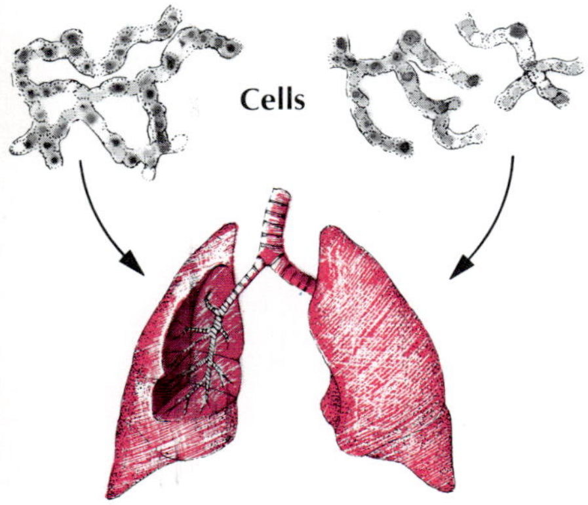

An organ: the lung

Seven different kinds of cells are shown in the picture across the page. ■ These cells are from different organs of plants and animals. Roots and leaves are organs in a plant. The brain, the intestine, the hand, are organs in an animal. All the organs are made of cells.

Cells have different sizes and shapes. But they are alike in this way: *They have a nucleus, cytoplasm, and a cell membrane*. Most cells you will see have these three parts. All the cells in the picture across the page have these parts.

●
Cells as Building Blocks

Cells are everywhere in your body. Are they lying about every which way? Not at all. Cells are found in organs. Each organ is made up of different kinds of cells. ● As you shall see, each kind of cell has a different function.

Your lungs are organs. Your lungs are made up of different kinds of cells. Each of your organs is *organized* (OR·gun·izd). This means each cell has its own special place in an organ. Cells are the building blocks of your organs.

You are organized. Organs are organized to make *you*. Each of your organs has a special place in your body.

Cells in a Team

Each organ of your body also has a function. Each organ does a special job. The different kinds of cells in an organ work together. The cells work together as a team.

Look at your hand with its fingers. With your hand you can touch. You can write. You can work. ▲ Your hand has many kinds of cells working together. Their functions allow your hand to move, touch, and write. Their functions allow you to work.

Does the brain have a function? Yes. The brain has many functions. One of its functions is to think. The eye is another organ. The eyes and brain have the function
▲ of seeing. Organs have functions.

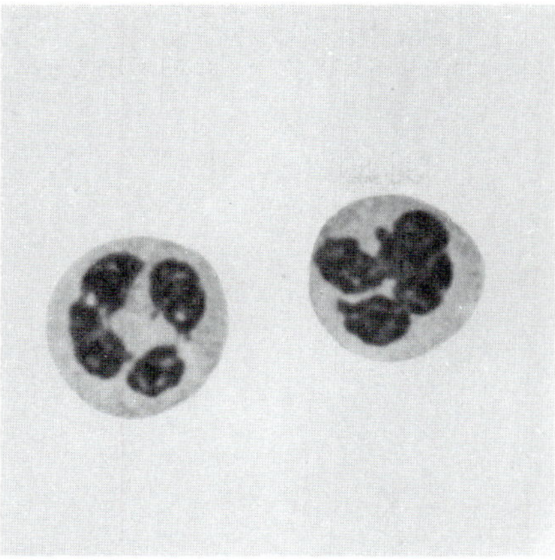
White blood cells

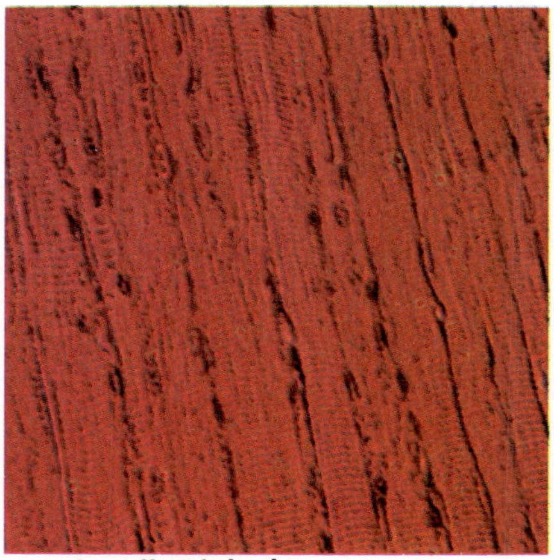
Muscle cells of the legs

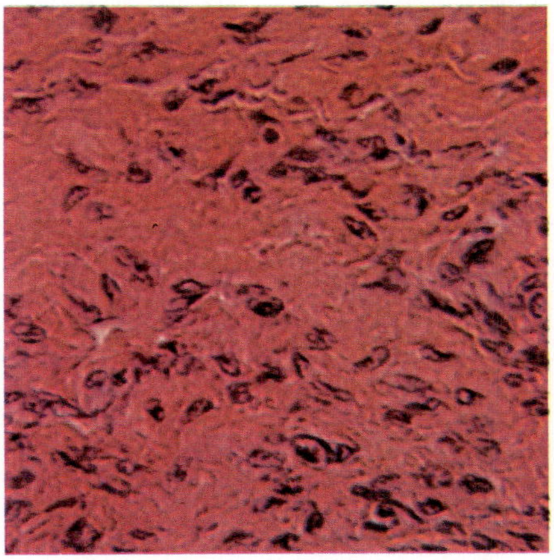
Muscle cells of the intestine

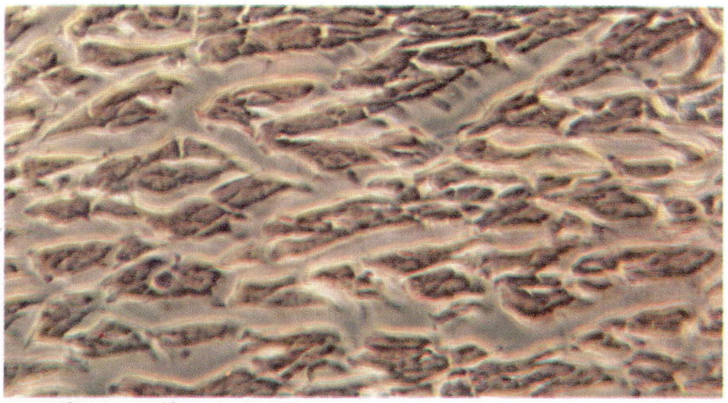
Cartilage cells

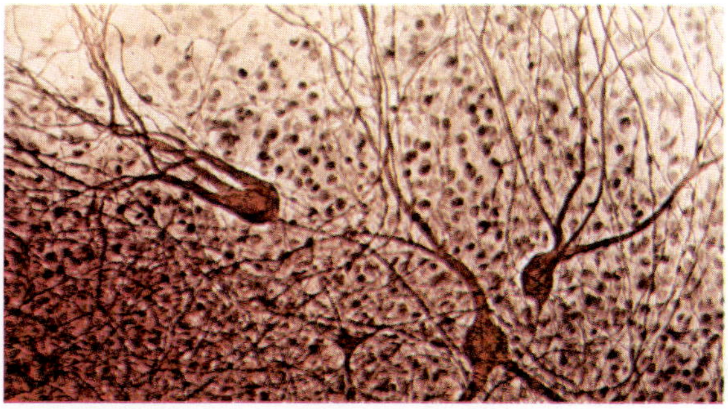
Nerve cells

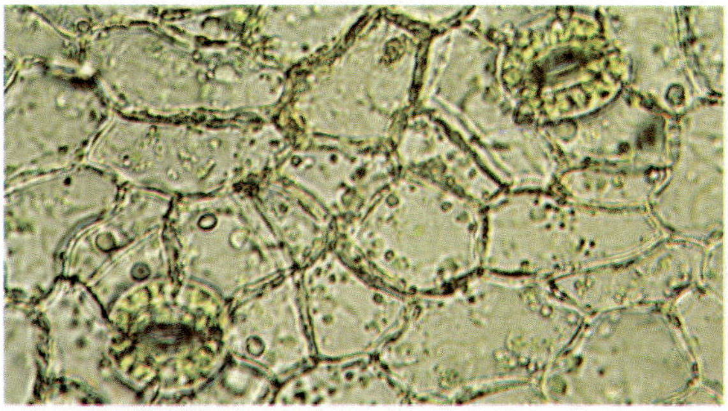
Green leaf cells

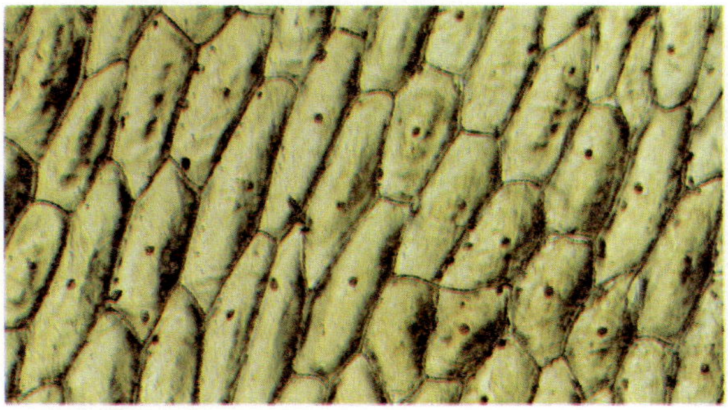
Onion skin cells

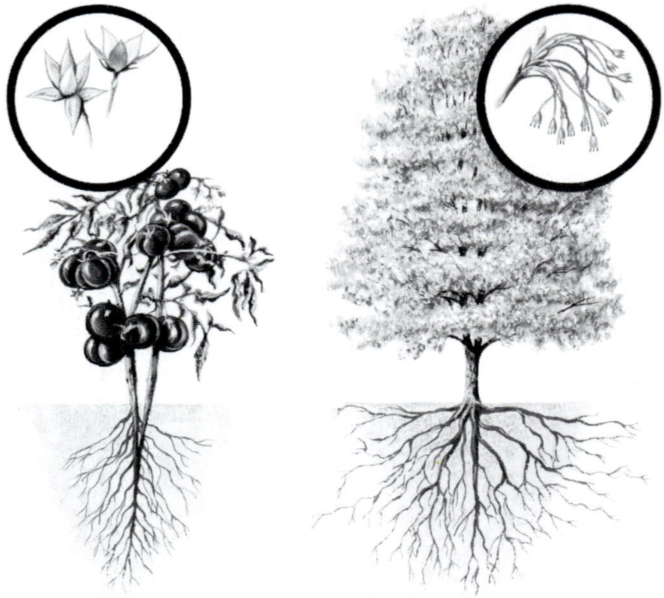

Tomato plant

Maple tree

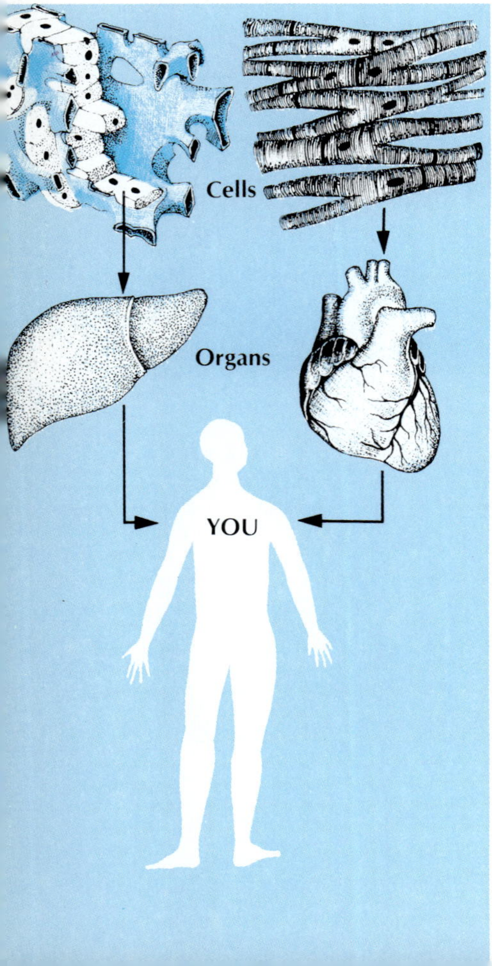

Plants have organs, too. What are some of the organs of a plant? *Roots* are organs. *Leaves* are organs. You probably know that *roots* have different functions than *leaves*.

You will study the cells of different organs under the microscope. You will see the many wonderful kinds of cells. For the time being, let us say this: *A living thing is made up of cells.*

Your cells are organized into organs. The cells have a special place in each organ. Now we can say that *cells* make up *organs*. *Organs* make up the *body*.

Let's think this way. Cells are the building blocks of your body.

A Try at Prediction

Animals and plants are organized. They are made up of cells.

Suppose a boy shows you a "magic green plant." He says it never needs water. Is it a real plant? How could you find out?

4. THE CELL—THE CENTER OF IT ALL

What do you find in plants and animals? Cells. There are cells in muscles and in blood. And there are cells in the brain. Cells are everywhere in animals.

Cells are also everywhere in plants. They make up the leaf, the root, and the stem. *The cell is the center of life.*

Why should you try to eat more than once a day? Because, in time, you would die if you didn't have enough food. Food is needed for energy and for growth. This is true.

Now you have a better way to say this. Without food, cells would not have what they need to build up ATP. If you did not eat, cells would have no ATP. Without ATP, cells could not do their work. The heart could not pump. Food could not be digested. The cells could not grow. A living thing dies when its cells die. The cell is the center of it all.

The cell is the center of life.

The Cells as Units

The word *unit* (YOO·nit) is a very useful one. You are a unit in your class. You and your classmates are the units that make up your class. Without each of you, as units, there would be no class.

Your class is organized. It is made up of units. It is made up of each of you. So it is with cells. Cells are organized.

The cell is the unit of order in you.▲ A cell is the smallest unit that makes a living thing like you. Or a living thing like an oak tree. Or like a whale. Or like a yeast cell or a cat. *The cell is the smallest unit of living things.* The cell is the building block of living things.

Organs are made up of units—cells. Cells make up the liver or the heart or any organ.

All the organs work together. An oak tree, a whale, an ant and you are made of organs. The organs work together.

The organs are organized. That is why scientists call a living thing an **organism** (OR·gun·iz·um). They call a

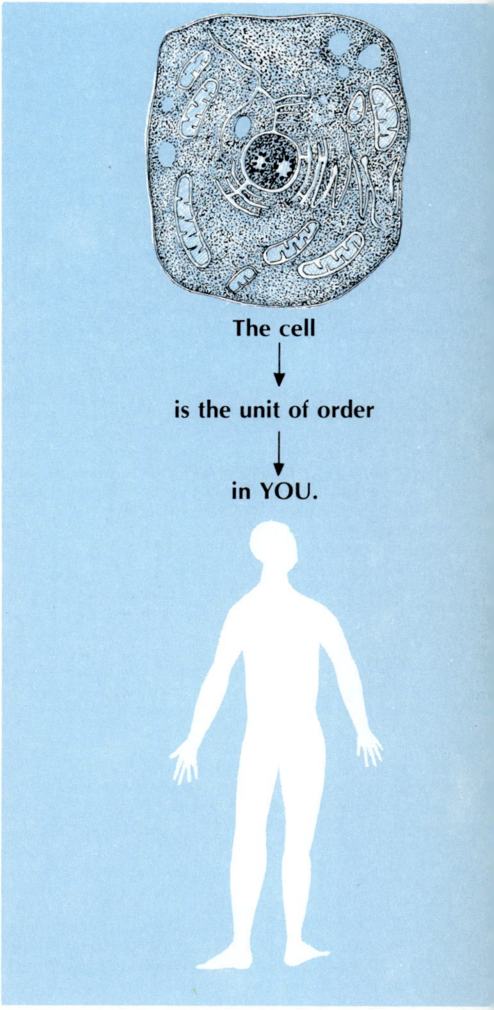

▲

living thing an organism because it is organized. Plants are organisms. Animals are organisms.

An organism may have many organs—all working together. Or an organism may be only one cell, like yeast cells. Or like the one-celled animals, paramecia (par·ah·MEE·she·ah). They are organized inside the one cell. The cell is the smallest unit of which living things are made. A living thing may be one cell, or many, many cells.

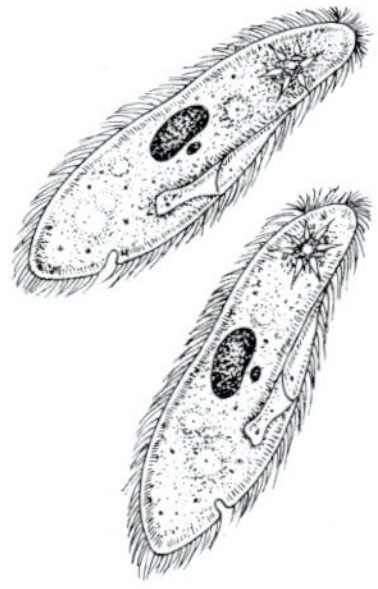

Paramecia

Cells as Units of the Body

Living things are made up of cells. Your lungs, liver, and heart are made up of cells. So are the lungs, livers, and hearts of animals like you. Cells are the units which make up the bodies of animals.

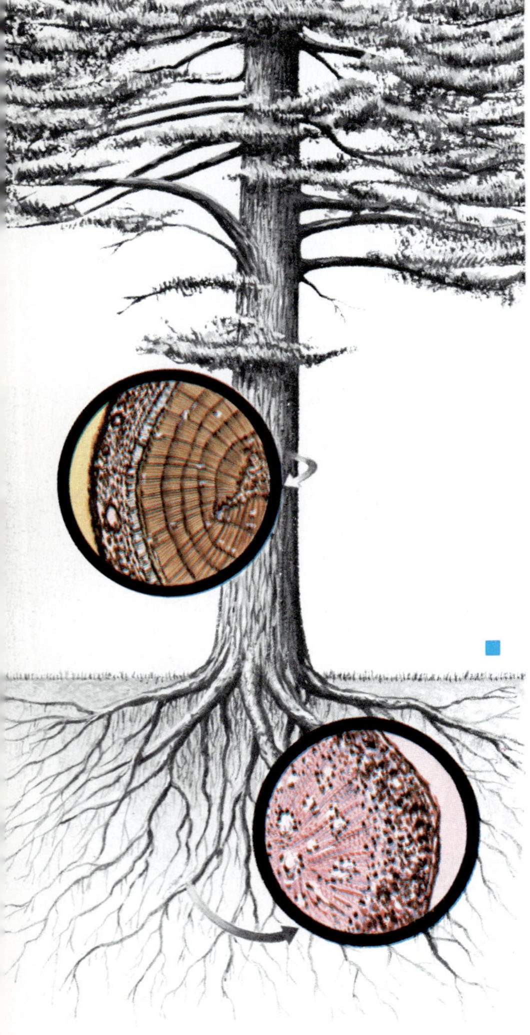

The way a plant is built depends on its cells. The form the cells make or build gives the plant shape. Cells build the leaves, stems, and roots. Cells *form* these organs. Cells are the units which make up the parts of the bodies of plants.

Cells are the building blocks of organs. Cells are the building blocks of living things.

Cells as Units of Function

Do you remember what *function* means? The workings of the different parts of the body are functions—or uses. Cells may have different shapes. But all cells are at work. You are breathing. Cells are at work. When you digest food or walk, your cells are at work.

The cell is the center of activity of the body. That is, *the cell is the unit of function.* Function is what the cell does in an organ. Function means action. Cells are the centers of action in living things. They are the centers of the functions that make life possible.

Breathing is a function. Digesting and moving are also functions. Remember the activities of living things you learned about in Unit I? They were

movement	growth
food-getting	reproduction
response	

These are all *functions*. These are the activities, or functions, of living things.

Find the smallest unit in an organ. It is the cell. The cell is the unit of function in the body.

You Add to Concepts of Life

Here are two concepts of living things, or organisms. These are the result of your study:

The cell is the building block of the body. You may say: **The cell is a unit of form.**

The cell is also the unit of the body's activity. You may say: **The cell is a unit of function** (action).

Now put the two together. What do you have? You have an important concept of living things: **The cell is the unit of form and function in living things.**

Is this concept true? Yes. Living things are made up of cells.

If an organism is only one cell, this cell is the unit of form and function. Paramecia have only one cell.● But the cell has a form. It has the functions of a living thing.

If an organism has *many* cells, the cells are organized. But the smallest unit is always the cell.

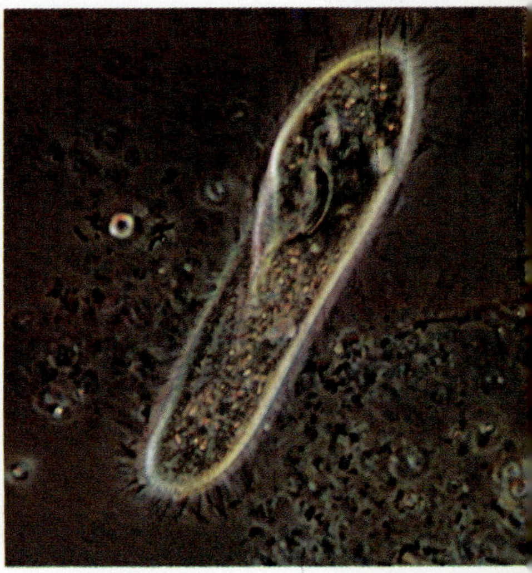

The Storehouse of Energy

3

INTRODUCTION

Teamwork in Your Food Tube

All these animals are getting food. All these animals are feeding. ■ The starfish turns its stomach inside out. This is the way it feeds on the clam.

The cow feeds on grass. It swallows a mouthful of grass. The grass goes into part of the stomach. Later, the cow can "cough" up a little grass at a time. Then the grass is slowly chewed.

You can say the baby is also feeding. And the fish is feeding too. Where does all this food go? What happens to food when it gets inside an animal?

You know the food must get to the cells. Food is broken down. It is broken down so it can get to the cells. This happens in the starfish. It happens in the cow. And it also happens in you.

Let's begin by learning what food has in it. You have to know this first. Many things must happen before food can get to cells. Then you can learn about teams of *chemicals* (KEM·ih·kulz). There are teams of chemicals at work in your food tube. The teams of chemicals work to break down food. They break food down into simple parts. Then these simple parts can go to the cells.

Let's begin.

Milk

Carbohydrates	++
Fats	+
Minerals	++++
Proteins	+++
Vitamins	+
Water	+++++

Steak

Carbohydrates	
Fats	++++
Minerals	++++
Proteins	++++
Vitamins	+++
Water	++++

Broccoli and apple

Carbohydrates	++
Fats	
Minerals	++
Proteins	++
Vitamins	+++
Water	+++++

Potato and bread with butter

Carbohydrates	+++++
Fats	+++
Minerals	+++++
Proteins	+++
Vitamins	
Water	++++

+ means **a very small supply,** ++ means **a small supply,** +++ means **a fair supply,** ++++ means **a good supply,** +++++ means **a very good supply.**

1. THE STOREHOUSE OF ENERGY: FOOD

Have you had lunch? Which foods in your lunch might have sugar? Ice cream and cake have sugar. Many foods have sugar.

Some foods you eat have starch. Fried potatoes and bread have starch. Your body can *change* this *starch to sugar* for energy and growth.

An apple is food. So is a hamburger. Each kind of food is made up of parts you need for energy and growth. These parts of food are called **nutrients** (NEW·tree·ents).

Food is made up of many nutrients. Your body needs nutrients for all cell functions. Let's look more closely at some of the nutrients in food.

For example, you must have heard of **vitamins** (VI·tah·minz). An apple—a food—has vitamins in it. An apple has *vitamin C.* Vitamins are parts of food. Vitamins are nutrients you need for good health.

There are other nutrients you need besides vitamins. You need **proteins** (PRO·teenz). Proteins are a necessary part of your cells. Proteins help build new cells in your body. That is, proteins are necessary for your **growth.** Proteins also help build up cells that are worn down.

Many foods have proteins. Meat, milk, and cheese, for example. Eggs and beans also have proteins. Proteins are nutrients.

Sugars and **starches** are nutrients. There is another word for sugars and starches. They are called **carbohydrates** (kar·bow·HY·drates). The sugars and starches you find in foods are carbohydrates. ▲ Carbohydrates are used by your cells to get energy. Carbohydrates are the nutrients most easily used by cells to get energy.

▲ 

Your body needs a lot of **water.** Cells have a lot of water in them. Water is also a main part of blood. Blood carries food to the cells. Blood also carries wastes from cells. Water also is a nutrient.

Your cells need **minerals** (MIN·er·ulz) like iron. Minerals are needed for the work of the cells. *Proteins, sugars, starches, water, minerals, and vitamins are all nutrients.*

There is still another nutrient you find in food. Butter is made up mainly of this nutrient. Meat has some of it. That nutrient is **fat.**

▲ 

Your body uses all the nutrients in the foods you eat. How many nutrients are there?

carbohydrates	*minerals*	*vitamins*
(sugars and starches)	*proteins*	*water*
fats		

These are the nutrients. *Cells need all these nutrients.*

The Nutrients in Food

Of course, you do not really eat these nutrients. You eat foods. The nutrients are in the foods you eat. These foods are made of different kinds of nutrients. This hamburger, for example, has many kinds of nutrients. ◆

◆

Hamburger	
Carbohydrates	
Fats	+++++
Minerals	++++
Proteins	+++++
Vitamins	+++
Water	++++

+ means **a very small supply,** ++ means **a small supply,** +++ means **a fair supply,** ++++ means **a good supply,** +++++ means **a very good supply.**

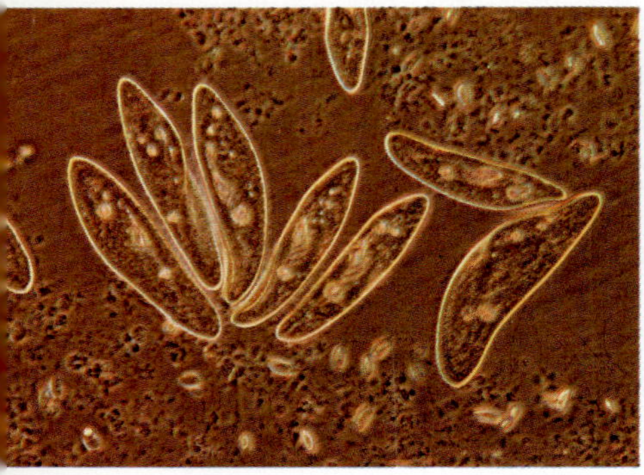

■

Most foods have more than one kind of nutrient. Some are more rich in one kind of nutrient than in the other kinds. For example, some foods are rich in proteins. You may think of meat or fish or cheese or eggs. These give you many proteins.

If you want carbohydrates, what foods would you look for? You might say apples or oranges. Or you could say cake or potatoes or beets. These foods are rich in carbohydrates. Your cells can easily use carbohydrates for energy.

Fruits and some vegetables (VEJ·tah·bulls) have the carbohydrates you need. They also have many vitamins and minerals. The next time you want candy, try eating an apple. Or an orange. You will get minerals and vitamins as well as sugar.

You need all kinds of food. You need food, and so do other living things. All plants and animals need food.

Users of Food

Plants as small as yeast cells need food. Animals do, too. Animals as small as paramecia or as large as a whale use food. Some kinds of whales are the largest animals alive.

●

▲

Look at the paramecia in a drop of water.■ Here, they are busy feeding on **bacteria** (bak·TIR·ee·uh) in the water. Bacteria are one-celled plants. The bacteria are food for the paramecia. Paramecia will break down the bacteria into nutrients.

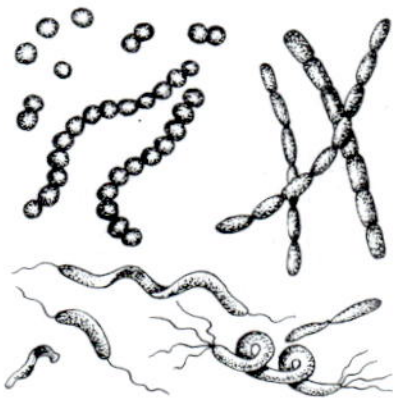

Bacteria

Paramecia can get their nutrients from bacteria. They get proteins, sugars, and starches from the bacteria. They get fats, minerals, and vitamins as well. This is how paramecia get energy to live.

Do grasshoppers need food for energy?● Or beetles? What do cotton boll weevils eat?

Here is what these beetles eat.▲

Do beetles also need nutrients? Do they need sugar, proteins, and vitamins just as you do? Of course, you know they do.

What Foods to Eat?

Some scientists study the way the body uses nutrients. They can tell us how much of each nutrient we need each day. They tell us we need some proteins each day. And some minerals and some vitamins. We should have some of *all* of the nutrients each day.

See if you can name a food rich in proteins. Think of something you like to eat. Study the chart of foods on pages 74–75 carefully. ◆ Is the food you named on the chart?

Most foods give us *more* than one nutrient. Look again at the food chart. Find a food you like that has proteins *and* minerals. Can you find a food rich in proteins? *And* one that also has *B vitamins?* Beef, liver, and pork are rich in proteins. And they have minerals and B vitamins.

What kinds of foods are rich in fats? Fats can also be used to give you energy. Name some foods rich in fats. Then check the chart.

Now, pretend you are planning the dinner tonight. What will you pick to eat? Pick out the foods that will give you many of the nutrients you need. You can pick meat, fish, eggs, or beans for proteins. All of these also have some fats. And supplies of B vitamins and minerals.

What special nutrients would vegetables and fruits supply? Study the chart on pages 74–75. ◆ Which vegetables and fruits are rich in *vitamin A?* Can you say which are rich in vitamin C? Some vegetables and fruits also have B vitamins.

What other nutrient do many fruits and vegetables supply? That's right, carbohydrates. Many fruits and vegetables have sugars and starches. Carbohydrates are needed to give you energy. Now pick two vegetables you like. Add these to your dinner.

What nutrient haven't we talked about? Water. If you eat some uncooked fruits or vegetables each day, you will get water. You also should drink water each day.

What do you think of when you plan a meal? You need some foods rich in proteins. You also need some to supply carbohydrates *or* fats. And you need minerals, vitamins, and water daily.

FOOD	NUTRIENT	SUPPLY
Apples (raw)	Carbohydrates	+++
	Fats	
	Minerals	+
	Proteins	
	Vitamin A	++
	B vitamins	+
	Vitamin C	+
Beef (cooked)	Carbohydrates	
	Fats	++++
	Minerals	++++
	Proteins	++++
	Vitamin A	+
	B vitamins	+++++
	Vitamin C	
Bread	Carbohydrates	+++++
	Fats	+
	Minerals	+++
	Proteins	+++
	Vitamin A	
	B vitamins	++++
	Vitamin C	
Broccoli (cooked)	Carbohydrates	+
	Fats	
	Minerals	+++
	Proteins	++
	Vitamin A	++++
	B vitamins	+++
	Vitamin C	+++++
Butter or margarine	Carbohydrates	
	Fats	+++++
	Minerals	+
	Proteins	
	Vitamin A	+++++
	B vitamins	
	Vitamin C	
Cabbage (cooked)	Carbohydrates	+
	Fats	
	Minerals	++
	Proteins	+
	Vitamin A	++
	B vitamins	+
	Vitamin C	++++
Cantaloupe and casaba (raw)	Carbohydrates	++
	Fats	
	Minerals	+
	Proteins	
	Vitamin A	++++
	B vitamins	++
	Vitamin C	+++
Carrots (raw)	Carbohydrates	+++
	Fats	
	Minerals	++
	Proteins	+
	Vitamin A	+++++
	B vitamins	++
	Vitamin C	++
Cereals (dry)	Carbohydrates	+++++
	Fats	
	Minerals	+++
	Proteins	+++
	Vitamin A	
	B vitamins	+++++
	Vitamin C	

FOOD	NUTRIENT	SUPPLY
Cheese (American)	Carbohydrates	+
	Fats	+++++
	Minerals	+++++
	Proteins	+++++
	Vitamin A	++++
	B vitamins	++
	Vitamin C	
Chicken (cooked)	Carbohydrates	
	Fats	+
	Minerals	+++++
	Proteins	+++++
	Vitamin A	++
	B vitamins	+++++
	Vitamin C	
Collards (cooked)	Carbohydrates	++
	Fats	
	Minerals	++++
	Proteins	++
	Vitamin A	+++++
	B vitamins	++++
	Vitamin C	++++
Dried fruit	Carbohydrates	+++++
	Fats	
	Minerals	+++
	Proteins	++
	Vitamin A	+++
	B vitamins	+++++
	Vitamin C	+
Eggs (cooked)	Carbohydrates	
	Fats	+++
	Minerals	++++
	Proteins	++++
	Vitamin A	++++
	B vitamins	++
	Vitamin C	
Fish (cooked)	Carbohydrates	
	Fats	++
	Minerals	+++++
	Proteins	+++++
	Vitamin A	++
	B vitamins	+++++
	Vitamin C	
Frankfurters (cooked)	Carbohydrates	+
	Fats	++++
	Minerals	+++
	Proteins	++++
	Vitamin A	
	B vitamins	++++
	Vitamin C	
Garbanzos (cooked)	Carbohydrates	+++++
	Fats	++
	Minerals	+++++
	Proteins	+++++
	Vitamin A	+
	B vitamins	++++
	Vitamin C	
Green peppers (raw)	Carbohydrates	++
	Fats	
	Minerals	+
	Proteins	+
	Vitamin A	+++
	B vitamins	++
	Vitamin C	+++++

The +'s in the chart tell you the supply of nutrients in foods. + means **a very small supply,** ++ means **a small supply,** +++ means **a fair supply,** ++++ means **a good supply,** +++++ means **a very good supply.**

FOOD	NUTRIENT	SUPPLY
Ice cream	Carbohydrates	++++
	Fats	+++
	Minerals	++++
	Proteins	+++
	Vitamin A	+++
	B vitamins	+
	Vitamin C	+
Lamb (cooked)	Carbohydrates	
	Fats	++++
	Minerals	+++
	Proteins	++++
	Vitamin A	
	B vitamins	+++++
	Vitamin C	
Lettuce (raw)	Carbohydrates	+
	Fats	
	Minerals	++
	Proteins	+
	Vitamin A	++++
	B vitamins	++
	Vitamin C	+++
Liver (cooked)	Carbohydrates	++
	Fats	+++
	Minerals	+++++
	Proteins	+++++
	Vitamin A	+++++
	B vitamins	+++++
	Vitamin C	++++
Lunch meat	Carbohydrates	+
	Fats	++++
	Minerals	++++
	Proteins	++++
	Vitamin A	++++
	B vitamins	++++
	Vitamin C	
Milk	Carbohydrates	++
	Fats	+
	Minerals	++++
	Proteins	+++
	Vitamin A	++
	B vitamins	+
	Vitamin C	+
Noodles (cooked)	Carbohydrates	++++
	Fats	+
	Minerals	++
	Proteins	+++
	Vitamin A	+
	B vitamins	+++
	Vitamin C	
Oranges (raw)	Carbohydrates	+++
	Fats	
	Minerals	++
	Proteins	+
	Vitamin A	+++
	B vitamins	++
	Vitamin C	+++++
Pancakes	Carbohydrates	+++++
	Fats	++
	Minerals	+++++
	Proteins	+++
	Vitamin A	+++
	B vitamins	+++
	Vitamin C	

FOOD	NUTRIENT	SUPPLY
Peanut butter	Carbohydrates	+++
	Fats	+++++
	Minerals	+++++
	Proteins	+++++
	Vitamin A	
	B vitamins	++++
	Vitamin C	
Pizza (with cheese)	Carbohydrates	++++
	Fats	++
	Minerals	+++++
	Proteins	++++
	Vitamin A	+++
	B vitamins	+++
	Vitamin C	++
Plantains (cooked)	Carbohydrates	+++++
	Fats	
	Minerals	+
	Proteins	+
	Vitamin A	+++
	B vitamins	++
	Vitamin C	++
Pork (cooked)	Carbohydrates	
	Fats	+++++
	Minerals	++++
	Proteins	+++++
	Vitamin A	
	B vitamins	+++++
	Vitamin C	
Rice (cooked)	Carbohydrates	++++
	Fats	
	Minerals	+
	Proteins	++
	Vitamin A	
	B vitamins	+++
	Vitamin C	
Spaghetti (cooked)	Carbohydrates	++++
	Fats	
	Minerals	++
	Proteins	+++
	Vitamin A	
	B vitamins	++++
	Vitamin C	
Sweet potatoes (cooked)	Carbohydrates	+++++
	Fats	
	Minerals	+++
	Proteins	++
	Vitamin A	+++++
	B vitamins	+++
	Vitamin C	+++
Tomatoes (raw)	Carbohydrates	++
	Fats	
	Minerals	+
	Proteins	+
	Vitamin A	++++
	B vitamins	+++
	Vitamin C	+++
Turnips (cooked)	Carbohydrates	++
	Fats	
	Minerals	++
	Proteins	
	Vitamin A	
	B vitamins	+
	Vitamin C	+++

How Are Nutrients Used?

Now let's go on to find what happens to the nutrients. What happens to the nutrients inside you?

The nutrients in the food you eat get to your cells. Some of the nutrients are used to get *energy*. Your cells use some of the energy to make more *ATP*. Cells use the energy in nutrients to make ATP.

For example, the cells use sugar to get energy. This energy is used to build ATP. Extra sugar is stored in the body. Some of this extra sugar may be changed to fat.

Try this. Close your fist tight. Now open it. Your muscle cells are doing work. They need energy to do this work. Where does this energy come from? You know the answer to this one. ■

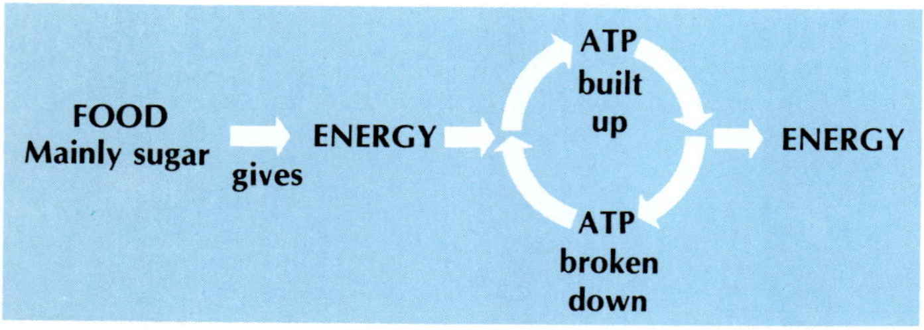

■

The energy comes from ATP. ATP breaks down to give energy. And cells use this energy to do their work. Your cells are using ATP now. They are working.

Do All Cells Do Work?

Yes. It is true that *all* cells do work. The cells in an insect or a tree do work. The cells in an earthworm or in a robin do work. The cells of these animals need the same things as yours do to do work. Their cells need nutrients.

Do paramecia have the same needs that you have? Do yeast cells have the same needs? Do they all need nutrients—just as you do?

They do. The insect, the earthworm, and the tree all need the same nutrients. They need these nutrients for energy to work and grow. They need these nutrients to live. So do you. You are using these nutrients now.

●

A Try at Prediction

A duck hatches out of a duck's egg. All its food is in the egg. Can you say what nutrients must be in the egg?

Would a snake's egg have the same kinds of nutrients?

Would an insect's egg have the same kinds of nutrients?●

Look at this tiny plant growing from the peanut.▲ A peanut plant is growing from the peanut. What nutrients do you think are in the peanut?

▲

An Exercise in Explanation

1. Suppose your friend picks cake and orange drink for lunch. You pick a meat sandwich and milk. Who has the more healthy lunch—you or your friend? Why?

2. Why is it that fish, eggs, or cheese can be eaten instead of meat at a meal? Check the chart on pages 74–75 before you answer.

What fruits or vegetables have some vitamin A, vitamin C, and some B vitamins? Check the chart on pages 74–75 before you answer. Why would it be good to pick them for a meal?

3. What makes *you* like some foods? And dislike others? Do you think your early training has anything to do with this?

Look at the charts on pages 74–75. Which foods have you never tasted? Why?

AN INDEPENDENT STUDY

A Need for Energy— A Likeness in Cells

All of you are made of cells. You might say you are all walking bundles of cells.

A whale is a swimming bundle of cells. ■ A bird is a flying bundle of cells. All are made of millions of cells too small to see.

This next part you can do on your own. Read each question on this page. Then write your answer *on your answer sheet.* Check each answer by looking across the page.

Check each answer before you start the next question.

1

Your hand is made up of very small parts called c----.

(Write your answer down. Put it next to number 1 on your answer sheet. Now, look across the page to box 1 for the answer. Are you right?)

2

Some of your cells are muscle cells. These cells help you move and lift things. They help you walk and work.

These cells are called m----- cells.

(Write your answer down by number 2 on your answer sheet. Check across the page to see if you are right.)

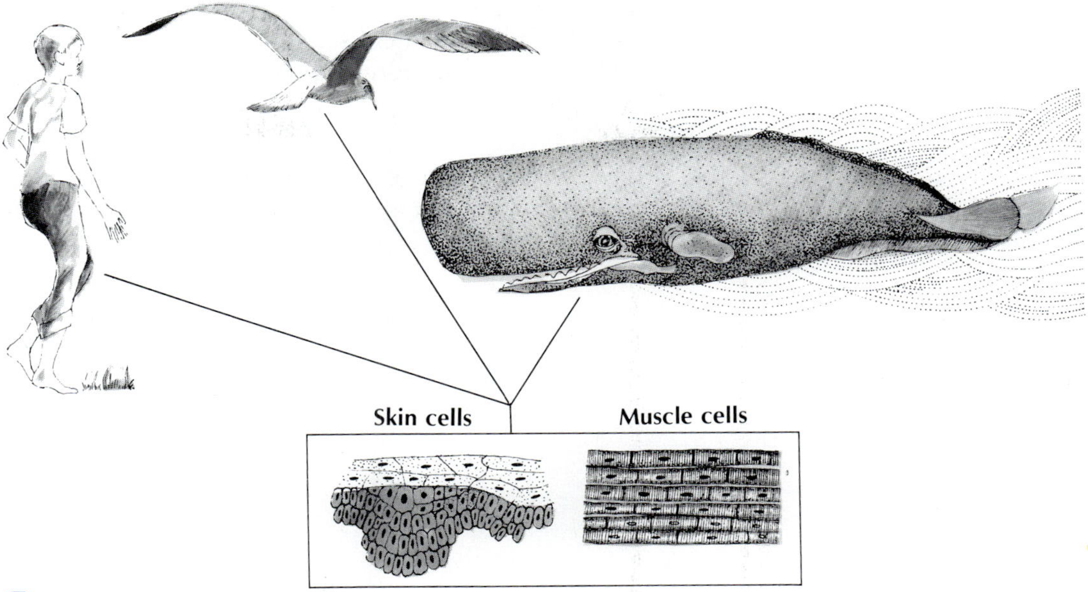

1

cells

Living things are made up of cells.

(Go on to question 2.)

2

muscle

(Now turn the page to question 3.)

3

Yes. Muscle cells. A bird can fly because it has muscle cells. A whale can swim because it has ______________ ______________.

(Write your answer down. Then check across the page to see if you are right.)

4

These muscle cells need energy. Any cell needs energy. You are more than walking bundles of cells. You are bundles of cells that need energy.

Where do cells get this energy—from *food* or from *air?*

(Write your answer on your answer sheet. Then, check across the page to box 4 to see if you are right.)

5

You know that energy is in sugar. And that many foods have sugars in them. Which food would give you more energy? an *apple* or a *lettuce leaf? Which* do you think?

(Write your answer on your answer sheet. Now check your answer across the page in box 5. Were you right?)

3

muscle cells

(Go on to question 4.)

4

from food

Yes. You know this. Some foods give you more energy than others.

(Go on to question 5.)

5

an apple

(Now turn the page to question 6.)

6

Why would an apple give more energy than a lettuce leaf? Because one apple has more sugar than a whole head of lettuce.

Why is sugar important? Because there is energy in the nutrient sugar. Sugar is a nutrient rich in ------.

7

Energy is needed to build up the chemical ATP in cells. Which nutrient do cells use for energy to build up ATP—*sugars* or *minerals?*

(Write your answer on your answer sheet. Then check across the page to box 7 to see if you are right.)

8

Then ATP breaks down to give energy. You need this energy for everything you do. You move a lot. Your heart beats. Your blood moves. All this takes energy. Scientists say you should eat breakfast. It is the meal that gives you energy to start the day.

Why should you eat breakfast? For ------.

(Write your answer on your answer sheet. Then check across the page to box 8 to see if you are right.)

6

energy

(Go on to question 7.)

7

sugars

Sugars have energy. This energy is used to build up ATP in your cells.

(Go on to question 8.)

8

energy

(Now turn the page to question 9.)

9

A whale needs energy just as you do. So does a bird. And every other living thing you can name. A whale grows. And so does a bird. And every other living thing grows, or makes new cells.

Most foods have sugars, starches, and proteins. The sugars and starches are mainly for ------. The proteins are mainly needed for making new c----.

(Write both your answers on your answer sheet. Check across to box 9 to see if you are right.)

10

Without proteins new cells cannot be made. The nutrient protein helps you to grow. Your body uses proteins to build new cells. The main nutrient you need for growth is --------.

11

We have been talking about animal cells—whales, birds, and you.

You have seen grass growing. Grass is also living. Grass is a plant. Does *grass* use sugars, starches, and proteins? And vitamins and minerals, too? *Yes* or *no?*

(Write your answer on your answer sheet. Then check across the page to box 11.)

9

energy
cells

(Go on to question 10.)

10

proteins

(Go on to question 11.)

11

yes

Grass is a living plant. Both plants and animals use nutrients to live.

■

●

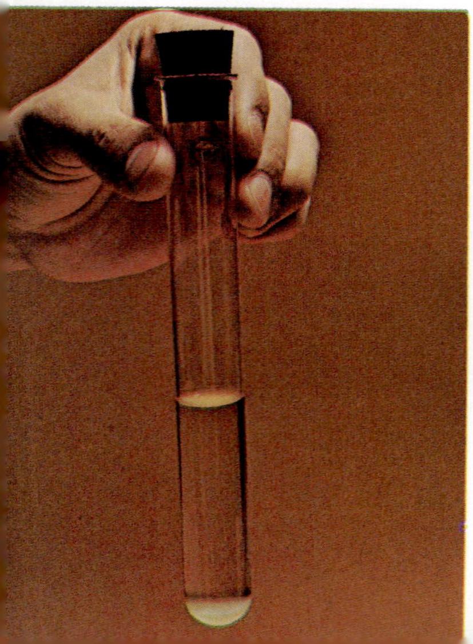

▲

2. WITHIN THE BODY: A BIT OF EGG WHITE

When you break open an egg, you find part of it is yellow. And part of it is white. The *white part* of an egg is *protein*.

Where does the chewed egg white go when you swallow it? It goes down to the *stomach*.

What happens to the egg white in your stomach? Two boys in Michigan tried to find out. They carried out their own **investigation** (in·VES·tih·gay·shun). Here is what they did.

What Happens to Egg White?

First, the two boys boiled an egg. Then they took off the shell. They peeled off the white part and chopped it. Then they put some in a test tube. ■

Next, they added a mix. This mix is almost like the juice in *your stomach*. The juice in your stomach is a water-like mix, called **gastric** (GAS·trik) **juice.** It is made of several things. It is made of **pepsin** (PEP·sin). It also has water. And it has a little acid (AS·id), called **hydrochloric** (hy·druh·KLOR·ik) **acid.** Pepsin, water, and hydrochloric acid are all chemicals.

The juice in your stomach is mainly made of: *pepsin, water,* and *hydrochloric acid.*

The boys made gastric juice by filling the test tube half full of water. To this, they added a pinch of pepsin powder. Then they added two or three drops of weak hydrochloric acid. They then had a mix something like the gastric juice in the stomach.

They poured the juice over the chopped egg white. Next they put a stopper in the test tube. ●

After a few days, the boys checked the test tube. They found the chopped egg white was gone. And the water was cloudy. ▲

This was quite a change. What do you think had happened? The juice had broken down the pieces of hard-boiled egg white. This breaking down of any food is called **digestion** (die·JES·chun). The egg white was digested. The egg white was changed, or broken down. This is what happens to proteins in your stomach. Proteins are changed in your stomach.

But we are getting ahead of the story. We need to ask some questions. And we need to find some answers. So let's go on.

Which Does the Job?

Which part of the juice breaks down the proteins in the egg white? Which part of the juice *digests,* or changes, the egg white?

Maybe it is the *pepsin?* Or is it the *weak acid?* It may be the *water.*

You can find out for yourself. You can make gastric juice.

Perhaps you will work alone. Or you may work in small groups to do this investigation. Here is what you can do.

Put the *same* amount of chopped egg white into eight test tubes. Next add about an inch of water to each test tube. Now you have test tubes of chopped egg white and water. Leave test tubes 1, 2, 3, and 4 alone now.♦

Mix some *pepsin* in water in a jar. Stir it so it mixes well. Add an inch of the pepsin mix to test tubes 5, 6, 7, and 8.♦

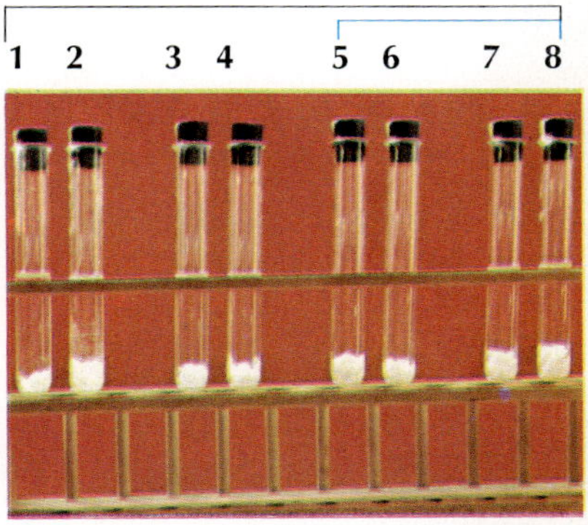

Now leave all the test tubes alone except 7 and 8. To tubes 7 and 8, add two or three drops of weak *hydrochloric acid.*★

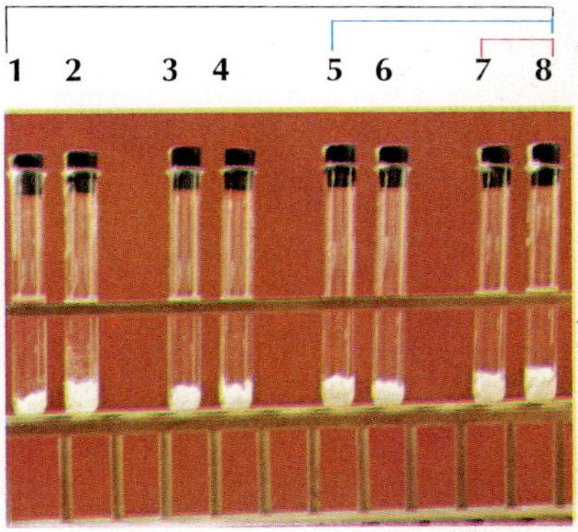

Now test tubes 7 and 8 are the only ones that have egg white + water + pepsin + weak hydrochloric acid. A mix of water, pepsin, and weak hydrochloric acid is like gastric juice. Test tubes 7 and 8 have a mix like gastric juice and egg white.

Let us now go back to test tubes 1, 2, 3, and 4. They have only water and egg white. Add two or three drops of weak hydrochloric acid *only* to tubes *3 and 4.*◇ These two tubes have only egg white, water, and hydrochloric acid. Test tubes 1 and 2 have only egg white and water.

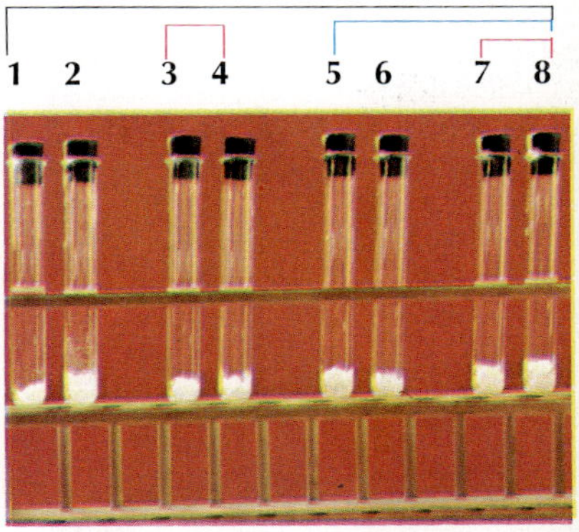

Now cork each test tube. Keep the tubes corked. Put labels on all the test tubes. List on the label what is in each test tube. You can keep a chart of what happens. Perhaps the chart shown is one you could use. ■

TEST TUBE	WHAT HAPPENED
1	
2	
3	
4	
5	
6	
7	
8	

■

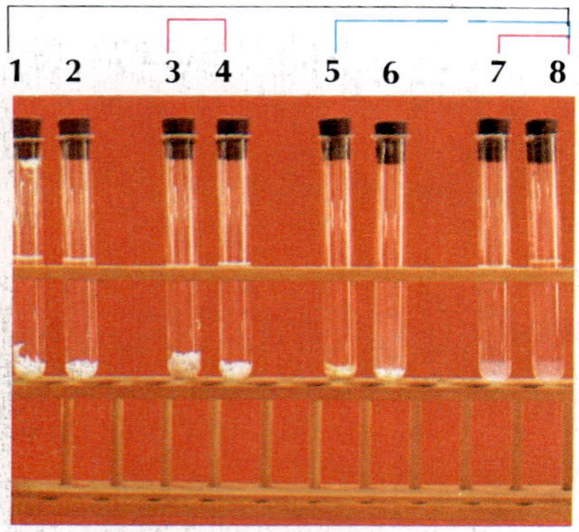

Keep all the test tubes in a warm place. After two days, look at what you have. Which part in gastric juice breaks down egg white? List the changes on the chart next to the numbers of the test tubes.

Let us see what happens in each test tube. ●

Does water alone change the egg white? (Look at test tubes 1 and 2.)

Does the hydrochloric acid with water change the egg white? (Look at test tubes 3 and 4.)

Does pepsin with water change the egg white? (Look at test tubes 5 and 6.)

Is the egg white changed by pepsin, hydrochloric acid, and water together? (Look at test tubes 7 and 8.)

What do you find? Do others in your class get the same results? (Be sure to write on your chart what you find in the tubes.)

Pepsin Digests Egg White

The changes you can see are in test tubes 5 and 6, and in test tubes 7 and 8.

Where is *most* of the change? Most of the change you see is in tubes 7 and 8. These are the tubes that have *both* pepsin and hydrochloric acid.

Now look again at test tubes 5 and 6. These are the test tubes where only *pepsin* was added. Pepsin alone digests, or changes, egg white a little, doesn't it? You can see the egg white is changed by pepsin. Look again at tubes 5 and 6.

Chemicals That Help

Pepsin is an **enzyme** (EN·zime). An enzyme is a chemical that helps a change take place.

In your study, pepsin is the enzyme. Pepsin changes the egg white. Pepsin makes the breakdown of egg white (protein) go faster. Pepsin digests egg white.

You also found that the enzyme pepsin works even faster with hydrochloric acid.

Look again at tubes 7 and 8. The acid in your gastric juice helps digestion go on better.

If you could see inside your stomach, you would see this change happen in a few hours.▲ Some proteins are digested in your stomach in a few hours. In the test tube, it takes a few days.

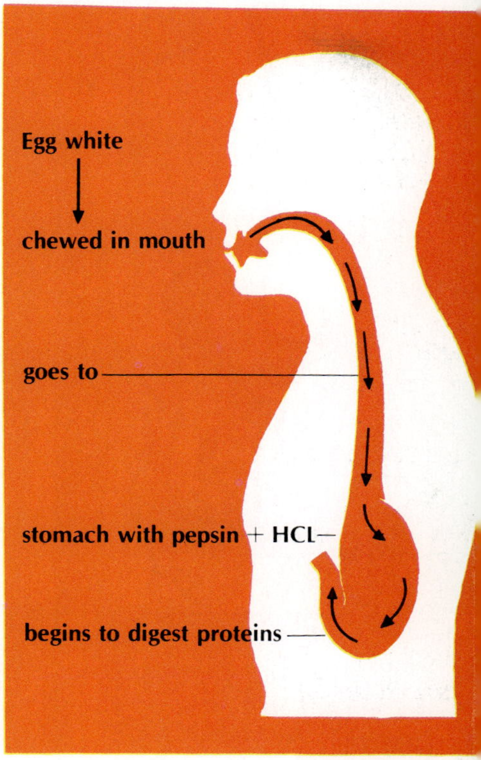

▲

A Try at Prediction

1. Pepsin is an *enzyme* that digests proteins, like egg white.

Would you expect pepsin to digest meat?

What is your "educated guess"? Add pepsin (with some hydrochloric acid) to a little chopped meat in a test tube. What do you find?

2. Some people can live well after the stomach has been removed. How can this be? Where are proteins digested if a person has no stomach? (Hint: Find out where food goes after leaving the stomach.)

An Exercise in Explanation

When fresh pineapple juice is added to chopped hard-boiled egg white, the egg white is digested. It is broken down. Can you say what kind of an enzyme is in the pineapple juice?

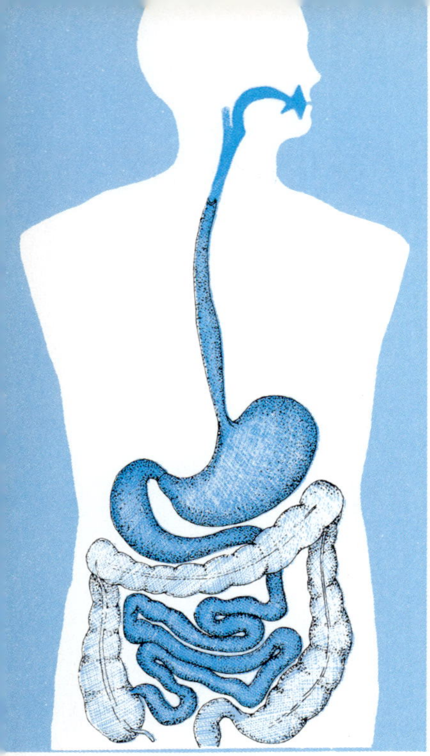

■

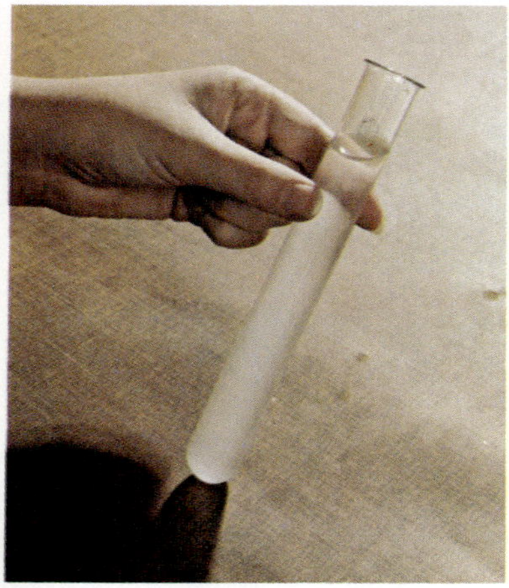

▲

◆

3. WITHIN THE BODY: A BIT OF STARCH

Look at it this way. All the food you eat goes into a long food tube. Here is a picture of the food tube. ■ The food you swallow goes into this food tube. The food does you no good if it stays there. It must get to your cells. Why?

Remember how the egg white was broken down? It became a cloud of digested proteins. Many of the nutrients you eat need to be digested. They need to be much smaller.

The foods you eat must be made very small. So small that they can pass through your food tube and get into your cells. This is why most nutrients need to be digested.

What nutrients must be digested? You eat foods that have fats and proteins. These fats and proteins must be broken down by enzymes. Carbohydrates like starches and sugars must be digested. They must be broken down. By what? By enzymes.

You saw what happens to proteins in the *enzyme pepsin*. Pepsin breaks down proteins like egg white. What happens to *starch* in your food tube? Does the same thing happen to starch as to proteins? Now see for yourself.

Starch and an Enzyme in You

To begin with, you will need some starch. You will also want to make a chart like the one on the page opposite. ●

For starch, you can use arrowroot (AR·oh·root). Arrowroot is a starch used for cooking. Mix a little of the starch in cold water. Then add more water. This makes a very thin water-like mix. Heat a little. Pour some of this *very thin starch mix* into a tall test tube. ▲ Add a drop or two of *very, very* weak iodine. Do you see a blue color? ◆ Iodine is used to test for starch. If there is a starch in a mix, it turns blue when you add iodine. *Save* this mix. (See A on your chart.)

Let's go on.

IN TEST TUBE	COLOR CHANGE	EXPLANATION
A–Starch paste and iodine		
B–Starch paste, iodine, and saliva		
C–Starch paste, with membrane, put in jar of water Iodine added to water (test for starch)		
D–Syrup, with membrane, in jar of water Benedict's liquid added to water and heated (test for simple sugar)		
E–Starch paste and saliva, with membrane, in jar of water Iodine added to water to test for starch Benedict's liquid added to water and heated (test for simple sugar)		
F—Starch paste and Benedict's liquid and heated		
G—Saliva and Benedict's liquid and heated		

●

Now collect some **saliva** (sa·LIE·vah), the liquid in your mouth. Put the saliva in a small test tube. ★ Now pour the saliva into the tube of light blue starch and iodine. ◊ Cork the tube and shake it upside down a few times. Is the blue color gone? ⊙ (See B on the chart.)

If there is no blue color now, what has happened to the starch? Is the starch gone? Where does this change happen in us? Why does the starch have to be changed? You'll see.

★

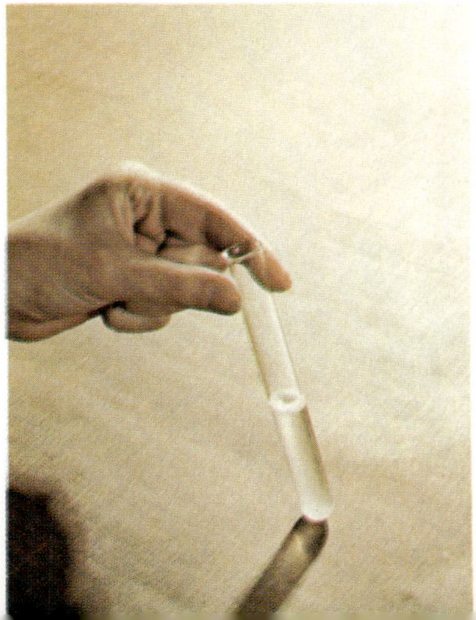

◊

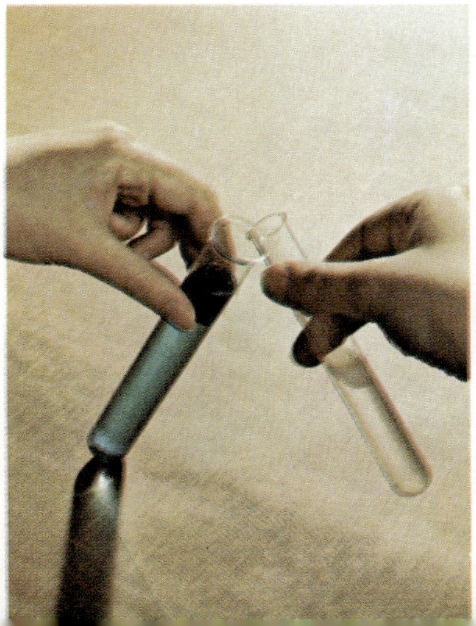

⊙

Starch and Sugar and a Membrane

Again mix a little starch in cold water. Add more water to make the mix very thin. Heat a little. Pour some of the mix into a test tube.

This time, cover the mouth of the test tube. Your teacher will give you a thin skin called a *membrane*. This membrane is like the lining of your mouth and food tube.

Wet the membrane. Cover the mouth of the test tube with it. ■ Use a rubber band to keep the membrane over the mouth of the tube.

Now, turn the tube upside down. Be sure the starch paste does not run out. Wash off the outside of the tube, too. Next, put the tube into a small jar. Pour about an inch of water into the small jar. Does your setup look like this? ● (See C on your chart.)

■

Now fix another test tube the same way. This time use a kind of sugar, called *simple sugar*. Simple sugars are in most syrups. Pour some syrup mixed in water into a test tube. Cover the test tube with a wet membrane. Wash off any syrup from the outside of the test tube. Fill another jar with an inch of water. Turn the tube of syrup upside down into this jar. ▲ (See D on your chart.)

Now for the third and last jar. Put about one inch of thin starch mix into a test tube. Now add about an inch of your saliva. Cover the mouth of the tube with a wet membrane as you did before. Shake the test tube to mix the starch and saliva. Put the test tube upside down into the jar with an inch of water. The setup should look like this. ♦ (See E on your chart.)

Now you have three setups.

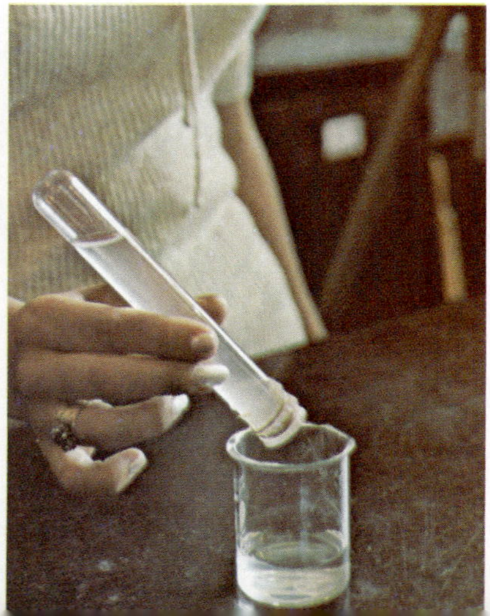

●

▲

Which Goes Through the Membrane: Starches or Simple Sugar?

Let's think together. You know that the food you swallow goes into your food tube. This food must be digested. It must be small enough to get to your cells. How does food get to your cells?

Think of the test tubes you just put in place. Think of them as very much like your food tube. The membrane over each tube is like the thin skin inside your food tube.

Can the starch go through this membrane into the water in the jar? What do you think? Can *starch* get out of your food tube?

There is syrup in another test tube. The syrup is made of simple sugar. Can this *sugar* go through the membrane into the water?

Is the sugar made of small enough **molecules** (MOL·uh·kyoolz)? *Molecules are small bits of matter.* Molecules are the *smallest* bits of any nutrient you eat.

Can these starch and sugar molecules go through the membrane? That is, can they both leave your food tube without being digested?

Let's see.

Most of the food you eat must be *digested.* It must be broken down into smaller molecules. Food must be broken down into smaller bits after you eat it. Why? ★

Nutrients have to be in small enough molecules to go through *membranes* of your food tube. From the food tube they go to your cells. They also must be small enough to get into your cells. That is where the nutrients are needed.

Much of the food in your food tube must be digested to get out. Then the very small molecules can go to the billions of cells that make up *you.*

★

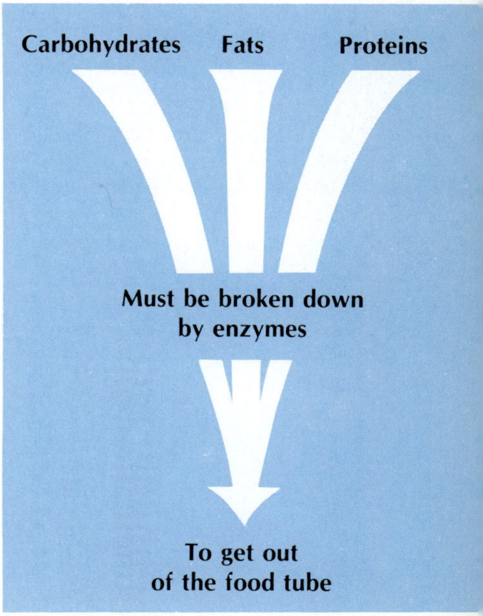

Through the Membrane

Which molecules—starch or these sugar molecules—can go through the membrane? Look at the test tube with starch. If *starch* can go through the membrane, starch should be in the water in the jar. You know this.

■

●

▲

Add a few drops of iodine to the water. The picture shows you how. ■ Does the water turn blue? No, the water does *not* turn blue. The water is the color of the iodine. This tells us there is *no* starch in the water. Starch is a nutrient that can't go through a membrane. (See C on your chart.)

Now to the sugar in the syrup. Do these sugar molecules go through the membrane? There is a special way to find out. You can use Benedict's (BEN·ih·dikt's) liquid as a test. Benedict's liquid is a test for simple sugar. You can use it. It can test for the kind of sugar in the syrup.

Pour a little of the water from the jar that has the syrup setup into a clean test tube. Add a little Benedict's liquid to this. Next heat the test tube carefully over a Bunsen (BUN·sin) burner. Be sure to point the test tube away from you as you heat it. ●

What happens in the test tube? Does the blue color of the Benedict's liquid change? Yes, it does. It becomes green or orange. ▲ What does this color change show you? This is a test for simple sugar. The green or orange color shows that sugar molecules are in the water in the jar.

If only a *little* sugar is in the water you test, Benedict's liquid turns *green*. If there is a lot of sugar, then it turns *orange*. Can you use this to tell how much sugar has gone through the membrane? How? (See D on your chart.)

Simple sugars do go through membranes. That is clear enough. Sugars like those in some kinds of pancake syrups go right through the food tube. They pass through the membrane of the food tube. The starch in the pancakes, however, stays in your food tube for awhile. What happens to it?

You already know that saliva breaks down starch.

What Happens to Starch?

Remember that you have a test tube with *starch and some of your saliva*. Has something passed through the membrane in this tube? Let's find out.

Use *iodine* to test the water in the jar *for starch*. And

use *Benedict's liquid* to test for *simple sugar*. Here is the way you can make your tests.

Take a sample of water from the jar. Pour the water into a test tube. Add iodine to the water. ♦ If there is starch, the water will turn blue. The water does not become blue. No *starch* seems to be present. Did you expect that? Starch does not go through the membrane. (See E in your chart.)

Take some more water from the jar. Pour it into a clean test tube. Now test it *for simple sugar*. Add some blue *Benedict's liquid*. Heat the test tube carefully. What happens to the water? Does the blue change color? Yes. The water turns a green or an orange color. ★ What does this show you? There is now a simple sugar in the water. (See E in your chart.)

Well, this becomes interesting. Something has changed. There wasn't any sugar in the starch paste when you started.

If you want to check this, you can. Just add some Benedict's liquid to the first thin starch mix you had. Heat it carefully. No, there is no simple sugar in the starch mix. (See F in your chart.)

Do the same with your saliva. Add some Benedict's liquid. Heat this in a test tube. You should not find sugar in your saliva. (See G in your chart.)

This is the story. ◊ Saliva changes, or breaks down, starch into simpler bits. That is, your saliva begins to *digest starch*.

Starch is broken down into small molecules of simple sugar. Simple sugar is not trapped in the food tube. *Simple sugar does not need to be digested. It can go through thin membranes. It can go through the thin membranes of your food tube.*

The starch you eat is digested. It is broken down into molecules of a simple sugar. It is broken down by enzymes in saliva and by other enzymes in the body. These molecules of simple sugar can then pass through the membranes of your food tube.

You have a few enzymes that digest starch. These enzymes are found not only in the mouth, but also in other parts of the food tube.

♦

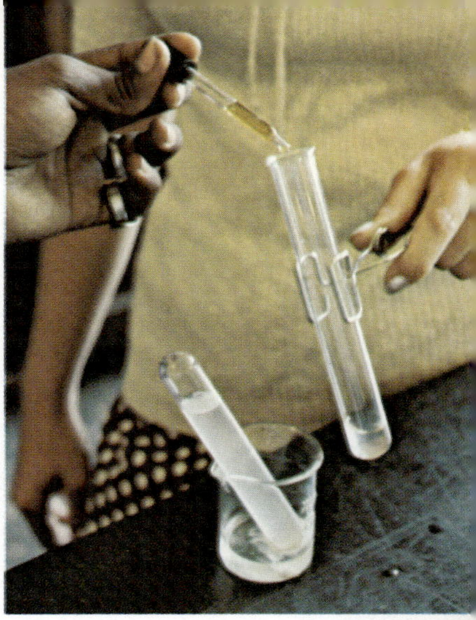

★

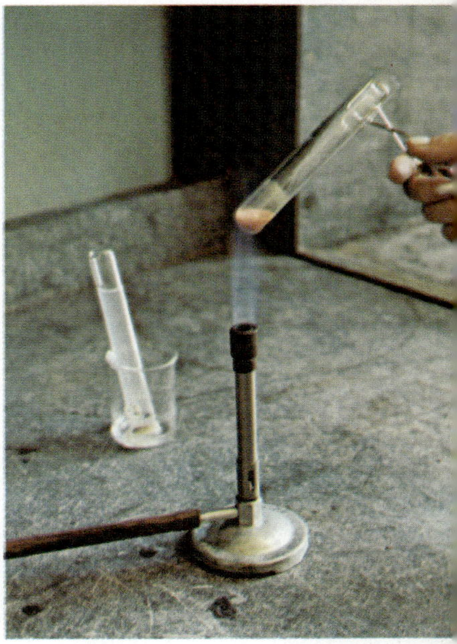

◊

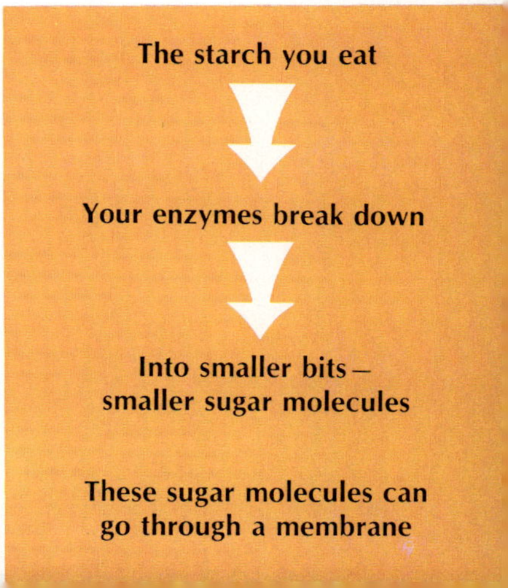

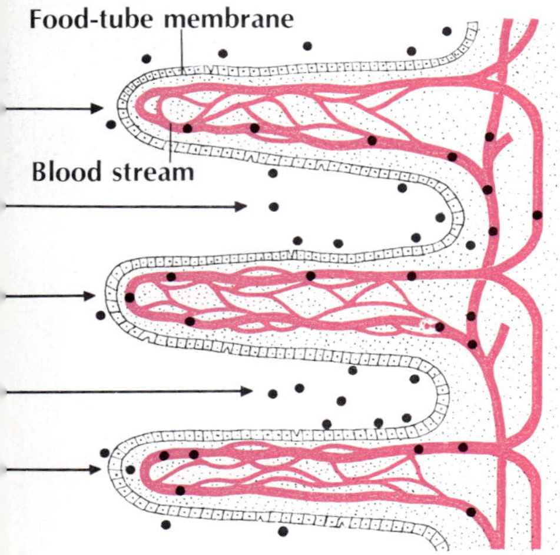

Out of the Food Tube into the Blood Stream

All the food an animal eats goes into its food tube. The food could be steak for you. The food could be insects for a woodpecker. ■

In the food tube of an animal, most of the food is digested. That means it is broken down into smaller bits. These smaller bits, or molecules, can go through the thin membrane of the food tube.

What happens to your food after it is digested?

The digested nutrients go through the food-tube membrane into the blood stream. ● The blood stream carries the digested nutrients to cells. The cells then have the nutrients to keep you alive. This is also how the woodpecker stays alive. Its cells get digested nutrients.

An Exercise in Prediction

1. A lion may eat a rabbit. The rabbit meat is protein.

Can you predict what happens to the meat in the lion's stomach?

What likeness do you think there is in a lion's food tube and in your food tube? Why is digestion necessary?

2. A tiny plant like bread mold lives on bread or other starch foods. Starch in the bread cannot get through a membrane. This means starch in the bread cannot get into the bread mold.

How do you think the bread mold gets its food?

3. Did you know that some plants eat insects? These plants eat insects. ▲ They use the proteins in the insects.

What enzyme do you think these plants have a lot of? Why do the plants need this enzyme?

4. A PATTERN—LIKENESSES IN ANIMALS

There is a pattern in many animals. There is a likeness in what happens to the food they eat. Animals eat food; the food goes into the food tube. The food is digested into smaller bits. Then the blood stream takes these molecules of nutrients to the cells.

What parts of the body make all this possible? In other words, where does the food go?

Let's go on and see.

Looking Inside

Soon you will *dissect* (die·SEKT) a frog.♦ This means you will open a frog. And you will look inside it. Then you will understand a lot more about the workings of your own body.

Think of what you might see inside the frog. The **food tube,** of course. And other organs, like the **liver** and **pancreas** (PAN·kree·us). These organs help digest food. They send juices into the food tube.

You will also see the pump that is so important. The pump is the **heart.** The heart pumps the blood around the body.

Look for the tiny tubes that carry the blood around the body. These are **blood vessels.** Look to see how they branch out. They go to all the *organs* of the frog.

Each organ has cells. The cells need the life-giving food molecules that the blood carries. Each organ has cells that must have nutrients.

Do you have your frog ready?

AN INVESTIGATION:

Dissecting A Frog

Place the frog on its back in a waxed pan. Pin down its front legs. Then pin down the hind legs in the wax.

Use tweezers (TWEE·zerz) to lift the skin between the hind legs. Then begin to cut through the skin. The picture shows you where to cut. Begin with the cutting line marked (1) in the picture. ■ Now you have cut the skin of the belly of the frog up to the chin. Next, cut as shown along lines (2) and (3). Then cut lines (4) and (5). You have now cut the skin in both pairs of legs. Now, pin back the skin. ■

Notice how thin the skin is. Look on the inside of the skin. ● Can you see the many small *blood vessels?* Frogs do most of their breathing through this thin, wet skin.

Oxygen can go through thin, wet skin. Frogs have very, very small lungs. They get most of the oxygen they need through their wet skin. This is how they get along with small lungs.

Under the skin you find shiny white **muscles.** ● What is under the muscles? All the organs.

Start again to cut the muscle between the hind legs of the frog. Use the tweezers to lift up the muscle. Make a small cut in the muscle. Use scissors (SIZ·erz) to do this. Cut very slowly. Take the tweezers and lift the muscles. Be sure to lift up the muscle as you cut. This is so you *do not* cut the organs underneath.

Cut up to the front legs. ▲ Watch as you get near the front legs. You will hit a bone. Carefully cut through this bone up to the chin. The heart is under this bone. Now take your thumbs and spread apart the front legs.

Can you see the *heart* in the thin bag? ♦ Look at the large blood vessels around the heart.

Now you may have to hunt a bit. Under the heart there are two thin brown bags. They are the **lungs.**

There is one very large organ in the body. Which organ is it? It is the *liver*. Can you see the dark liver? ♦ It is divided into several parts. Now turn the liver over. You will see a big green bag, the **gall bladder** (GALL·BLAD·er).

If your frog is a female, it *may* have lots of eggs. ♦ Each mass of eggs makes up part of an **ovary** (OH·vah·ree). An ovary makes eggs. If there are eggs, take them away. Then you can see more of the organs.

■

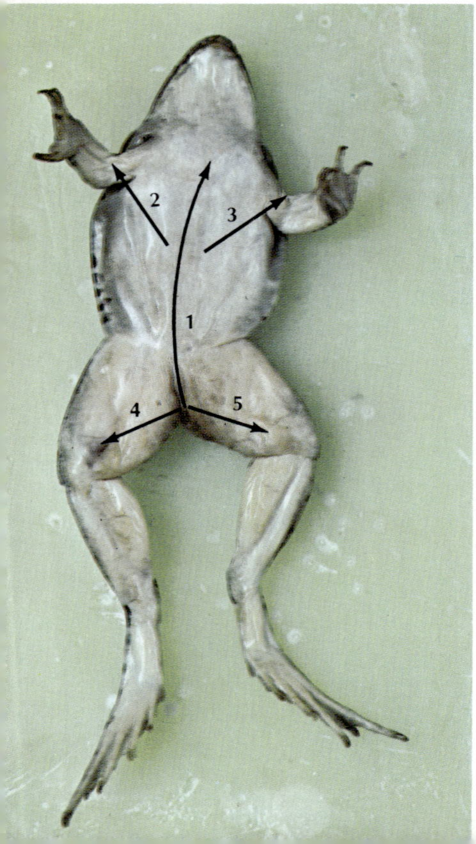

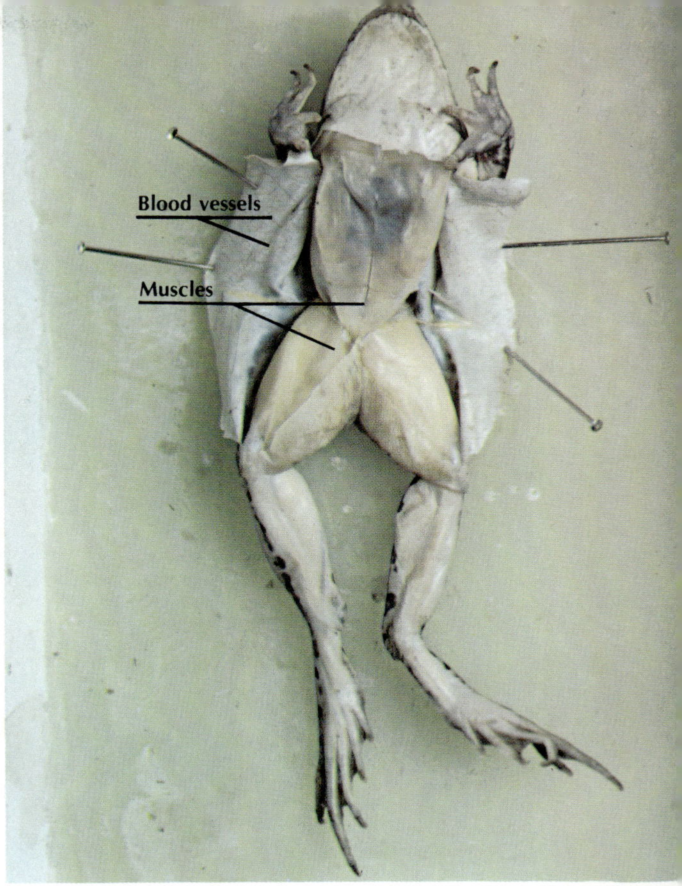
Blood vessels
Muscles

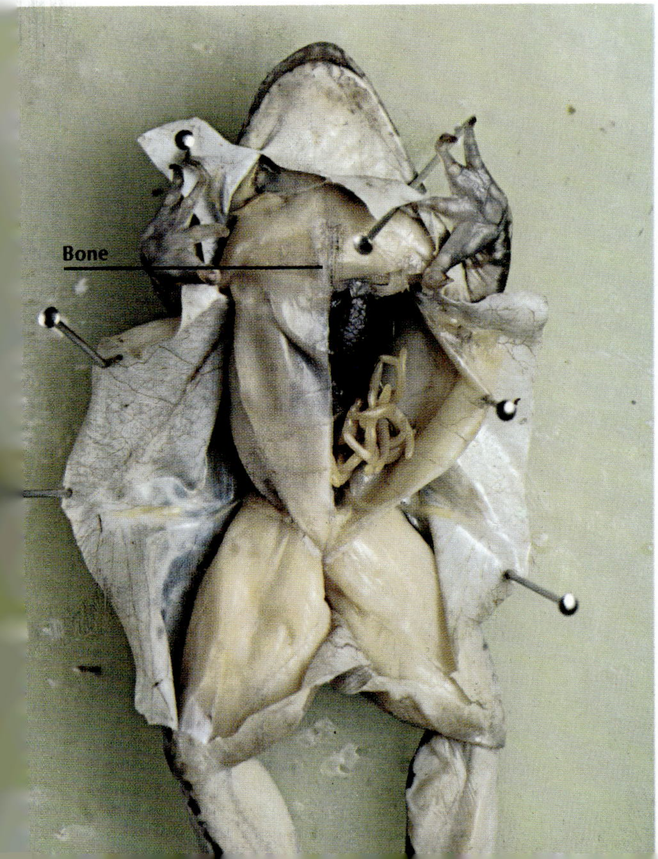
Bone

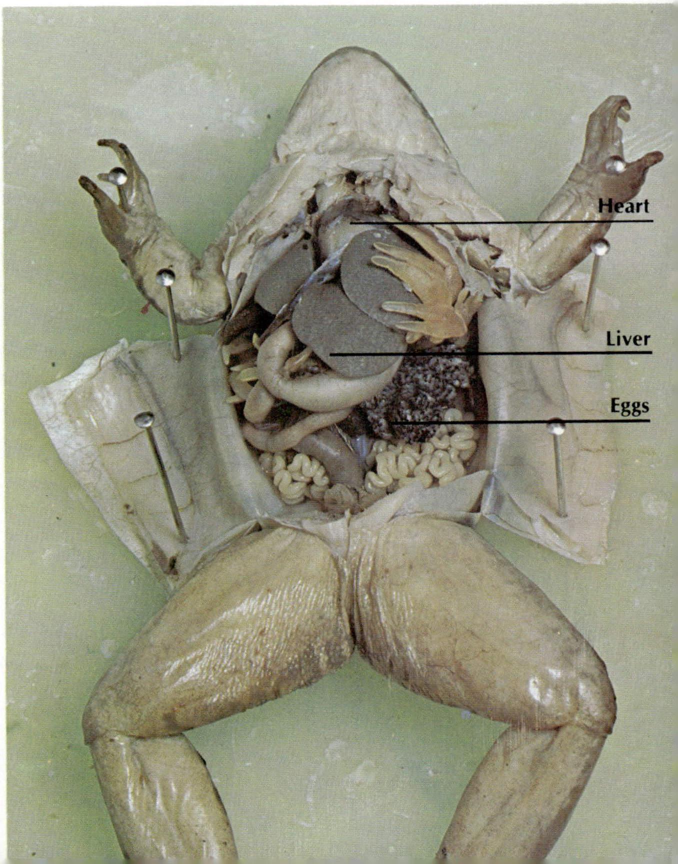
Heart
Liver
Eggs

Now push the liver aside. You can easily see the *gall-bladder* now. ■ And you can then find a white organ, the **stomach.** ■ Look at the long thin tube coming from the stomach. This is the **intestine** (in·TES·tin). ■

There is a membrane between the stomach and the intestine. Look at this membrane. See if you can find a pinkish-brown organ in it. This is the *pancreas.* ■ The pancreas is near the end of the stomach.

Do you have a female frog? Can you tell the difference between the intestine and the egg tubes? The egg tubes in a female frog are long and coiled. ■ So is the intestine. Can you tell the difference?

Now cut out the food tube. To do this, cut the **gullet** (GUL·it). The gullet is the tube that passes food from the mouth to the stomach. It lies ahead of the stomach. Make another cut at the **rectum** (REK·tum). ■ The rectum is at the end of the intestine.

Now you can lift out the whole food tube. Place it in the pan next to the frog. ●

Notice that the pancreas comes out with the food tube. ● The pancreas helps digest food. Next, cut out the liver. It also helps digest food.

Now place the liver in the pan too. ● Can you place these organs as they are found in the frog? Both send juices to the stomach to digest food.

If you couldn't find the *lungs* before, you can now. The lungs are on each side of the heart. ●

If you have a *male* frog, you can see two yellow organs. These are the **testes** (TES·teez). ● The testes make **sperm** (SPURM) **cells** for reproduction.

You saw the *ovaries* (OH·vah·reez) with eggs, if you had a female frog. The eggs are needed for reproduction.

Look near the backbone of the frog. ● You can see the long, flat deep red **kidneys** (KID·neez). ●

One kidney has been moved. Now you can see **nerves** (NERVZ) underneath. ● A **spinal cord** is inside the *backbone.* ● Some of the nerves from the spinal cord go to the muscles of the legs.

You have now dissected a frog. You see that a frog is organized. You have seen its many organs. You know much about the frog. Let's find out more.

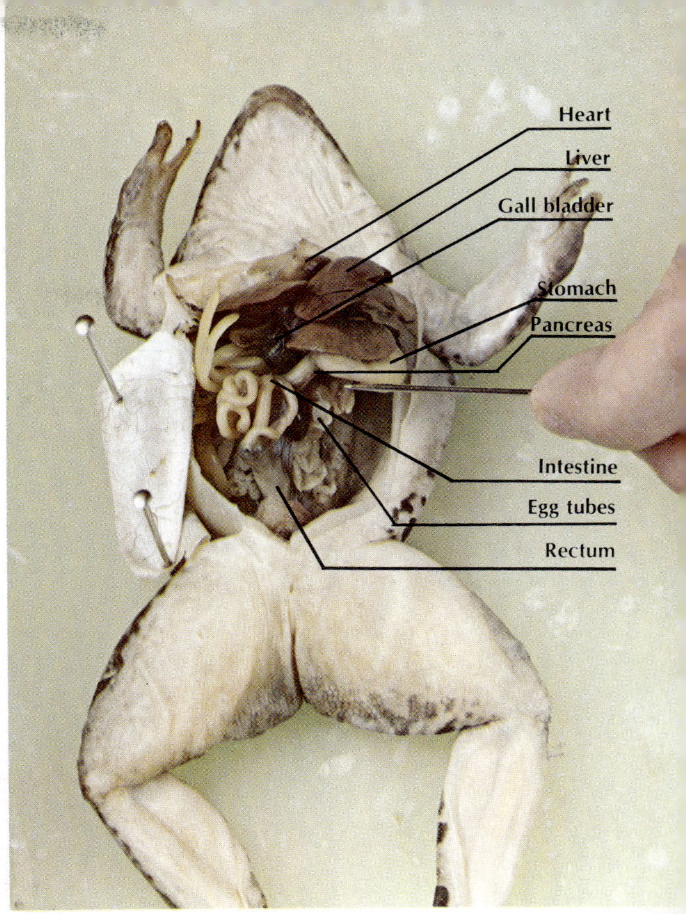
Heart
Liver
Gall bladder
Stomach
Pancreas
Intestine
Egg tubes
Rectum

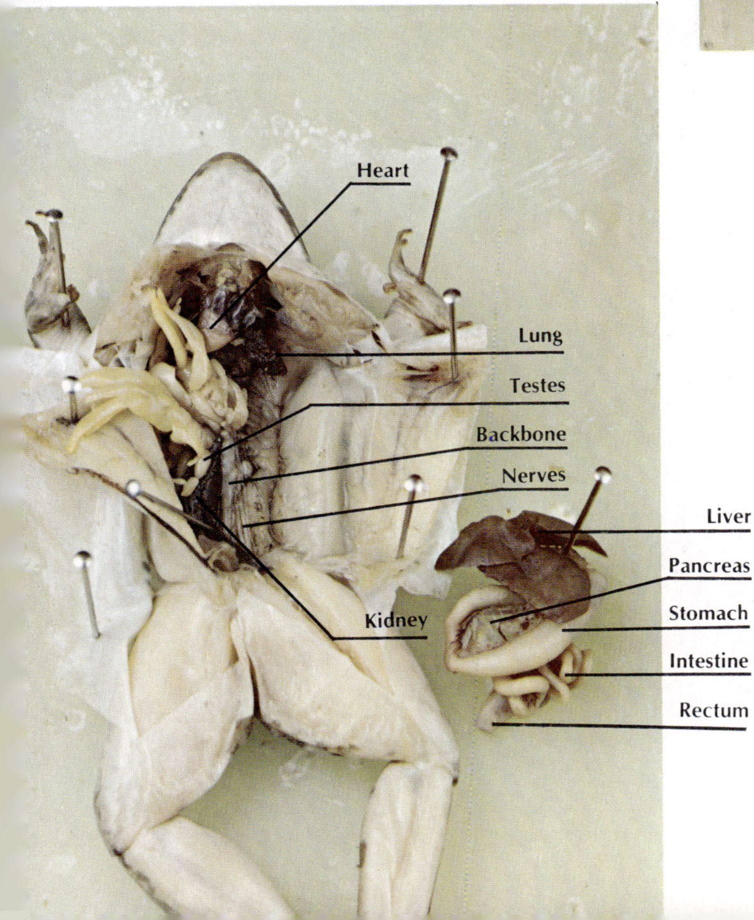
Heart
Lung
Testes
Backbone
Nerves
Kidney
Liver
Pancreas
Stomach
Intestine
Rectum

The Frog Is Organized

You have looked at the insides of a frog. The organs are organized to make up the frog. The frog is organized. You can now say that a frog is an *organism*. You saw many of the organs that a frog has.

These organs carry on the life activities of frogs:

The **skin** and **lungs** are the organs where gases go in and out.

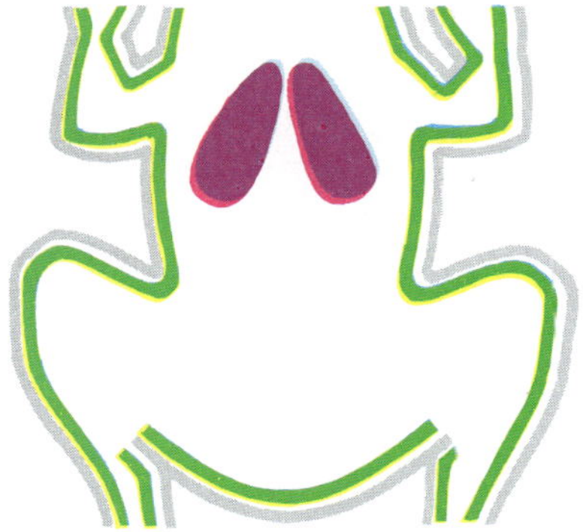

The **heart** is the pumping organ.

The **blood vessels** carry blood around the body.

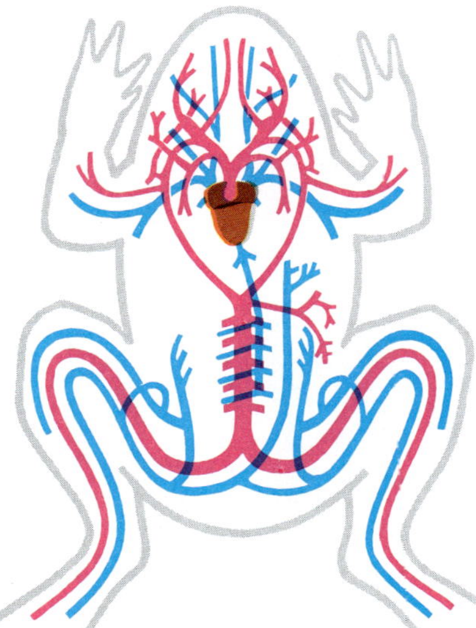

The **testes** are the organs in the male where sperm cells are made.

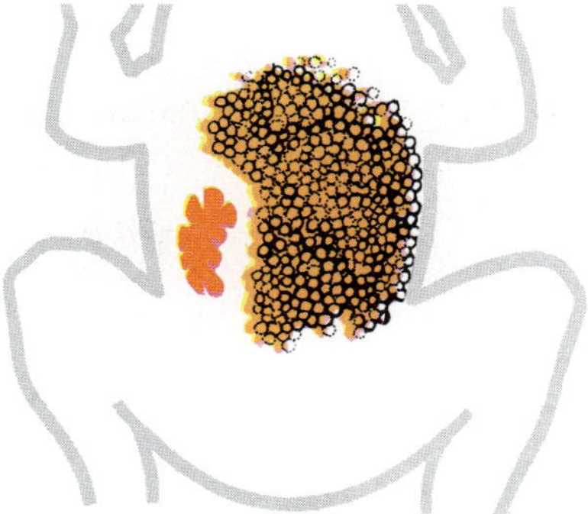

The **ovaries** are the organs in the female where the masses of egg cells are made.

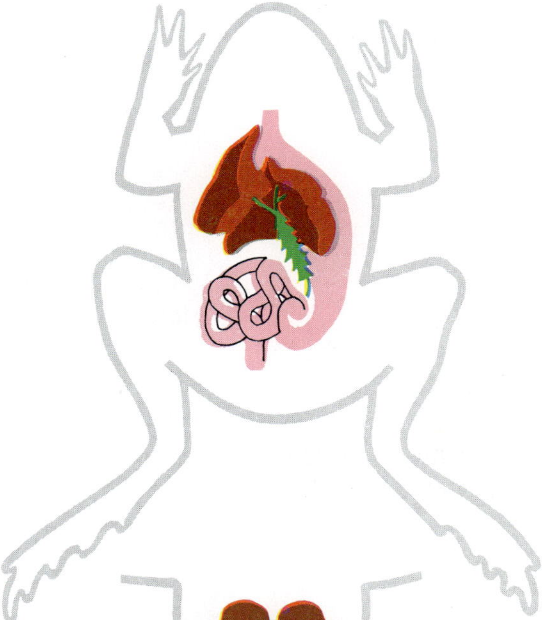

The **gullet, stomach, liver, pancreas,** and **intestine** help the frog digest food.

The **kidneys** are the organs that rid the body of some wastes.

Now you know how a frog is organized. That is, you know something of the way the frog's organs work together. Your organs have the same pattern. Your organs work together, too.

You might ask, "But what are these organs made of?"

You know.

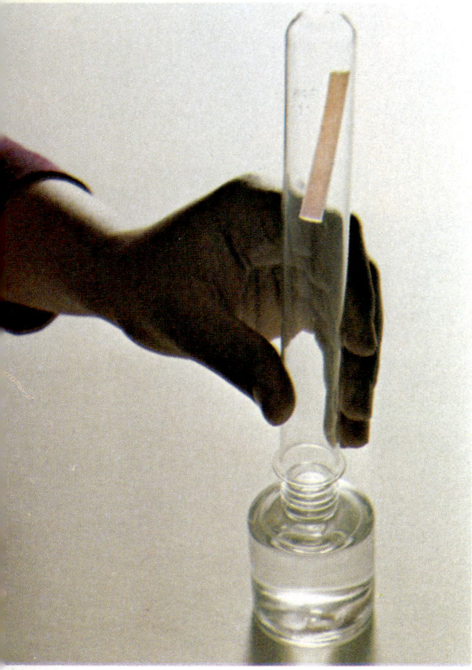

5. AT LAST—TO EVERY CELL

Can you smell things you can't see? You can smell something baking or cooking. That's a good smell.

You can smell a burning campfire. You can smell perfume in the air. And you can smell gasoline. These are all *things you can't see* but you know are there.

How do you know they are there? Small bits come from each thing you smell. These small bits are *molecules*. Molecules, you remember, are very small. You can't see them. Molecules of perfume mix in the air. And so do molecules from the burning wood of a fire.

You learned about molecules that can go through a membrane. When food is digested, it is made into smaller molecules.

It is not easy to show you molecules. They are smaller than your eye can see. They may move very fast. But sometimes you can see their tracks.

■ Moving Molecules

Here is a way to follow the track of molecules.

First you need a big test tube. Put a wet piece of red litmus (LIT·mus) paper into the tube. Turn the test tube upside down. Put it over an open bottle of ammonia (uh·MOAN·yuh). ■ Ammonia is a gas mixed in water. Ammonia has no color, and it has a strong smell.

How quickly the litmus paper changes color! ● The ammonia molecules move quickly. Ammonia molecules change the color of the litmus paper.

How does ammonia gas get to the litmus? Molecules of ammonia gas move up and around. Ammonia liquid changes to ammonia gas. Ammonia molecules quickly reach the paper in the test tube.

Membranes and Molecules

What would happen if you put a membrane in the way of the moving ammonia molecules? Watch.

Put some water in a test tube. Now let's use red litmus liquid. Red litmus liquid is what is used to make litmus paper. Add a little of the liquid to the water in the test ● tube.

Now cover the mouth of the test tube with a wet membrane. Hold the membrane in place with a rubber band.

Turn the test tube upside down. Open the bottle of ammonia carefully. Put the test tube over the open bottle of ammonia.▲ Now watch. Can you see a sort of blue stream?♦ Do you see it rising in the test tube?

You probably know what is happening. Can you say what is moving up the test tube? Yes, molecules of ammonia gas. From where? They go from the bottle of ammonia, *through the membrane* at the mouth of the test tube.

When the molecules of ammonia touch the red litmus, a blue color shows. The litmus changes from red to blue. The ammonia causes this change. You are watching tracks of ammonia molecules moving up the test tube. Now you have an idea of how some kinds of molecules can pass through a membrane.

Can molecules pass through your food tube? What is your "educated guess"?

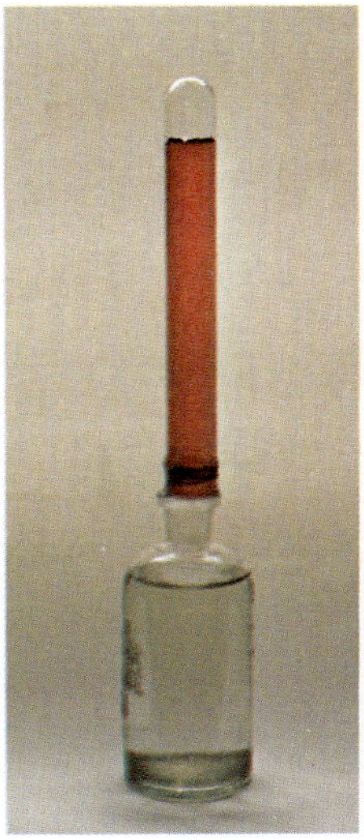

▲

♦

To the Cells

You have seen a drop of blood. You will say you did not see food in it. But food is there. The nutrients are there. Tiny molecules of digested nutrients are there. There are digested proteins, sugars, starches, and fats in your blood. There are also minerals and vitamins in your blood.

Digestion is a part of your life. Digestion goes on all your life. Can you say that the food you eat will be changed? Yes, it will be digested.

Digested foods go through the membranes of the food tube. They get into your blood stream. From the blood stream, the molecules enter the cells. How do the molecules enter the cells? They enter through the *membrane of each cell.*

These digested foods will be part of a store of nutrients. The cells draw on these nutrients. These nutrients enter cells. Here they will be used.

For what? For energy! For growth! The nutrients are turned into more of you.

The Capture of Energy

4

INTRODUCTION

Green and White Seedlings

Have you ever seen corn seedlings? Here are some green corn seedlings. And some white ones. ■

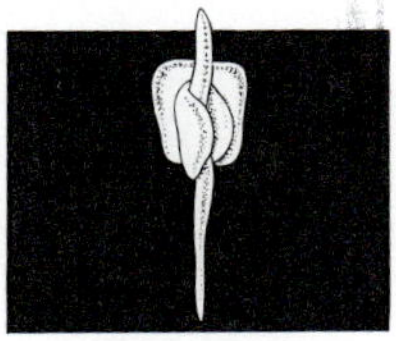

Albino seedling

The white ones are called **albino** (al·BY·no) seedlings. ■ Albino means without *normal* or *usual* color. An albino plant is a plant that is not its usual green color. Its leaves are white, not green.

Two girls in Florida grew some of these albino plants. From their teacher they got some corn grains. ● They left the grains in water overnight. Then they planted them in damp sand. ▲

In a few weeks, the girls found the corn grains had grown into seedlings. They found some normal green corn seedlings. ■ And they also found some white, or albino, seedlings.

Can *both* kinds of seedlings live? No. The albino plants cannot stay alive. Why can't albino plants stay alive? What makes these two kinds of plants so different?

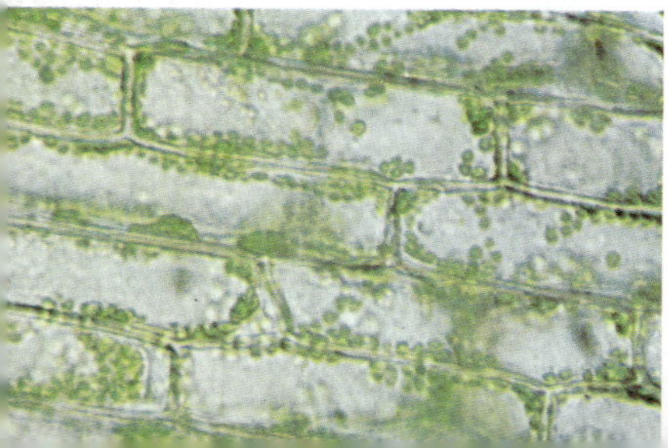

1. THE COLOR GREEN

You have seen green leaves and green grass. You may have wondered why these plants are green. They are green because they have **chlorophyll** (KLOR·oh·fill). Chlorophyll gives green plants their green color. They have chlorophyll in their leaves.

You can see where the chlorophyll is in green plant cells. You will need a leaf of a fresh water plant like elodea (eh·LOAD·ee·uh). ■ Elodea is a common green plant that grows in a lake or a fish tank.

Put a drop of water from a fish tank on a slide. ■ Put one green leaf of elodea in the water. Flatten the leaf carefully with a cover-slip. ● Use the low power of the microscope. **Focus** (FOE·kus) on your slide. Focus means to move the lens of the microscope so you can see the object more clearly.

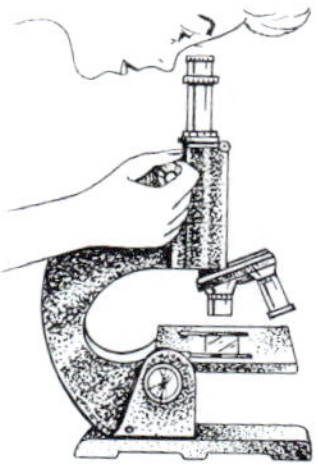
Focus

Here is what one girl saw. ▲ You can see what makes up the leaf. It is made up of box-like cells. Can you also see tiny green bodies inside the cells? ◆ These green bodies are called **chloroplasts** (KLOR·oh·plasts). The chloroplasts are green because of the chlorophyll in them. There are a lot of chloroplasts in these cells.

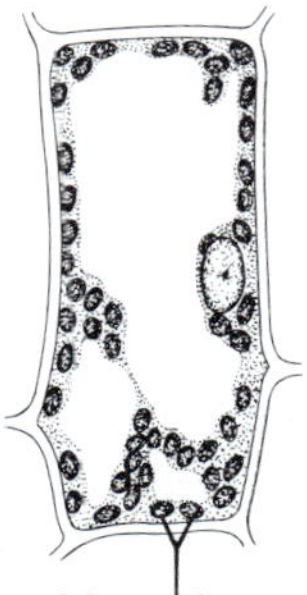

Green plants make sugar and starch in the chloroplasts. Green plants can make sugar and starch because they have chlorophyll.

Albino plants cannot make their own sugar and starch. They have no chlorophyll. That is why their leaves are white instead of green. And that is why they cannot stay alive.

Green plants have chlorophyll. Sugar and starch are made in plants with chlorophyll.

★

Chlorophyll and Food-making

Without chlorophyll you would not have food. There would be no vegetables, fruits, or seeds. The meat or milk you like would also be missing. There would be no cattle. Cattle need grass.

No chlorophyll—then no beef and no milk.

Why is chlorophyll so important? Can we show that chlorophyll is needed to make the food we eat? Is chlorophyll needed to make sugar and starch? You can see for yourself.

◆

Study this picture carefully.★ This is a silver-edged geranium (ji·RAY·nee·um) plant. Each leaf is mostly green, with a white rim.◆

The question is, do both parts of each leaf make sugar and starch? Can the white part make them? And can the green part make them? You can find out.

In green plants, sugar is made first. In most plants, sugar is quickly changed to starch. Starch is the matter most green plants store.

You can easily test for starch. How? Iodine turns starch blue or black.⊙

⊙

⊙

Where Is Starch Found in Leaves?

Here is what you need to find out:

Is starch made in the white part of leaves?

Is starch made only in cells with chloroplasts (in the green part of leaves)?

Two students in California tested several leaves from silver-edged geranium plants.◆ You can do this, too. Follow the steps they used.

■

●

▲

They heated the leaves in warm water. ■ This was to break open cells. Next, they put the soft leaves in warm alcohol (AL·kuh·hall). ●

What happened when the leaves were put in warm alcohol? The alcohol turned green. Why? Alcohol takes the chlorophyll out of the leaves.

Then, they took the leaves out of the alcohol. You can see the leaves lost color. ▲ You cannot see the rim of white that was on the edge of each leaf. The green part of the leaves lost chlorophyll. All parts of the leaves are now white. ▲

The green color is out. And now you can see where the starch is made.

Starch turns blue or black in iodine. Now you can see why the chlorophyll is taken out of the leaves first. It is easier to see if the color changes when the iodine is added. You will be able to find where the starch is in each leaf.

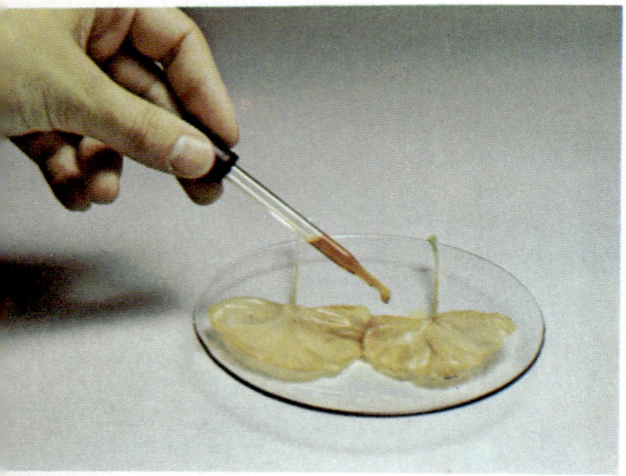

♦

Spread out the white leaves in a dish. Add some iodine to these leaves. ♦ What do you see? Can you see the outside rim of each leaf now? The rim of each leaf does *not* turn black. That part stays iodine-color. ★

Why does this happen? Look at the rest of each leaf. Most of each leaf is black. This part of the leaf has starch.

Which part of each leaf makes starch—the *green* part or the *white* part? Look at your leaves. Look at the picture of the leaf again.

★

What does the green part of the leaf have that the white part does not? *Chlorophyll.*

A Try at Prediction

1. What will happen to seedlings with albino leaves? If you can get some grains of albino corn plants, you can see for yourself.

2. How do the cells in the white rim of the geranium leaves get nutrients? You know starch does not go through a cell membrane. Starch can't pass into these cells in the white rim.

Which substance could go from cell to cell? Simple sugar can, but not starch. Cells in the green part of leaves make sugar. How can sugar get to the white part?

2. THE CAPTURE OF LIGHT

No one would put a green plant in a dark closet. The plant could not make sugar and starch. It would soon die. Green plants need light to begin to make sugar and starch.

A Simple Investigation

You can show that geranium leaves need light energy to begin to make sugar and starch. To do this you need to ask if leaves can make starch without light.

Try this investigation.

You will need a healthy geranium plant. Cover some leaves of the plant with black paper or silver foil. ◆ Use pins to hold the paper around some leaves. Now the rest of the leaves on the plant are in light.

The geranium plant now has some green leaves in the light. And some green leaves do *not* get light. These leaves are in the dark. *In which leaves will starch be found?*

Keep the plant near the window, where it gets sunlight. Or keep it near an electric light.

After two days, you can test the leaves. Which leaves have starch? Test the leaves the same way you tested the silver-edged geranium leaves. (See page 110.)

◆ 

■

●

▲ Leaves that were in light — Leaves that were not in light

Remove the black paper or foil from the leaves. Cut off the leaves that were *not in light* from the plant. Put these leaves in one beaker of water. ■

Now cut some leaves that were *in light* from the plant. ● Put the leaves that were *in light* in a *different* beaker of water.

Heat both beakers until the leaves are soft. Then pick up the leaves with tweezers. Put the leaves that were *in light* in warm alcohol. Put the leaves that were *not in light* in *another* beaker of warm alcohol. Be sure the leaves are kept apart. ▲ You should know which leaves were in the light. And which were not.

◆

Now look closely. What happens in each beaker of alcohol? Does the alcohol turn green? Yes. And the leaves lose their chlorophyll and turn white.

Why do you want the leaves white? To test for starch. If the leaves are white, you can see what happens. You can see what happens when iodine is added. Iodine is used to test for starch. You can see what happens when the leaves are white instead of green. How can you tell which leaves have starch?

Spread the leaves on two dishes. Add iodine to the leaves on both dishes. Look closely at the leaves. ◆

◆

Which leaves have starch? The ones that turn almost black. These are the leaves that were *in the light*. The leaves that were *not in light* stay the color of iodine. They have no starch. Leaves cannot make starch if they don't have some light.

How do you know light is needed to make sugar and starch? Only the green leaves that have been *in the light* show starch. So light is needed. After they have had light, green leaves make sugar. Then the sugar is changed to starch. You tested for starch. You found starch in the green leaves that had light.

What Does Light Do?

You have seen that green plants make sugar and starch. They do this after they have had light.

The way this all happens is called **photosynthesis** (foh·toh·SIN·thuh·sis). Photo means light. And synthesis means putting together, or making something. So photosynthesis is the name for the way green plants make sugar and starch. Green plants *need light* to start photosynthesis. ★

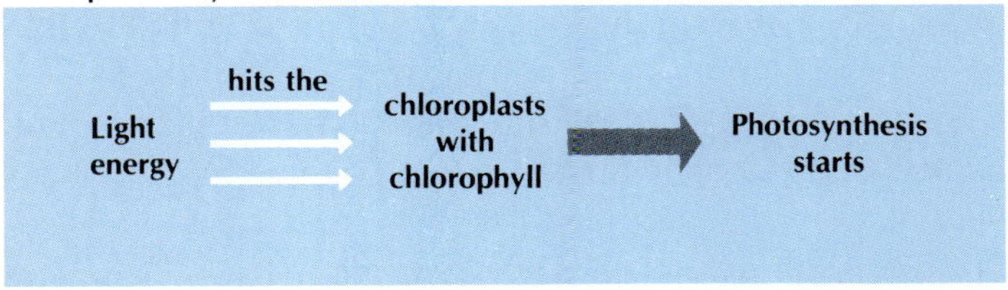

★

What does a green plant make?
Sugar and starch.
What kind of plants carry on photosynthesis?
Green plants.
Why can *only* green plants carry on photosynthesis?
Green plants have chlorophyll. Chlorophyll makes chloroplasts green.

And all these chloroplasts get energy from light. That is, they capture energy from the sun or energy from an electric light bulb. Chloroplasts capture the energy from light.

First the chloroplasts get energy from light. Then the cells can begin to put together sugar and starch. But this is only the start of the story.

It has taken more than 50 years to find out what goes on in chloroplasts when light energy is captured. There are still some questions. Scientists still cannot explain everything about how this happens. But they do know this. Light energy is captured by the chloroplasts. Then the energy from light is used to split water.

When molecules of water are split, energy is given off. Some of this energy goes into making ATP in the cells. This ATP then runs the rest of photosynthesis. The ATP is needed to put together sugar molecules.

The Making of Sugar

Scientists have a way to write simple sugar. You learned about simple sugars before. This is simple sugar: $\mathbf{C_6H_{12}O_6}$. **C** stands for carbon. **H** stands for hydrogen (HY·druh·jun). **O** stands for oxygen. Sugar has six parts of carbon, twelve of hydrogen, and six of oxygen.

Notice the hydrogen in the sugar. This H_{12} means there are twelve parts of hydrogen in a simple sugar molecule. Where does the hydrogen come from?

The hydrogen comes from *water*. Cells of plants need water. Cells need water to make sugar. The scientist's way of writing water is H_2O.

Green plants need H_2O to make *sugar*. You can think of the *first* step in the making of sugar this way:

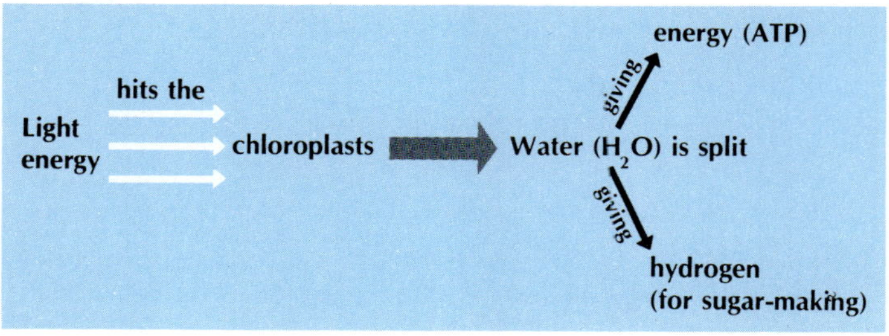

What will be made in chloroplasts? Simple sugar, $C_6H_{12}O_6$.

Where does the hydrogen part of sugar come from? From water.

What is the next part of the story? You may ask where the C's and the O's in sugar come from.

You know a gas that has both C and O. Of course, it's CO_2, carbon dioxide. Green plants need CO_2 to add to the H. Then sugar, $C_6H_{12}O_6$, can be made.

Now let's look at the gases green plant cells take in and give off.

3. THE USES OF A GAS

Do you remember what happens when you breathe through a straw into brom blue? Let's try it again.

You only need to breathe a few times before the brom blue turns yellow. ▲ The large amount of carbon dioxide in your exhaled breath turns the brom blue to yellow.

Three students in Minnesota had an idea. They breathed into jars of brom blue and turned it yellow. Next, they put green plants, like elodea, into the jars of brom yellow. ◆

When they looked at the jars the next day, *the liquid was blue again.* ★ This is what they thought would happen. What causes the brom yellow to turn back to blue? What would you say?

Here is what these students thought. They had exhaled carbon dioxide into the first brom blue. It turned yellow. Brom blue turned yellow because there was a large amount of carbon dioxide in it. Now why does this yellow liquid turn blue again? *The carbon dioxide must be gone!*

What took the carbon dioxide out of the brom yellow? Find out for yourself.

▲

◆

★

◇

Do Green Plants Take In Carbon Dioxide?

Begin, perhaps, this way.

Pour brom yellow into six test tubes. ◇ Put elodea plants into four test tubes of brom yellow. ◇ Leave two test tubes with *only* brom yellow in them—no plants. Then cork all the tubes. Now nothing can get out of the tubes. Let them stand for one or two days in light.

What do you think will happen? *Do not* turn the page until you think this out. Will the brom yellow turn back to blue in any of the tubes? What if it only turns in the tubes *with green plants?* What will this tell you?

Suppose the brom yellow turns back to blue only in the tubes *without* green plants. What will this tell you?

Now turn the page. Check yourself by the picture on the next page. Then check your test tubes after two days. Do you find the same as in the picture?

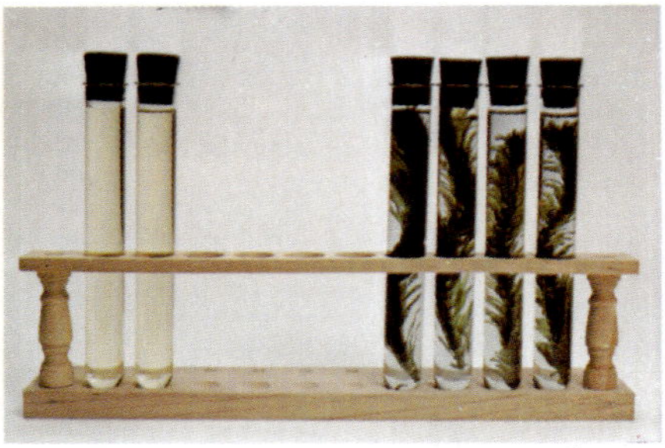

■

The brom yellow turned back to blue in those tubes *with the green plants.* ■ The green plants took the carbon dioxide out of the brom yellow. That's why the brom yellow turned back to blue.

When do green plants take in carbon dioxide? You can answer this. Where did you keep your test tubes? In light? Yes. Green plants take in carbon dioxide when they have been in light. If your test tubes were in light, they should look like this. ■

What do they use carbon dioxide for? Green plants use carbon dioxide to make sugar and starch. You already know this. The chloroplasts in green plants use carbon dioxide to make sugar and starch. Only a very small amount of carbon dioxide is found in air. But this is enough for green plants. With it they can make sugar and starch.

●

What Do Green Plants Do After They Have Been in Light?

In light, green plant cells split water. They make ATP. And they take in carbon dioxide. This is how green plant cells get hydrogen, carbon, and oxygen. And, this is how they can begin to make sugar and starch. We can say this in another way. In the chloroplasts of green plants:

$$\underset{H_2O}{\text{WATER}} + \underset{CO_2}{\text{CARBON DIOXIDE}} \xrightarrow[\text{gives}]{\text{in LIGHT}} \underset{C_6H_{12}O_6}{\text{SUGAR}} + \ ?$$

Green plants make sugar. And they store sugar as starch. Notice the last + sign? Do green plants make something else? Let's find out now.

▲

A Stream of Bubbles

Here are elodea plants in water. ● A strong light is near the elodea. The green plants are making sugar and starch.

Look closely at the plants. ▲ You can see bubbles coming out of the elodea plants.

In the next picture, you can see the light has been moved away from the plants.◆ It is now a few feet from the elodea. The plants do not get much light. Look closely at this picture.★ There are no bubbles coming from the plants.

You can see that green plants give off bubbles.▲ And that light seems to be needed to make the bubbles start. What are these bubbles? They are a gas. What gas do plants give off? What gas is given off as plants make sugar and starch?

Look at this setup.◇ Suppose you could put the gas from the plants into a test tube.◇

How could you find out what the gas is? First, light a piece of wood with a match. Then put out the flame so the wood only glows. Now put this glowing piece of wood into the tube of gas.⊙ See how the wood burns brighter.⬟ This is the test for oxygen gas. The elodea plants are giving off *oxygen*. They give off oxygen when they have had light.

◆

★

◇

⊙

⬟
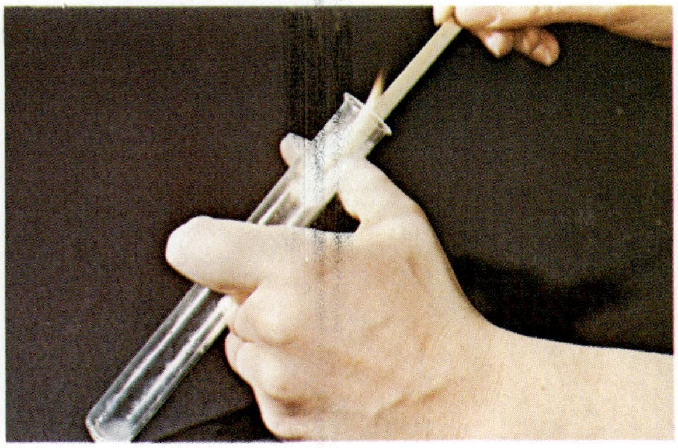

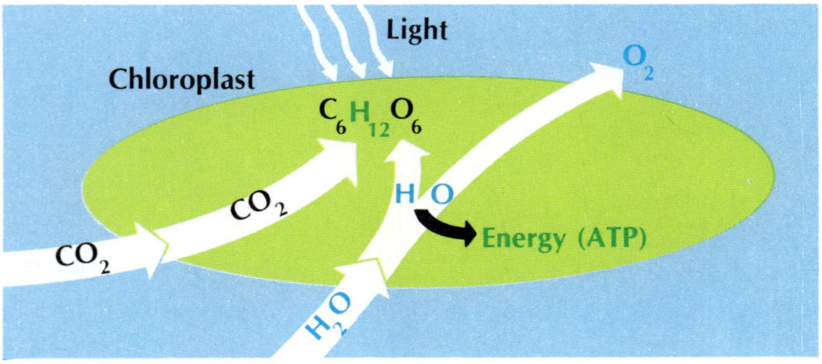

You may recall that energy from light splits water in the chloroplasts. The hydrogen is used to make sugar. What is left over? Oxygen from the water (H_2O). This is given off.

Green plants take in carbon dioxide and give off oxygen. They do this when they have had some light. They do this as they make sugar and starch. Green plants make sugar and starch in the chloroplasts of their cells.

Can you go back and finish what we started before? In the chloroplasts of green plants:

WATER	+	CARBON DIOXIDE	in LIGHT →	SUGAR	+	OXYGEN
H_2O		CO_2	gives	$C_6H_{12}O_6$		O_2
						given off

Green plants make sugar. The sugar is stored as starch. And they give off oxygen, too.

What happens when the light is moved away from the elodea plants? The bubbles slow down. Then they stop. Less oxygen is given off. Then no oxygen is given off. And sugar and starch making is slowed down in the cells of the plants.

Green plants need some light. And they need carbon dioxide to make sugar and starch in their cells. They also give off oxygen.

Green plants can do something animals cannot do. Green plants use carbon dioxide and give off oxygen. When? In making sugar and starch.

There is a way green plants are like animals. They both take in oxygen. When? What for? You'll see.

4. CYCLES OF GASES

You know now that green plants make sugar. They use water and carbon dioxide. They give off oxygen.

You know green plants supply food. And green plants put oxygen into the air. No animal can do this. Animals cannot make their own sugar. And they don't give off oxygen.

Green Plants Use the Sugar They Make

Green plants need energy to grow. They need energy to carry on their life activities. Just as animals need energy. A bird and a worm get energy from sugar. An elephant and you get energy from sugar. And green plants also get energy from sugar.

Green plants can break down the sugar they make. They can use it to get energy. They use oxygen in this job of breaking down sugar. This is what animals do too. Both green plants and animals break down sugar for energy. They both take in oxygen to do this job.

$$\underset{C_6H_{12}O_6}{\text{SUGAR}} + \underset{O_2}{\text{OXYGEN}} \xrightarrow{\text{gives}} \text{ENERGY} + \underset{CO_2}{\text{CARBON DIOXIDE}} + \underset{H_2O}{\text{WATER}}$$

You can say that green plants and animals are alike in this. They both use oxygen to break down sugar. Energy is given off. Green plants and animals take in oxygen and give off carbon dioxide. When? While sugar is broken down to get energy.

What Gases Go In and Out of Cells?

This may surprise you. *Green plants take in carbon dioxide and give off oxygen.* They do this when they have had light. They do this as they make sugar.

But *green plants also use oxygen.* And *they give off carbon dioxide.* When? All the time. They take in oxygen and give off carbon dioxide. Just as animals do. They do this all the time. They do this as sugar is broken down to give energy.

■

●

You know that brom blue turns yellow when you blow into it. Why does it turn yellow? It turns yellow because you exhale carbon dioxide.

Green plants also give off carbon dioxide. And you can show they change brom blue to yellow. Just as you do. But you need to keep the green plants out of light to do this.

Put some growing elodea plants in test tubes of brom blue. ■ Set up some test tubes of brom blue without plants. Keep all the test tubes out of the light. Or cover the tubes with silver foil. ■

After a day or two look at the tubes. Brom blue has changed to yellow in the tubes with elodea. ● What must have happened? The green plants gave off carbon dioxide. Just as you do when you exhale.

Can you say why the elodea plants had to be kept without light to show this? Why do they have to be in the dark to turn brom blue to yellow? Here is the answer.

After plants have light, they take in carbon dioxide. They take in the carbon dioxide fast. After plants have had light, they take in a lot of carbon dioxide. So the carbon dioxide does not stay in the liquid in the tubes. You need to stop sugar-making. Then you can see that green plants give off another gas. The gas is carbon dioxide. This is why you need to keep the green plants without light. This way you can show they give off carbon dioxide. Just as you do.

Picture the cycle of gases in *animal cells* this way:

OXYGEN (O_2)
↓ is used when
SUGAR
↓ is broken down to give
CARBON DIOXIDE (CO_2) + ENERGY

In animal cells, oxygen goes in. Carbon dioxide goes out. Here is what happens in *green plant cells:*

OXYGEN (O_2)
↓ is used when
SUGAR
↓ is broken down to give
CARBON DIOXIDE (CO_2) + ENERGY

So far, the cycle is the *same* in animals and plants. Sugar is broken down to give energy. Oxygen is needed in plants and animals for this. Carbon dioxide and energy are given off.

But green plants also make sugar. They do this when they have had light. So you have to *add another cycle of gases:*

CARBON DIOXIDE (CO_2)
↓ is used to make
SUGAR
↓ and
OXYGEN (O_2)
is given off

Can you show the two gas cycles in *plants* together now?

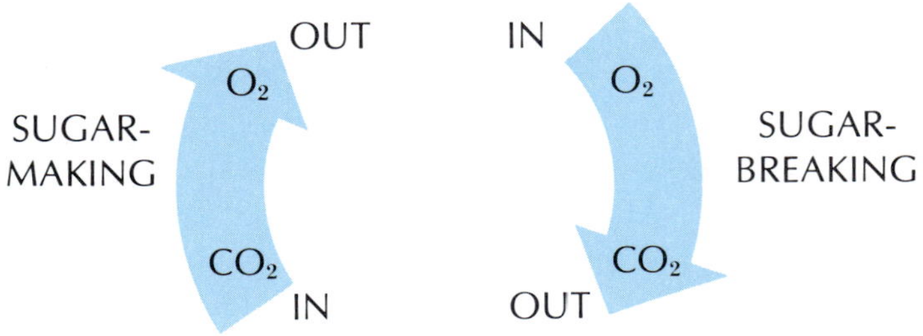

Animals use one gas in the air. They use oxygen. Green plants use *two gases* in the air. They use oxygen and carbon dioxide.

Can you say which living things put oxygen back into the air? You are right if you say green plants return oxygen to the air. It is a good thing. Otherwise all the oxygen supply might be used up.

What does this mean? Oxygen and carbon dioxide are used over and over. These are the cycles of oxygen and carbon dioxide.

Where Gases Get In and Out

Look at another green plant. Look at cells in a lettuce leaf. This is a green *land* plant. The leaves use gases from the air. You need a thin layer of lettuce to see the cells.

First take a leaf of lettuce. Tear the leaf from the middle to the outside edge.■ Can you see the thin layer at the torn edge?● Peel off the thin layer from the underside of the leaf.

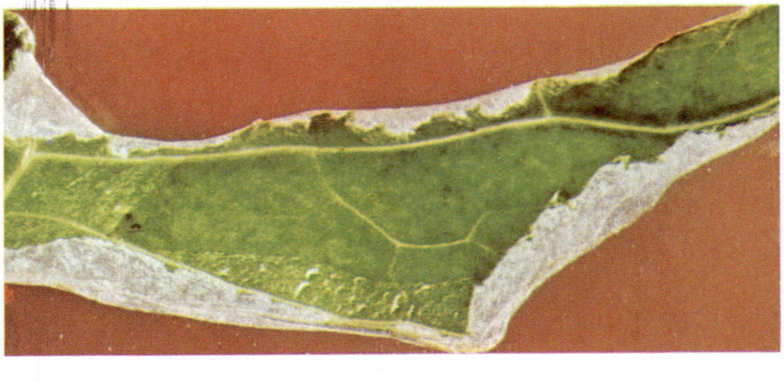

Put a small bit of this thin layer in a drop of water on a slide.▲ This is a thin layer of cells. Look at it under the low power of your microscope.♦ Then look at the same cells under high power.★ You can now see the cells better.

Now try to find tiny openings in the leaf. Like these.★ There are millions of openings in leaves of lettuce. And in all land plants. Here is where gases go in and out of leaves. Oxygen and carbon dioxide go in and out of leaves.

In water plants, gases pass in and out of the membranes of cells. All cells are taking in gases and giving off gases.

▲

A Try at Prediction

A boy plans to keep some goldfish alive in a fish tank. The fish need oxygen. What living things could he add to the fish tank? He wants to make sure the fish will get oxygen.

Predict whether the fish tank can be kept in a dark part of a room. Why or why not?

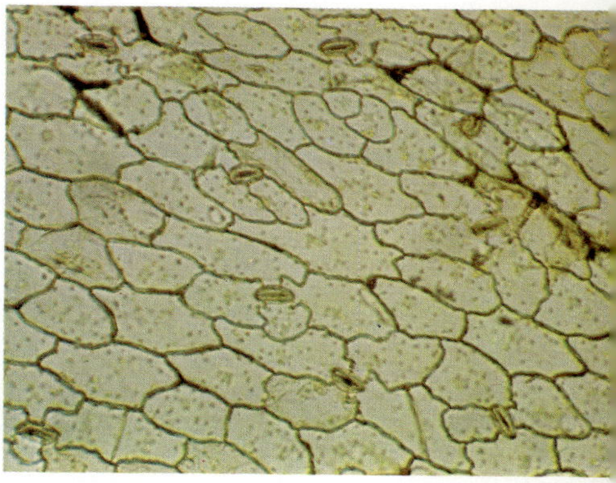

♦

A Try at Explanation

1. Let's suppose a living thing changed brom yellow to blue.

a. Was the living thing an *animal*, or a *green plant*? Explain.

b. When could a living thing change the brom yellow back to blue?

c. What kinds of living things can change brom blue to yellow? When?

2. Imagine a forest with many tall pine trees. And other kinds of very tall trees. Only a little light reaches the floor of the forest.

Can you explain why the young seedlings of tomato plants cannot live in the forest? After all, the seedlings are green. There is water. And there is carbon dioxide in the air. Why can't they live?

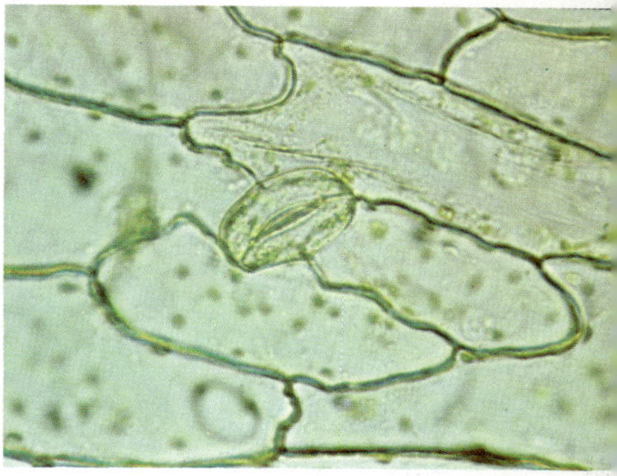

★

AN INDEPENDENT STUDY

Green Plants: The Capture and Use of Energy

In the next pages you can learn on your own. You can find out what you know. And what you do not know. The pages are like this one. You will find questions on the left-hand page. Place a piece of paper over the page on the right. This will be your answer sheet. When you have the answer, write it on your answer sheet.

How can you find out if your answer is correct? You should know how to do that now. You can check your answer with the answer in the box on the right-hand page. Check your answer before you go on to the next question.

Now go through the study and you will find out for yourself. You will find out what you know. And what you do not know. You will remember about the use of energy in green plants. You will also see that green plants and animals have another likeness. They both need energy for the life of their cells.

1

Here are things you know. When *you* breathe into brom blue, it turns yellow. And a green plant can give off a gas that turns brom blue to yellow. Look at the pictures on the next page. ■ You and green plants can turn brom blue to yellow.

You give off carbon dioxide.

Plants give off a gas, too.

The gas the green plant gives off is ____________ ____________.

2

You give off carbon dioxide all the time. Green plants are like you.

All the time, green plants give off the gas ____________ ____________.

1

carbon dioxide
CO_2

Both you and green plants give off carbon dioxide.

(Go on to question 2.)

2

carbon dioxide
CO_2

Plants and animals give off carbon dioxide all the time.

(Turn the page to question 3.)

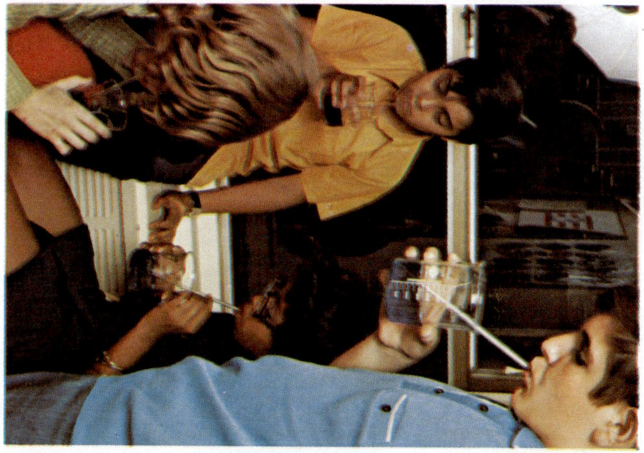

3

Both you and green plants take in a gas. This is the gas you need to carry on all life activities. Green plants need it, too.

This gas is ______.

4

If this is so, all the oxygen in the air should be used up. You and millions of people and animals are *all* taking in oxygen.

Plants are also taking in ____________ for life activities.

5

What do green plants give off after they have had light? They give off a gas that can make a match burn brightly. Remember the burning wood put into the test tube? (See page 117.) The test tube had gas given off by the plants under water. The match burned brighter inside the test tube.

This gas that the green plants are giving off is ____________.

(Write your answer down. Check yourself across the page.)

6

You can give off only one gas, carbon dioxide.

Green plants give off the same gas you do. It is called ____________ ________________. They give off this gas all the time.

3

oxygen

O_2

You and green plants take in oxygen.

(Go on to question 4.)

4

oxygen

O_2

(Go on to question 5.)

5

oxygen

O_2

Green plants give off oxygen when they have had *light*.

(Go on to question 6.)

6

carbon dioxide

CO_2

Green plants give off carbon dioxide all the time.

(Turn the page to question 7.)

7

After green plants have been in light, they do something animals cannot do. They can give off more than carbon dioxide. They can also give off another gas.

This gas they give off when they have had light is ------.

8

When do green plants give off oxygen?

Green plants make sugar and give off oxygen when they have had some light.

In photosynthesis green plants make sugar and give off the gas ------.

9

If green plants were kept without light for a long time, they would stop making sugar and starch.

They would stop giving off the gas you use and other animals and plants use.

This gas is ____________.

10

One gas cycle is made up of oxygen taken in and carbon dioxide given off.

Where does it happen? in all living things? or *in green plants only?* Which?

7

oxygen

O_2

(Go on to question 8.)

8

oxygen

O_2

(Go on to question 9.)

9

oxygen

O_2

(Go on to question 10.)

10

in all living things

All living things take in oxygen and give off carbon dioxide.

(Turn the page to question 11.)

11

In some living things there is a special cycle. Carbon dioxide is taken *in* and oxygen is given *off*. Where does it happen? *in all living things?* or *in green plants only?* Which?

12

When does this cycle take place in green plants? When is carbon dioxide taken in and oxygen given off? Does this happen *all the time?* or *after the plant has had light?*

13

You have the concept now. Green plants have two cycles.

One all the time:

Oxygen in ⟶ *carbon dioxide* out.

And a special one that happens after the plant has had light:

Carbon dioxide in ⟶ *oxygen* out.

What nutrients do green plants make after they have had light?

Green plants can make ____________ and ____________ after they have had light.

14

Green plants make sugar first: $C_6H_{12}O_6$.

The *C* part (or carbon) and the *O* part (or oxygen) in sugar come from the gas called ____________ ________________.

11

in green plants only

(Go on to question 12.)

12

after the plant has had light

Green plants have a special gas cycle in light.

(Go on to question 13.)

13

sugar and starch

Green plants make sugar and starch if they have had light.

(Go on to question 14.)

14

carbon dioxide
CO_2

Carbon dioxide gives the carbon (or *C*) and the oxygen (or *O*) for sugar.

(Turn the page to question 15.)

15

Where does the *H* part (or hydrogen) in the $C_6H_{12}O_6$ come from? Does the *H* come *from water* (H_2O)? or does it come *from carbon dioxide* (CO_2)? Which?

16

How do you know the *H* part in sugar comes from water? The way we write water is H_2O. Water is made up of hydrogen *(H)* and oxygen *(O)*. Green plants use the hydrogen to make sugar. What is left over? What is the gas given off in photosynthesis? It is ____________.

17

Sugar is made in green plant cells. These cells put together carbon dioxide and hydrogen in a certain way. They need water and light to do this.

This is the way green plants make sugar ($C_6H_{12}O_6$). This process is called photosynthesis.

What living things will use the sugar? *plants only?* or *plants and all animals?* Which?

15

from water

H_2O

(Go on to question 16.)

16

oxygen

O_2

(Go on to question 17.)

17

plants and all animals

5. WITHOUT CHLOROPHYLL

How many leaves are on a tree? There are hundreds. Many hundreds. What happens to all these leaves? There have been trees on the earth for about 300 million years. Leaves have been falling off trees since then. Where do they all go?

You know that dead leaves rot. They **decay** (di·KAY). Decay means to rot. The leaves break down, or decay, after they fall to the ground.

Leaves, after all, are made up of cells. These cells, you remember, have sugars and starches in them. They also have in them other nutrients such as proteins, vitamins, minerals, and water.

Leaves have in them the matter that animals can feed upon. You are not surprised. The Koala (koh·AH·luh) of Australia lives on the living leaves of a tree. Cows live on the living leaves of grass plants. These animals live on the nutrients in the leaves. The leaves are food.

▲

◆

But leaves do fall from a tree. Grass does often turn brown. Then what happens? What would happen if leaves fell to the ground year after year? Suppose they just piled up.

The earth would be choked with leaves. But this does not happen. The leaves decay, or are broken down.

Plants, Animals, and Bacteria

What causes leaves to decay? ▲ *Bacteria* (bak·TIR·ee·uh) in the soil. Millions of tiny bacteria cause decay. These bacteria of decay also cause dead animals to rot.

Soil that has decaying leaves and animals is rich soil. It is rich in minerals. Minerals come from decaying leaves and animals. This rich soil is called *humus* (HEW·mus). Humus is added to poor soil in order to grow more healthy plants.

Plants That Use Ready-made Nutrients

Mushroom

Look at the floor of the forest. You can see a fern, moss, and a kind of **mushroom** (MUSH·room). ◆ Ferns and mosses are green. They make their own food. But mushrooms are plants without chlorophyll. Without chlorophyll they *cannot* make sugar and starch.

These plants are growing in soil rich in decaying plant and animal matter. What do the green ferns and green mosses take from this soil? To make their own food, they take water and minerals from the soil. They take in water and minerals through many root-like parts.

Let's look at the mushrooms. ■ Mushrooms are not green. They do not have chlorophyll. They may be orange, yellow, or black. But they are not green plants. They cannot make their own nutrients.

Mushrooms live on decaying plant and animal matter. Or in soil rich in this matter. They take in nutrients from this decaying matter.

Many plants live in the same way as mushrooms. That is, they live on matter that was once living. You know some of these plants. One is the mold plant. You have seen mold plants growing on bread and lemons. They even grow on leather.

In what way are bread, lemons, and leather like decaying plant and animal matter? *They are all made of plant or animal matter.* They are all made up of matter that was once living. Bread is made of ground-up grains that come from green plants. How can molds grow on leather? Leather comes from animal skins.

Some plants, like the molds, use ready-made nutrients. How does a mold plant grow? See for yourself.

AN INVESTIGATION

A Second Look at Bread Molds

You grew mold plants once before. (See page 28.) Now grow some molds again and look more closely. You know that mold plants cannot make their own nutrient supply. They do not have chlorophyll. They use nutrients made by green plants.

Let's see how the mold plants grow into the bread. These root-like parts are also used to take in the nutrients. ●

First sprinkle a few drops of water on a piece of cracker or bread. Then brush the bread along a dusty window sill. Now put this in a plastic box and cover it.

In a week—or sooner—you should see white fuzz with dark dots like this. ▲

Now use a hand lens. You can see more now. There are round black dots mixed in the white threads. ♦

Let's look even closer. What does bread mold look like under a microscope? Be careful not to open the box too much. You do not want to spread bread mold around the room. Take only a bit of mold to make a slide.

Put this mold in a drop of alcohol on your slide. Place a coverslip over the slide to spread the drop out on the slide. Now place your slide under low power on the microscope. Can you see threads with round black tops like these? ★

●

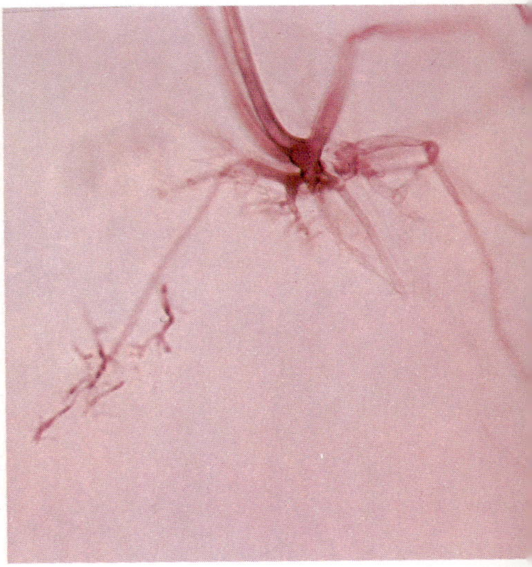

▲

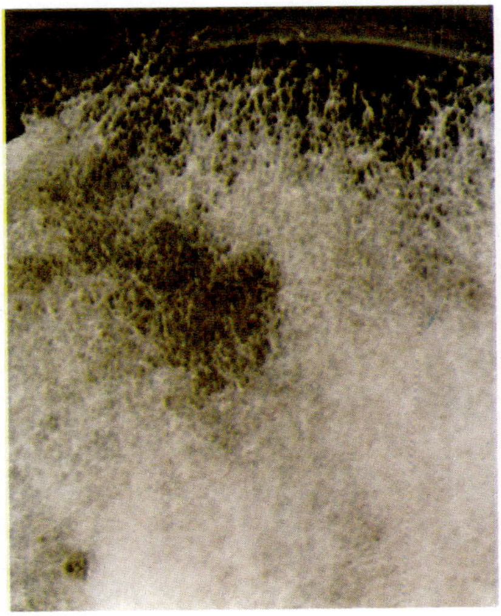

♦

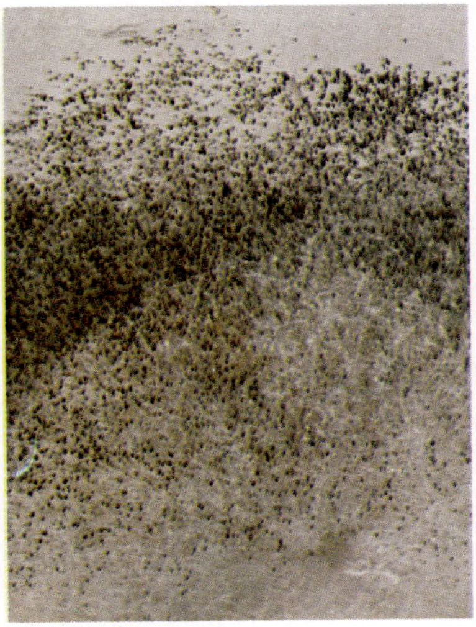

★

How Molds Get Nutrients from Bread

Bread is rich in starch. But starch cannot go through membranes of cells. Mold plants need nutrients like sugar. They need nutrients that can go through the membranes.

What do mold plants do to the starch in the bread? They break down the starch. They have *enzymes*. These enzymes go into the bread. They break down starch to sugar. The enzymes digest the starch. Then the sugar can go through the cell membranes into the cells of the mold.

Fungi: Plants That Are Consumers

Water fleas feed on green algae. Horses and sheep eat hay or grass. Water fleas, horses, and sheep cannot make their own nutrients. They are **consumers** (con·SOO·merz). Consumers cannot make their own nutrients.

Green plants capture the energy of light to make sugar and starch. These nutrients have energy from light locked in them. Green plants that make their own nutrients are called **producers** (pro·DOO·serz). Producers make their own nutrients. *Consumers feed on producers.*

Now you know that some *plants* are also consumers. Such as the molds that live on bread, fruit, or leather. And the mushrooms. Molds and animals are alike in the way they get their nutrients and energy. They do not make their own nutrients.

Can you name a plant that gets its energy like an animal? Bread mold. Or mushrooms. They are consumers, like man. They do not make their own nutrients.

Both molds and mushrooms are **fungi** (FUN·gee). Fungi are plants that do not have chlorophyll. Fungi feed on nutrients made by other living things. Just as animals do.

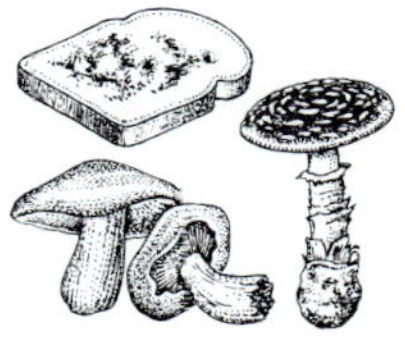

Fungi

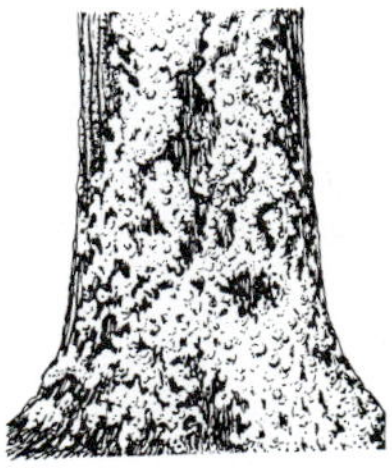

Algae

Look at all the organisms in the picture. ■ Can you see some whitish shelves on the tree? These are plants. They are fungi. Fungi grow on other plants and animals. The greenish growth on the tree is **algae** (AL·gee). These green algae are one-celled plants with chlorophyll.

Which one of these gets its nutrients from the tree—the whitish fungi or the green algae? Which sends out enzymes to break down starch? Which is the producer?

An Exercise in Prediction

What are the beetles, earthworms, termites, and chipmunks in the picture? ● Are they *producers* or *consumers?*

Plants and animals live together in this place. Could consumers live long without the producers?

A Try at Explanation

1. This earthworm you see swallows rich soil. ▲ The food, perhaps decaying leaves, is digested inside the worm. The digested matter can pass through membranes of the food tube in the worm. The rest of the soil passes out of the body. The digested food gets to all the cells in the worm.

What helps to break down the food so that the nutrients can get to all the cells of the worm?

■

●

▲

2. Look at these two pictures. Which picture has only producers?

Which picture has *both* producers and consumers? Name the producer and the consumer in this picture.

3. How can the mushrooms live in the dark in the soil in the cave?

Could roses or grass or corn plants grow in this cave? Why or why not?

6. THE MAIN CONCEPT: CHLOROPHYLL AND THE WORLD'S FOOD SUPPLY

Animals cannot live just anywhere. They cannot live alone on land or in the water. They need green plants. You need green plants. *Animals depend on chlorophyll.*

What does chlorophyll do? *The chlorophyll in green plants* is important for *photosynthesis*.

As you know, sugar and starch are made in photosynthesis. You may know that other nutrients such as proteins and fats are found in many seeds, and in fruits and vegetables. Plants need energy to make all these nutrients. They need the chemical energy ATP. If photosynthesis did not happen, there would be no ATP. And there would be no fats and no proteins in plants.

So without chlorophyll there would be no photosynthesis. There would be no nutrients. There would be no food. Animals could not live. The chlorophyll of green plants traps energy from the sun. Green plants make the world's food supply.

The First Part in the Chain

Green plants always seem at the beginning of the story. Almost all plant and animal life depends on the work of chlorophyll. This is true of almost all life.

There are some tiny kinds of cells that get their energy in another way. But this is a small part of the whole mass of plants and animals found on earth.

You depend for your very life on chlorophyll found in green plant cells.

True?

An Energy Cycle

If you understand what you have been studying, some concepts stand out.

Green plants capture energy from light.

They use this energy to make many nutrients.

Animals use the nutrients of green plants.

Some plants, such as fungi, use the nutrients of green plants.

Without green plants there would be no nutrients. There would be no food.

Almost all living things—green plants, animals, and many bacteria and fungi—depend on chlorophyll.

Chlorophyll traps energy from light. This begins an energy cycle. Plant cells with chlorophyll do special work. Their work is food-making. You—and all the world—depend on green plant cells for food and energy. You—and the world—depend on chlorophyll.

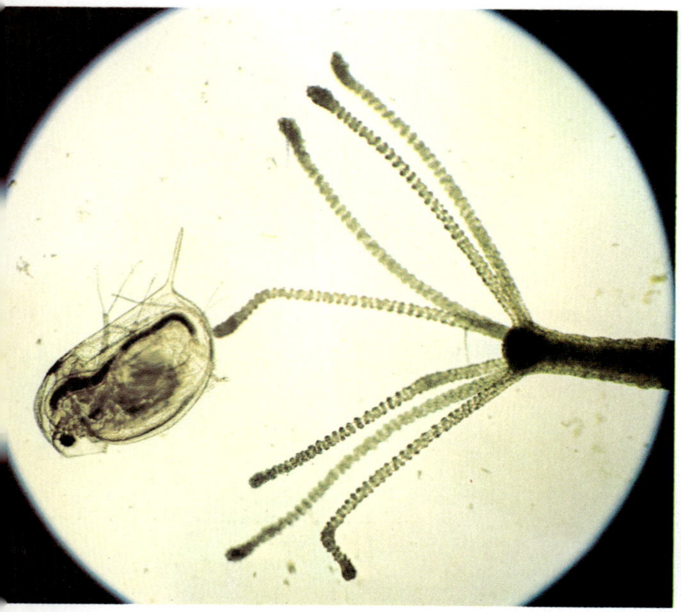

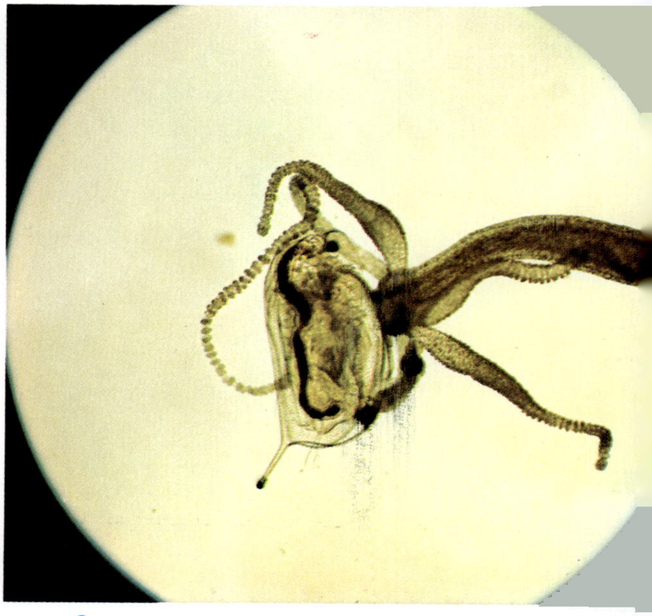

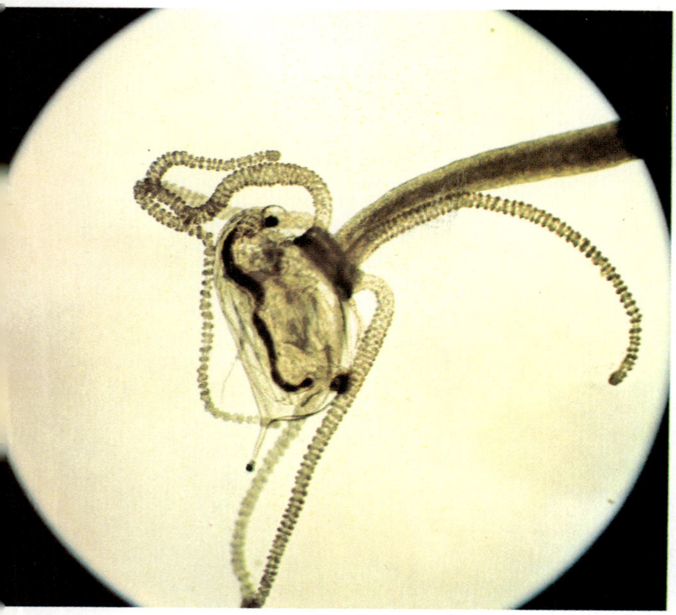

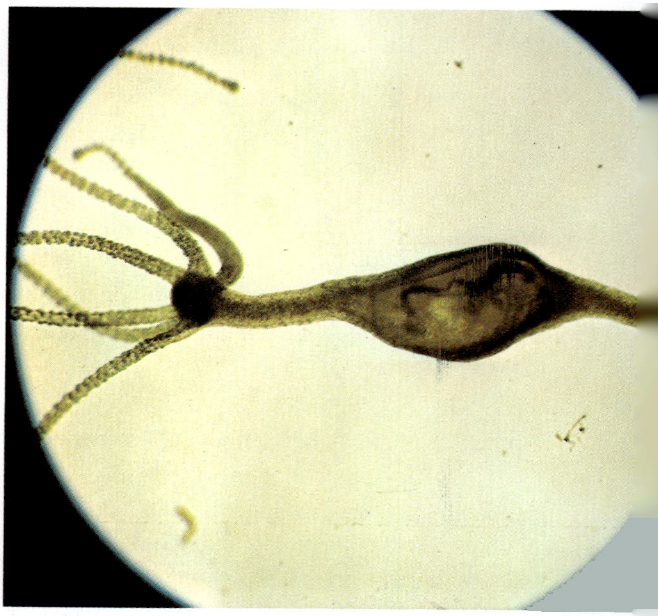

The Earth As Man's Home

5

INTRODUCTION

The Hydra And Its Prey

You are alive. An eagle, a lion, and an oak tree are also alive. And so is the hydra you see in the pictures on page 142.

Remember the hydra (page 31)? A hydra looks like a small piece of thread in the water. Most hydras are smaller than this line ___.

A hydra is so small you may be surprised at what it eats. It eats other animals. A hydra gets its food by feeding on other animals. That is, the hydra gets its energy for life by feeding on other animals.

The hydra lives in fresh water, like lakes and ponds. Another kind of animal, the water flea (FLEE), also lives in fresh water. Look what happens when a water flea comes near a hydra. ■ Look at the thread-like arms of the hydra.

As the water flea comes near the hydra, suddenly it stops—as if it has been shot. And this is what happened. The hydra shot a poison dart into the water flea. The dart stops the water flea. It may even kill the water flea.

Now the hydra uses its arms. ● It pushes the water flea into its mouth. ▲ Then the water flea goes inside the hydra. There it is digested. ♦

To live, the hydra needs the right surroundings

(sur·ROUND·ings). And it needs the right food. The hydra lives in fresh water. So do water fleas. The hydra eats water fleas. The hydra has the right surroundings. And it has the right food.

Each living thing needs the right food. Each living thing needs the right surroundings, or **environment** (en·VIE·run·ment).

What about you? You need the right food. You need the right environment, too. Your environment is everything around you. It is the air, the water, the soil, the plants, and all the animals around you. And more, too. In a sense, your environment is the earth, your home.

What happened when man went to the moon? He had to take part of the earth's environment with him. He needed to take the right environment to the moon.

Let's find out more about your environment. Let's look closely into your environment—and that of the hydra.

1. THE EARTH: FIT FOR LIFE

Man has landed on the moon. Will he land on other bodies in space? What do you think? What is out in space?

Way out in space are masses of clouds thick with stars and gases. Each mass is called a *galaxy* (GAL·ek·see). A galaxy is a large mass of stars and clouds and gases. There are many galaxies in space.

Galaxy

The galaxy that interests *us* is the Milky Way galaxy. Why? Our earth is part of this galaxy. Our sun and the earth are at one end of this galaxy.

See the arrow in the Milky Way galaxy? ■ It points to our sun. And to the earth and the other eight planets that go around our sun. This is our solar system.

There are other suns, bigger suns, in this galaxy. But this one star is *our* sun. Why is our sun important to us? This is the sun we depend on for light and heat. All plants and animals on earth depend on this sun, or star.

Energy from 93 Million Miles Away

You probably think of the earth as your whole world. Most of us think that way. But some people say our world goes way out into space. Is this so?

What do green plants depend on? Sunlight. That is true. Chlorophyll of green plants can only do its work if it gets light. It needs light from a sun that is 93 million miles away from the earth. In a way, we can say our world goes beyond earth and way out into space.

Life on earth depends on light from the sun. Yet only a small part of all the light from the sun gets to earth. The rest goes off into space in other directions. Only a small amount of the sun's light energy gets to earth. But this small amount of energy is enough for chlorophyll to do its work.

Chlorophyll picks up the light energy that gets to earth. This trapped light energy is then changed into a different kind of energy. What kind?

You know that green plants use light energy to make sugar. Sugar is made in the leaves of plants. Sugar is chemical energy. This is the chemical energy that cells use. Without this energy, cells would not be able to make ATP. Cells would die. There would be no life on earth.

Do you suppose that there are other worlds like ours out in space? Other worlds that depend, as we do, on a sun for energy? There is much out in space we are not sure of.

Here is another picture of a galaxy. ● Are there planets moving around the suns in this galaxy? Some of these planets may get energy from their sun. They may be fit for some kinds of life. We do not know.

●

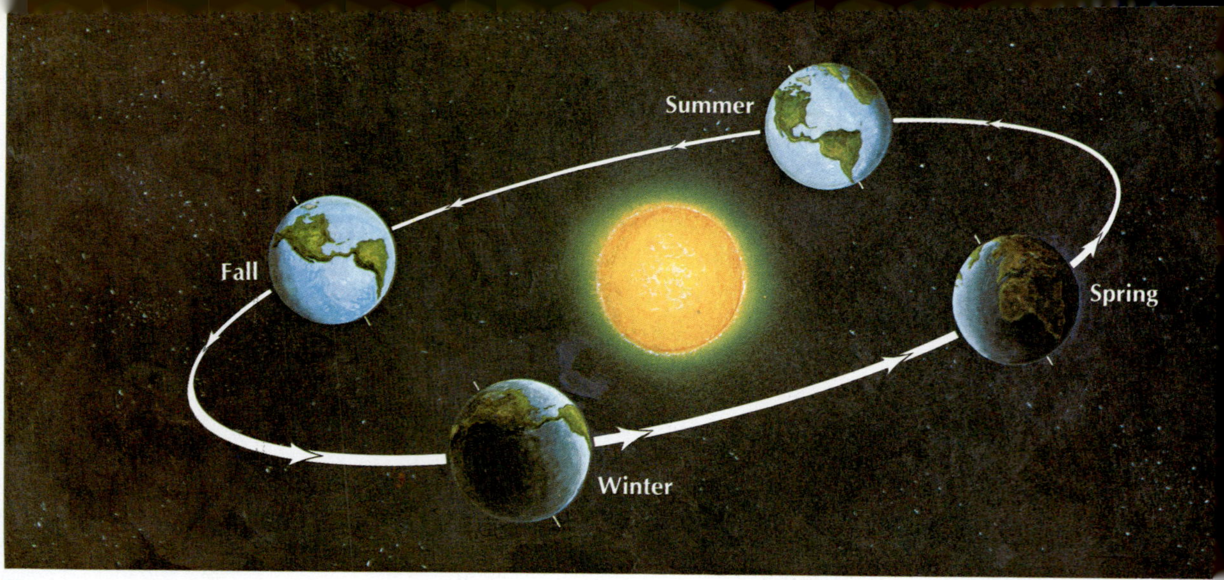

Our Earth: A Closer Look

Let's look closer at our earth and sun. Our earth spins around the sun. Our earth is rushing around the sun at a speed of 66,500 miles an hour. In one year, the earth makes one full circle around the sun.

The earth spins around the sun in a fixed path. As it moves around the sun, the earth goes through seasons. Spring, summer, fall, winter—all these seasons come at fixed times of the year. ■

The earth also spins like a top as it moves around the sun. It makes one full turn in 24 hours. It is this turning that gives us light and darkness. Day and night come as the earth turns on itself. ●

Our earth has different seasons. There are also day and night.

What is it like at any one place you might be on earth?

The Earth: Land and Water

Look at a globe. The globe shows our earth. Spin your globe the way the boy in the picture is spinning his globe. ▲ What do you see? You see a lot of water and some land. The land masses are called *continents* (CON·tih·nentz). Only about one fourth of our earth is land. About three fourths of the earth is water. ♦ This is not much land for all the people on earth. It is not much land for all the animals and the plants on earth.

Which are the largest oceans? Can you name them? (Check your answers with the names on the bottom of the page.*)

At the top and bottom of the globe are the poles. The North Pole is at the top. At the bottom is the South Pole. What is around the middle of the globe? The *equator* (ih·KWAY·tor), of course. The equator is a make-believe belt around the middle of the earth.

Look again at the globe. You can see *more land masses* lie above the equator than below it.

Which continents, or land masses, are found below the equator? Australia is sometimes called "down under." Why? It is not really under. No matter where you are on earth, you would not be under, or on top, of anything.

▲

♦

Places and Climates

At different places on the earth, the **climate** (KLY·mit) is different. Climate is the kind of weather a place has.

What makes climates different? The sun? Yes. The slant of the sun's rays. Look at the picture. ★ If the sun's rays were coming right to place *A* on the globe, it would be warm in place *A*. Places that get *direct rays* of the sun are warm. The climate in these places is warm. It is warm at the equator. The equator gets the direct rays of the sun.

Places like *B* and *C* would not be so warm. They are getting *slanted rays* of light.

* They are the Pacific Ocean and the Atlantic Ocean.

★

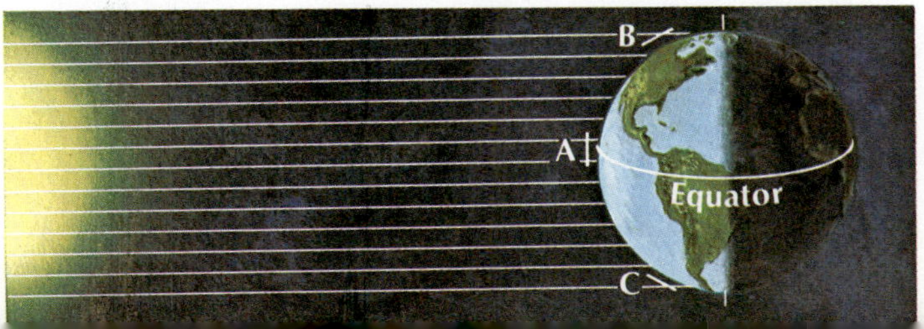

Find Australia on the globe. Predict what climate it should have. Why do the United States and England have much the same climate?

How *long a time* the sun's rays fall on a place also makes a difference in the climate. Some places on earth have a longer day than other places. The sun shines on the land a longer time. Other places have a short day. Which places would be warmer?

Places around the equator have a climate that is much the same all year. Why is this so? Around the equator the day is always as long as the night. There are 12 hours of light. And 12 hours of darkness. You could guess that it should be warmer at the equator than in the United States. Or warmer at the equator than in Russia or Japan.

Is the climate the same in most places all year long? You know there are different seasons. The climate at most places is not exactly the same all year round.

When we talk of the *kind* of climate in one place, we usually mean in a general way. We think of all the seasons taken together. For example, check the temperatures in some places over the earth. ■ You can see about how hot or cold it gets at these places.

CITY	WINTER	SUMMER
Bismarck, North Dakota	9°F	71°F
Detroit, Michigan	26°F	74°F
Albuquerque, New Mexico	35°F	78°F
Boise, Idaho	29°F	75°F
Miami, Florida	66°F	82°F
New York, New York	32°F	76°F
Madrid, Spain	40°F	74°F
Nairobi, Kenya	60°F	67°F
Tokyo, Japan	34°F	79°F
London, England	39°F	64°F
San Juan, Puerto Rico	75°F	80°F
Nassau, Bahama Islands	70°F	82°F

■

A Try at Prediction

1. One complete day is 24 hours. Part of the day is light, and part is dark. In one day the earth turns once on itself. It also moves in its trip around the sun. It goes a little farther along its path around the sun.

Find spot *B* near the equator in this picture.● Does spot *B* get the same amount of the sun's rays all year around? Here are temperatures during a year for spot *B*.▲

AVERAGE MONTHLY TEMPERATURES IN LIMA, PERU

January	74°F	July	62°F
February	75°F	August	61°F
March	74°F	September	62°F
April	71°F	October	64°F
May	67°F	November	67°F
June	63°F	December	70°F

▲

Which temperature shows this spot getting the most direct rays of the sun? Is your answer like ours? (Check at the bottom of the page.*)

2. Try again. What kind of clothes would you take with you in July to go to southern Chile? This is southern Chile in July.♦

Climates are different around the world. They are different because the earth spins around the sun. And they are different because of the slant of the earth.

* 75°F. Lima, Peru gets the most direct rays of the sun during the winter months. But the climate is much the same all year.

●

♦

Air Currents and Climates

But climates are different for other reasons, too. As the earth spins around, the *air moves* over the earth. We say there are air currents (KUR·entz). These air currents make the big winds that move over the earth.

Heat from the sun warms the land. The air is heated by the warm, dark land. This air rises up. As the earth spins, this air moves to other places. This moving air is a wind, or air current. ■

The air also can move from the land over the oceans. ■ As clouds of air move over the oceans, the air picks up water molecules. Warm air currents hold the water molecules.

The earth is not smooth like a marble. It has mountains. Sometimes high mountains slow down the moving air. Then this air rises up higher and slides over the mountains. When warm air *rises* near high mountains, the warm air gets *colder*. As air climbs up the mountain, rain or snow falls. ●

Can you say why it rains when the air cools? You have seen something like this happen. Have you ever put an ice-cold can of soda in a warm room? What happens? The can gets wet on the outside. Doesn't it? ▲

The warm air hits against the cold can. Warm air ● holds a lot of water molecules. But cold air cannot. Warm air from the room gets cooler when it hits the can. The cooler air cannot hold as many water molecules as the warm air. So water forms on the outside of the can. This water comes from the air around the cold can. This water is like rain. When warm air becomes cooler, water forms.

This is what happens up on the mountains. The warm air gets cooler as it rises over the mountains. Cold air cannot hold the water molecules. Rain or snow falls out of the air.

It rains or snows on one side of the mountain. Here there are lots of trees and plants. ◆ What do you think the land on the other side looks like? Why is it so dry on the other side? A dry place like this is called a **desert** ▲ (DEZ·ert). ★

The air currents slide down the mountains. This is a dry air mass. It is *dry* because it lost the water as it moved *up* one side of the mountain.

Find the Rocky Mountains on a map. What is the climate of Montana? What kind of climate is found in states like Wyoming? What kinds of plants live in places like Montana and Wyoming?

What kinds of living things can live in these different places on earth? What kinds of plants and animals can live in these different climates?

2. THE EARTH: DIFFERENT ENVIRONMENTS

Look at your globe again. Suppose you could cut the globe open. Then you could spread it out flat on paper. This is one way a map of the earth is made.

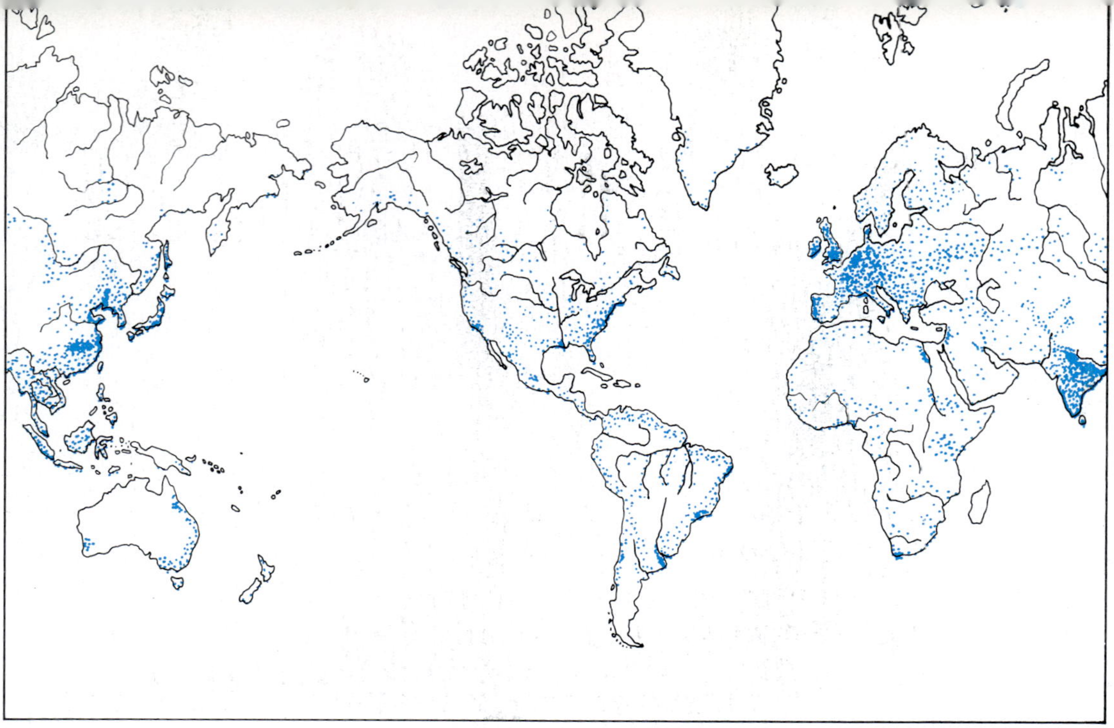

■

Look at this map. ■ You know that man lives on the continents. And plants and animals that are part of man's environment also live there.

But you can see that man does not live everywhere on this earth. People are not spread out evenly over the land. Neither are plants. Neither are animals.

What are some places where it is hard for man to live? On the tops of high mountains. In dry, hot deserts. At the North Pole. At the South Pole. In jungles.

You have learned that the kind of land and the climate seem to go together. This is a pattern. Now you will see another pattern. *The kind of living things also seems to go with the land and the climate.*

Kinds of Environments

There are many special environments on earth. These special environments are called **biomes** (BY·omes). Biomes are named after the kinds of plants that grow in them. The plants of a biome are easy to see. But, of course, animals also live in the biomes. As long as there are plants in a place, animals can go there. Animals need plants for their energy supply. Animals go where plants are, if they can live in the climate.

Here is an example of a biome. Picture in your mind a jungle. Does it look something like the one in this picture? There are many trees and vines. The plant growth is thick. You have to chop your way through to make a path. Everything is green and damp and warm.

Let's picture another plant environment, or biome. What is the climate like near the North Pole? It's cold, and there are frozen soil and snow. There are not a lot of tall trees and vines.

But it is not all snow and ice. Many places in this Arctic area have soil that stays frozen *most* of the year. But there is a short springtime. Small bunches of green plants appear. They are little plants like the bits of green you find on some rocks. Or like the ones on the sides of trees.

■

Another biome on earth is desert. See the plants with thick stems and spines? ■ These are cactus (KAK·tuss) plants. Or there are Joshua (JOSH·oo·ah) trees in some deserts.

Some places on earth have forests. Not all forests are the same. These forests are in Canada and the Northwest. ● This kind of forest grows high up on the mountains over the world. The trees stay green all year. These are evergreen trees. They are pine, spruce, and fir trees. You know some of them as Christmas trees.

Here is a different kind of forest. ▲ These trees lose their leaves in winter. These are oak, maple, walnut, and ash trees. They are the trees that have leaves that turn red, yellow, or orange in the fall. Then the leaves fall off the trees.

Other places on the earth have miles and miles of grassland. Here there are no tall trees. ♦

●

▲

♦

★

◆

⊙

The Right Environment

You see that different kinds of plants live in the special environments on earth. The plants live in the environments that are right for them. They live in the environments that they are *fitted* to live in.

Animals also live in the environments they are fitted to live in. Zebras and antelopes (AN·tuh·lopes) are found in Africa. ★ They are fitted to live in this environment. Tigers are found in Asia. ◆ The kangaroo is found in Australia. So is the wallaby (WALL·ah·bee). ⊙

These are Emperor (EM·purr·or) penguins (PEN·gwinz). ⬢ Emperor penguins are found only in the Antarctic. The polar (PO·ler) bear, on the other hand, is found only in the Arctic, or near the North Pole. It probably could live in the Antarctic, or near the South Pole. But it never traveled that far.

You know that orange trees do not grow in Washington, D.C. They do not grow in New York City or in Detroit. These places do not have the right environment for orange trees.

Spanish moss hangs from the trees in the South. ⬢ Not from the trees in Boston or San Francisco. The Joshua tree is fitted for living in the desert. It is not fitted for the Okeefenokee (oh·key·fih·NO·key) swamp in southern Georgia and northern Florida. Why?

⬢

⬢

A Special Case: The Environment of Andean Indians

Most groups of people have not made their homes on very high mountains. First of all, it is hard to breathe and work in these very high places. Also, it gets very cold at night.

But there are Indians who live high up in the Andes (AN·deez) Mountains in Peru. ■ They are able to live two to three miles higher up than sea level. Up that high there is about one half as much oxygen in the air as where you are. But the Andean Indians need the same amount of oxygen for their bodies as you do.

These people live all year around in these high places. It is cold at night, and it is warm in the daytime. But it is always hard to breathe.

How can these Indians live at 15,000 feet above sea level? They have larger chests and bigger lungs than other people. Their lungs hold more air. There are many small blood vessels in their lungs. Oxygen and carbon dioxide go back and forth through the blood vessels. Red blood cells carry oxygen to all the cells of the body.

Most people have about five quarts of blood. Andean Indians have about seven quarts of blood. They have more blood than other people and lots of blood vessels in their lungs. They have *more* red blood cells than other

■

people. And their red blood cells are bigger, too. A lot of blood goes back and forth to their hands and feet. Even in snow, their hands and feet are warm.

What pumps the blood? A large heart is needed to pump the blood. Andean Indians have larger hearts than other people. This larger heart has a slow beat. It beats slower than the heart of a person who lives at sea level.

How have these Indians become so well fitted for living on high mountains? How do they happen to have larger chests than other people? How do they happen to have larger hearts, bigger lungs, and more blood cells than other people?

These likenesses of the Andean Indians are the kind that are passed along from parents to children. These likenesses have been passed along in the Andean Indians for hundreds of years. We say likenesses that are passed along are **inherited** (in·HER·ih·ted).

Then we must ask another question. How did the ancestors of these Andean Indians become fitted to stay alive on the high mountains?

Andean Indians come from the Inca (IN·cah) peoples of Peru. Hundreds of years ago, the Incas fled high up on the mountains. They fled rather than become slaves of the Spanish armies that had invaded their land.

Some of the Indians who moved to the mountains had stocky builds. And they had big hearts and good lungs. They were able to live high up on the mountains. They were fitted to live there. The Indians who were not fit for such places could not stay there. They had to move somewhere else.

Young people learned from their fit parents how to stay alive there. They worked hard from childhood. They became used to living and working in these high places. ● They were fit to live where most people cannot live. They were **adapted** (a·DAP·ted) to their environment.

These Andean Indians are good examples to explain that people are fit for their special environment. People inherit certain likenesses that may fit them to the environment in which they live. If they are not fit, they have to leave or die.

A Try at Prediction

1. We have talked about different biomes, or environments, over the earth. Is a biome the same in all parts of the world? For example, here are two desert biomes.

The Gobi (GO·bee) Desert is in Outer Mongolia (mon·GO·lee·ah). ■ This is in the northern part of China.

The Mojave (mo·HAH·vee) Desert is in southern California. ●

These two deserts are great distances from each other. In what way are the Gobi and Mojave deserts alike? Yes. They are both hot and dry. There is little rainfall.

But plants and animals can live in both deserts. The animals and plants are not exactly the same kinds. But all are fitted to live in a desert.

Can you say in what ways?

■

●

2. Which picture shows a rain-forest environment?
Where is it? In a warm place? or a cold place?

Can you say why there are many, many kinds of plants and animals in the rain forest?

There are many special biomes on land. Are there special environments in the lakes and oceans on earth?

3. UNDER WATER

Today people can go skin diving far under water. However, they need to take certain things with them. They need to take part of their environment with them under water.

The men who go skin diving must take with them part of their environment. One part they must take is *air*. Of course, you know they take the air they need to breathe. What else do they need to take? What other parts of their environment do they need?

A Skin Diver's Environment

What kind of environment would a skin diver take with him? What would *you* need to stay under water?

First, you would need a skin-diver's suit. You would also need a mask that fits over your face. Why would you need these?

You would need the suit to stay warm. And you would need the mask to see clearly under water. The mask would also keep you from breathing water into your lungs.

■ Here you see a diver walking into a lake from the shore. ■ Air tanks are part of his equipment (ee·KWIP·ment). He uses the tanks to get air to breathe.

These divers are swimming under water. ● Can you see swim fins on their feet? They use the fins to make swimming easier.

●

▲

What if you were a diver? You could stay under water for a long time. You would bring your air with you in a tank. You could breathe under water. You would not be too cold, just warm enough. You would have brought part of your environment with you.

On the Bottom of a Lake

Let's suppose you are a diver on the bottom of a lake. What can you see if you look near the shore? The picture you see was made with an underwater camera. ▲

If you were on the bottom of a lake, you could see several kinds of fish. There are all kinds of small fish. Look closely around you. You can see them eating. They are feeding on smaller animals and plants.

In a way, we can say they are "catching" their *energy*. They get energy by feeding on other living things.

You can see many water plants in the lake. They are also getting energy. They get energy in a different way from animals. Green plants use energy from sunlight for food-making. Near the shore, the water is not deep. So sunlight can get through the water to the plants.

Look again at the water plants near you. You may see some small white threads on the stem of a plant. These white threads are hydras. They are the tiny killers of water fleas you read about on page 143.

■

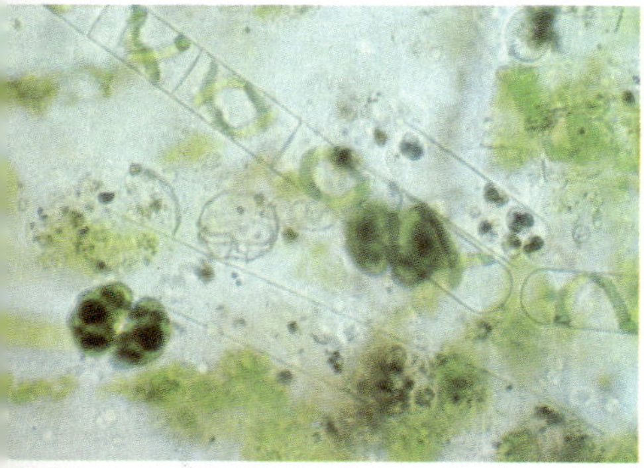

●

▲

A Fresh-water Environment

You know about part of *your* environment. What is the environment of hydras and green water plants? What makes up their environment? What plants and animals live together in *fresh water?* Perhaps you can find out.

Do you live near a pond or lake? Take a large plastic bag to the lake. Use the plastic bag to bring back to class part of the underwater environment.■ You can get a gallon of water from the lake for the whole class.

This picture shows you what might be in the bag of lake water you bring back.● Try to bring back some plants and, maybe, some snails and hydras.

Now you have a bag of lake water and plants and animals. What do you really have in this plastic bag? You have part of a fresh-water environment.

Skin Diving in the Ocean: A Salt-water Environment

Let's go to another environment—an *ocean*. We can go to the Atlantic Ocean or the Pacific Ocean. Or we could go to a faraway ocean, say the Indian Ocean.

Let's go skin diving with a diver. Here again, you need a skin-diving suit. You must also be a good swimmer. The waves in the ocean are often very strong. The water is salty.

You are swimming now in a **marine** (mah·REEN) environment. A marine environment is a *salt-water* environment. *Marine* means "of the sea or ocean."

You are in a marine environment if you swim in the ocean. How is this environment different from the fresh-water environment?

If you are in a warm, southern ocean, you can see fish all around you.▲ What colors! Even the plants, the seaweeds, have very bright colors and shapes.

These plants are not like the plants in a fresh-water lake. The animals are different, too. Here you can find a shrimp or a crab crawling along the bottom. Look carefully. You might see snails, too. The snails, shrimps, and crabs are different from those in fresh water.

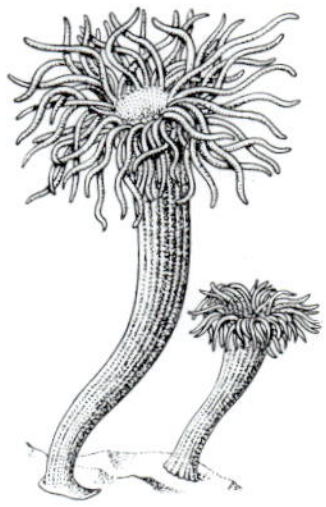

Sea anemone

In the ocean there is something like the hydra. It is the sea flower, or sea *anemone* (ah·NEM·uh·nee). ◆ The sea flower and the hydra don't look very much alike. But the sea flower shoots out poison darts the way the hydra does. It catches food the way the hydra does. It is like the hydra in many other ways, too.

◆

Suppose you fill a bag with sea water the way you did with fresh water from the lake. This salt water will have sea snails and plants. It may also have a sea flower in it.

How will this sea-water environment be different from the fresh-water environment?

A Look at Two Environments

What is the difference between fresh water and sea water? Here is a way to find out. You could boil some ocean water in one dish. In another dish you could boil some lake water.

The water will boil away in each dish. What is left behind? Which dish would you say has some salt left in it? ★

★

A Try at Prediction

Would it make any difference if you put animals and plants of the lake into salty ocean water? Or if you put sea-water animals and seaweeds into a fresh-water lake?

What you would be doing is *changing* the environments of the plants and animals. Would it make any difference? What would happen?

You are right if you think these living things could not live in their new environment. They would die. Can you say why?

A Try at Explanation

1. Why would fresh-water animals die in salt water? How close do you come to the explanation given on pages 165–66?

2. Why must astronauts take oxygen with them? Couldn't the astronauts "learn" to breathe some other gas and to live in it? Why? Why not?

3. Let us say that a small, unknown monkey is found in South America. Some scientists want to study the way it lives. They will take it to a laboratory in Boston.

What will the scientists need to do to *keep the monkey alive in Boston?* Can you tell why?

4. CELLS IN TWO ENVIRONMENTS

■

●

Could you live on the moon? How?

To live from day to day you need energy and oxygen. It would be great if living things could get energy and oxygen *anywhere*. But they can't.

Many living things can't move easily from one environment to another. A fish can't get its food on land. You cannot live under water the way a fish does. A fresh-water hydra can't live in the salty ocean.

Why must fresh-water life get its food in a fresh-water environment? Why can't an animal living in *fresh water* change easily to a *salt-water* environment?

See for yourself.

From Fresh to Salt Water

You will first need a piece of *elodea* (eh·LOAD·ee·ah). Elodea is a fresh-water plant. You will need a microscope, slide, coverslip, dropper, and a blotter. You will also need a *solution* (suh·LOO·shun) of table salt. A solution of table salt is a mix of water and salt.

Place a leaf of elodea in a drop of pond water, or water from a fish tank, on a slide. ■ This is the water elodea grows in. Flatten the leaf with a cover slip. ● Focus with the low power of the microscope. You looked at elodea cells before (page 108).

Now switch to high power. You can see the green chloroplasts. They are in the cytoplasm in the cells. Remember that elodea is a fresh-water plant. Now let's change the environment. Change the elodea cells from fresh water to *salt water*.

Take up some of the salt water in a dropper. Put a drop of the salt water at one side of the coverslip. At the other side of the coverslip put a small piece of blotter. ▲

Watch how the blotter draws the *fresh water*. It draws the water from under the coverslip. Now the *salt water* moves under the coverslip. It takes the place of the fresh water.

The cells of elodea are in a new environment. This is very much like putting a *fresh-water plant into the ocean*.

Study the cells after three minutes. What has happened to the cells? This is what was seen in one class. ◆ Do you find that everything inside the cells of your leaf seems to shrink?

In about 10 minutes the inside of each cell seems to be squeezed together in the middle of the cell. Only the stiff cell walls stay in place. ★

Remember the cytoplasm that fills up most of the cell? All cytoplasm has water in it. The cytoplasm of elodea is nearly all water. In most cells, cytoplasm is close to 90% water. Cells must have water to do their work.

When a cell is placed in salt water, the water from the cytoplasm goes out of the cell. When the water goes out, the cytoplasm of each cell shrinks.

A Problem on Your Own

What if you could bring the cytoplasm back to the right size? Can you return the cytoplasm back to its normal space in the cell? Why not try? You have some fresh water or pond water on hand.

How would you do it? Could you put the elodea back into a drop of pond water on a slide? Would the cells fill up again with water? Why not try it?

There is a point at which elodea cells do not return to normal. That is, sometimes cells lose so much water they die.

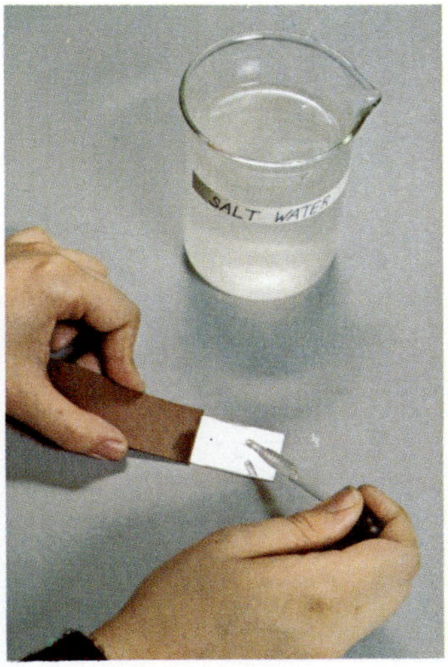

▲

◆

After 3 minutes

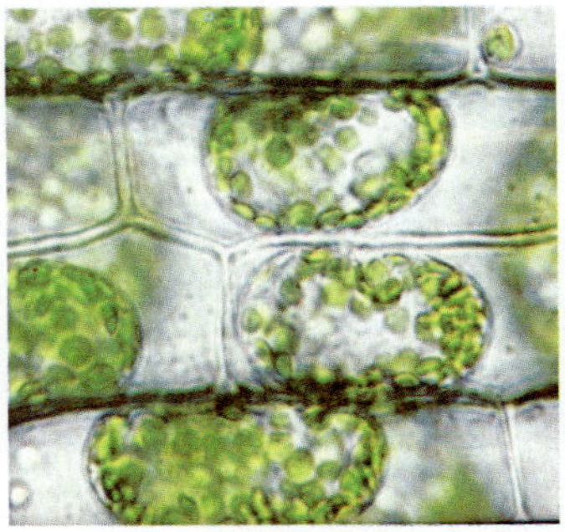
★

After 10 minutes

■

●

▲

◆

An Explanation

You have seen that water left the cells. When these fresh-water cells are placed in salt water, the cytoplasm seems to shrink.

Cells that live in fresh water cannot live in salt water. Their cells lose water when put in a salt-water environment.

Cells live best in their own special environment.

Potato Cells — Two Environments

You can find the same thing in other cells. Use slices of raw potatoes or raw carrots to find out. Put a few slices of potato or carrot into a jar of salt water. ■

Into another jar put some fresh water and some slices of the potato or carrot. ●

You now have potato cells in two different environments. Which environment is *right* for the potato cells?

Compare all the slices after one day. Then compare them again, after *two days*.

What do you find? What happens to the pieces in the salt water? Are these slices wrinkled (RING·kuld)? Did water leave the cells? Here is a picture of wrinkled potato slices in salt water. ▲ Do yours look like this?

Some of the water left the cells in the potato slices. In salt water the cells of the potato or carrot lose water. This is very much the same as what happened to the cells of elodea plants.

What about the potato slices in fresh water? Are they as wrinkled as the slices in salt water? No. The potato slices do not shrink in *fresh* water. ◆ Why? Let's see.

Cells Are Fit for Their Environment

Fresh water organisms live in fresh water. They cannot stay alive in a salt-water environment. The amount of water in cells of living things is important. It seems to fit the cells to their environment. *Cells are fitted for the environment in which they live.*

This is why living things in fresh water cannot easily be changed to salt water. The fresh-water cells that are put in salt water lose some water. Their cytoplasm shrinks. Then the cells die. When their cells die, living things die.

You must understand what we mean when we say a living thing is fitted to its environment. You can say a living thing gets along best in the place it lives. Salt water is right for salt-water plants and animals. Fresh water is right for fresh-water plants and animals.

Salt-water plants die in fresh water. The cells of a salt-water plant *swell up* in fresh water. Water goes into its cells. Then the cells swell up and burst. Salt water plants are not fit to live in fresh water.

Adding to the Concept

Your concept of the cell grows. First you learned the cell is the unit of *form* of a living thing. You also learned the cell is the unit of *function* of a living thing.

Now you can add another idea. *Cells of plants and animals are adapted to live in a special environment.* Adapted means fitted to live in a certain environment.

You are alive. But you are alive only in a very special environment. You are fitted, or adapted, to a certain way of living. You live on land. Your cells are adapted, or fitted, for life on land.

An elodea plant lives in fresh water. Its cells are adapted, or fitted, for life in fresh water. A hydra is adapted for life in fresh water, too. A salt-water fish lives in salt water. Its cells are adapted, or fitted, for life in a salt-water environment.

Each living thing has its own special environment.

Finding the Right Environment

Sometimes we can make the right environment for a plant or animal. For example, the polar bear lives in the Arctic. But it can live well in a zoo. How? Because we copy the environment for it. We copy the environment it is adapted to. ★

★

In a hothouse we can make the right climate for a palm tree. We make the dampness and temperature the same here as in the environment it came from.

AN INVESTIGATION

What is a Right Environment?

It takes time to find out what each plant or animal needs in its environment. You can take one living thing and find out what it needs.

You can use a white potato. It is easy to grow potato plants. What is the right environment for potato plants?

A white potato is really part of the stem of a plant. It is a stem that grows underground and stores starch. New plants grow out of the buds or "eyes" on the potato. This is a new plant that is starting to grow. ■

This is how farmers grow potatoes. They cut potatoes into many pieces. Each piece has two or three buds or "eyes." These pieces are planted in soil.

Does it make any difference whether potato buds get light? Do they need a special temperature? How much water do they need? Light, heat, and water are part of the environment of a living thing.

You can grow potatoes the same way a farmer does. Take a large white potato and cut it into six equal pieces. Be sure each piece has a few "eyes."

Weigh the six pieces to see if they are the same weight. If one weighs more than the others, you can cut off a little. Keep cutting off a little at a time until this piece weighs the same as the others.

Now you need six plastic cups or small pots. Put some washed sand in each pot. Then plant one piece of potato in each pot. Bury each potato piece about one inch under the washed sand.

Look at the chart across the page. ● It shows you what to do with the six pots of potato buds. See that five pots are watered. Do not water the sixth pot. Keep the potato piece in this pot in *dry* sand.

Put one pot in a very light place. Put the second pot or cup in a dark closet. And place the third one in a refrigerator. Be sure these pots have enough water.

What should be done to the fourth, fifth, and sixth pots? Look at the chart. Yes. Keep the fourth one at room temperature in a little light.

The fifth pot should have a *lot* of water. Give it more water than the rest of the pots.

Remember the sixth pot is *not* watered at all.

Check the plants at the times the chart shows.

▲ **After 6 days**

◆ **After 9 days**

★ **After 12 days**

Here are pictures of what several boys found. Which plants have grown after 6 days?▲ after 9 days?◆ after 12 days?★ after 15 days?◇ How do your results compare?

You can see that some of the potatoes grew. Which one looks normal? What do potato buds seem to need to grow well? The right temperature and the right amount of water. They also need light to turn green.

Temperature, water, and light are parts of the environment of the potato plants. You can say that they are three **conditions** (con·DISH·onz) of the environment.

Potato plants grow well if all three conditions are right. These are only three conditions in the environment you are studying. Water, light, and temperature.

If one of these three conditions is *not* just right, then something the plant needs is missing. This one condition that is not just right stops the growth of the plant.

◇ **After 15 days**

			OBSERVATIONS		
Planted potato pieces with buds	**What environment the potato buds get**	**After 6 days**	**After 9 days**	**After 12 days**	**After 15 days**
Pot 1	Room temperature in bright light (with water)				
Pot 2	Room temperature in a dark closet (with water)				
Pot 3	In a refrigerator (with water)				
Pot 4	Room temperature (with a little light and water)				
Pot 5	Room temperature with plenty of water (no way for water to drain out: cork the hole)				
Pot 6	Room temperature without water				

Suppose there is the right amount of light. Good. And the right temperature. Good. But suppose there is not enough water. The plants do not grow well without water.

Another time there may not be enough light for the living thing to grow. Living things need all the conditions that are right for them.

A Try at Explanation

Anthurium (an·THOOR·ee·um) is a Hawaiian plant. It has bright red heart-shaped leaves. Its environment is a tropical jungle. In the jungle the anthurium lives in rich soil in full shade. It does not grow flowers in full sunlight.

These plants need a special environment. They need

- a temperature that does not go below 50°F
- dampness
- rich soil
- shade

In Hawaii, people grow the plants under wide spreading ferns. They copy the jungle environment in every way.

What do these plants need to grow flowers? What other conditions have to be right for them to live well and to grow?

A Try at Prediction

1. A family from Illinois goes to the seashore for a vacation. They find some plants called beach plums. These plants grow wild in the sand near the beach. Father and son carefully dig up some of these plants. They take them home to Illinois. They put the plants in the soil of their own backyard.

What do you predict will happen? Why?

2. A boy raises some tropical fish called guppies (GUP·eez). He uses air pumps to pump air into the tanks for his fish. His guppies grow well and have large numbers of young guppies.

He gives a pair of guppies to a friend. This friend keeps his fish in a tank just like the one the first boy has. He does everything the same as his friend except for one thing. He does not have an air pump. His guppies grow up fast.

What do you think will happen in the tank? What will happen as the fish reproduce? What will happen when the tank has a lot of fish? Why?

Is it the same with us? Do people need special environments? Yes. We need special environments to grow and be healthy.

5. THREE LIVING THINGS

Animals catch their food. They eat what they need. This is how animals get energy and grow.

How does the hydra get its food? Remember how the hydra kills a water flea? The water flea is food. The hydra gets its energy from the water flea.

But what about the other water fleas? Some are *not* eaten by hydras. They live and grow and reproduce. They need food for energy. How do these water fleas get their food?

Both hydras and water fleas live in a fresh-water environment. They usually live in lakes or ponds. But they could live in a fish tank in your classroom.

Sometimes water in a fish tank turns green. Look at some of this green water. You will need a microscope to see what causes the water to turn green. You will see green cells in the water. These green cells are, for the most part, the plants called *algae*. There are many kinds of algae. (One cell is called an *alga* [AL·guh].)

If you could make a slide of this green water, you would probably find many algae. One kind you may find is chlorella (kluh·REL·uh). Chlorella is a fresh-water alga. Here are the round green cells of this alga.● You may also find some other kinds of algae.

If you can make this slide of green water, add some water fleas. Now you have water fleas on the slide of chlorella. Watch the water fleas feed on the chlorella.

●
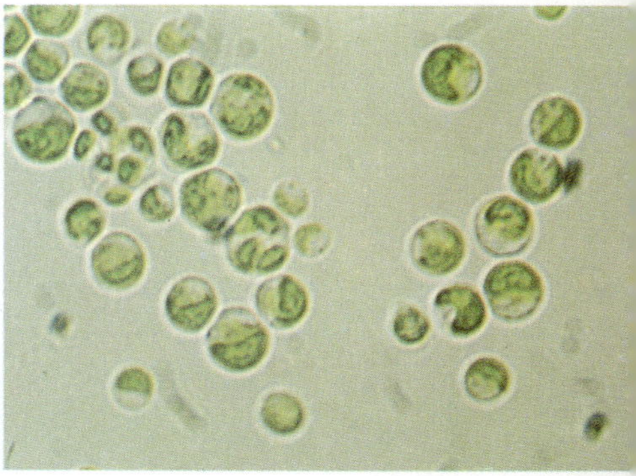

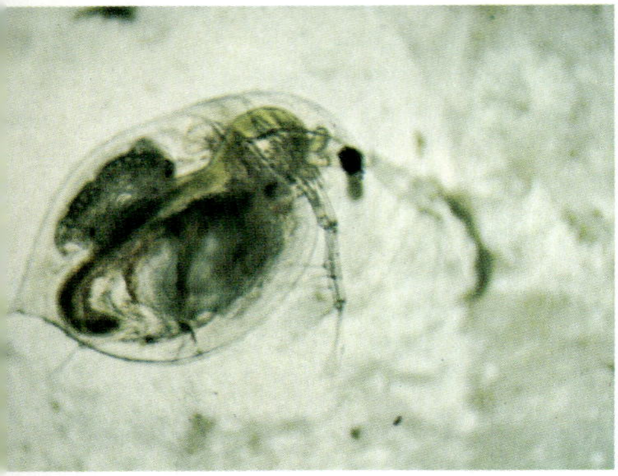

Watch the water fleas feed on the round green cells. Hundreds of these green cells go into the water fleas. In fact, you will soon see the whole food tube of each water flea turn green. ■

What happens to the hundreds of green cells trapped in the food tube? You are right if you think the green cells will be digested. They will be digested in the food tube of the water fleas.

You know this is what happens to the food *you* eat. This is also what happens in a grasshopper, in a horse, or in a zebra.

A Food Chain: A Chain of Energy

What would happen if the water fleas ate all the green cells? The water fleas would die off. They would have eaten up their food. The water fleas would not be able to get energy for life. The water fleas would have changed their environment. They would have no more food in their environment.

Hydras capture water fleas. Water fleas capture green algae. This is called a **food chain.** ● Green plants are eaten by water fleas. Water fleas are eaten by hydras.

What is at the start of this food chain? Green cells—the green algae. You can see that the green cells are needed for both the water fleas and the hydras. You may say that a hydra does not eat the green cells. That is true. Hydras eat the water fleas.

But what does the water flea eat? In this case, it eats the algae in the water. Water fleas may also eat tiny water animals. What do you think these tiny animals would feed on? Yes, algae or other tiny green plants or tiny animals.

What if there were no more green algae? Soon there would be no water fleas.

Would the hydras live?

●

No. In the case we are studying, hydras eat water fleas, mainly. When the water fleas died, the hydras would have no food. The food chain would be broken. *When a food chain is broken, one kind of animal has no food.*

One way or another, a *food chain goes back to a green plant.* Or so it seems. In this case, the hydras clearly need the water fleas or small animals like them. The water fleas need the green algae—or tiny water animals. Both the hydras and the water fleas are *consumers.*

Now you will want to guess what the tiny water animals feed on. Make an "educated guess."

Yes, it is the same story if the food chain is in the water or on land. *Consumers* depend on *producers* for their energy. What are the producers in this food chain in the fish tank? ▲

▲

6. ONE TIGER ON A HILL

There is an old Chinese saying that only one tiger can live on a hill. How can this be? Why can't a hundred tigers live on a hill? The answer, of course, goes back to *food.* What food does the tiger eat? Doesn't the tiger hunt another animal for food?

One tiger gets a lot of animals in a month or in a year. A hundred tigers on one hill would wipe out all the animals they feed on. They would wipe out their food.

There is a sort of **balance** (BAL·ens) needed. Balance, here, means the right number of each living thing in one place.

There must be enough animals to go around to feed the tigers. If there are only a few in one place, how many tigers could live in that place? How many tigers could live on a hill?

The triangle (TRY·an·gel) tells the story. ◆ Which animal is at the top? Yes, a tiger.

Which animals are shown in the next larger part? Grazing animals, mainly. Grazing animals are animals that eat grass. This space in the triangle is larger than the

◆

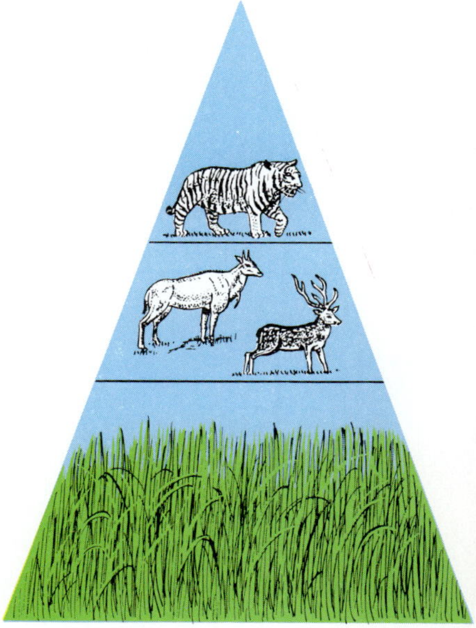

tiger's space. This means there should be more grazing animals than tigers in one place. Tigers need many animals for a food supply. One tiger needs a large number of the animals it feeds on. ■

Look at the bottom of the triangle. This space is the largest of all. What is in it? Grass. Why is the largest part of the triangle grass? A large amount of grass is needed as food for the grazing animals. These grazing animals, in turn, are food for one tiger. Again you see a food chain. *At one end of the food chain are green plants.*

Of course, the Chinese saying does not mean only *one* tiger can live on a hill. It means that only a certain number of animals like tigers can live in one place.

If too many tigers live in one place, they eat up all their food too fast. Then they lose their supply of food for energy. What would happen to them? With no way to get food for energy, the tigers would soon die out.

How Does This Food Chain Work?

Plants and animals seem to live in a *balance*. Finding out how they all live together is like solving a mystery (MISS·ter·ee). Let's look into this a bit further.

On a certain island off South America there live some birds called finches (FIN·chez). There are many kinds of finches on the island. Let's look at two kinds that live in the *same* place.

Both kinds eat seeds. Both kinds get along fine. They both get enough to eat. *How can they live together when they both look for seeds for food—for energy?* Are there enough seeds for *all* the finches?

If you watched these birds for long, you would soon see the answer. One kind of finch has a large beak. ● These finches eat large seeds.

What do you think the other kind of finch feeds on? Look at the picture. ● You can see this other kind has a *small* beak. Of course! These finches feed on very small seeds.

These two kinds of finches can live in the same place. They are both seed-eating birds. But they do not get in each other's way. They eat seeds of different sizes.

They live in a balance. They get their food for energy in different ways.

These finches do not fight with each other. They can live in the same place because they do not use the same food to stay alive.

Here again the food chain goes back to green plants. These birds all eat seeds. Seeds are food. Seeds come from green plants.

You and a Food Chain

Are you part of a food chain? Do you get your food from green plants? Perhaps you eat meat that comes from a cow. Cows eat grass for food. Grass is a green plant. Look at the food chain. ▲

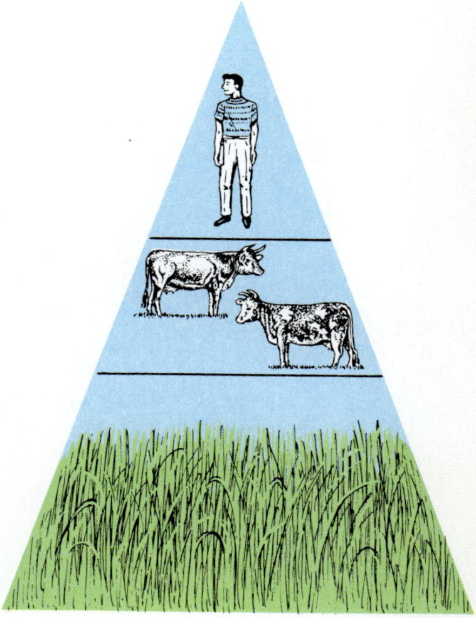

▲

Plankton

You may eat fish for food. ♦ Large fish may eat small fish for food. Small fish may feed on **plankton** (PLANK·ton). Plankton are masses of tiny animals and green plants that drift through water.

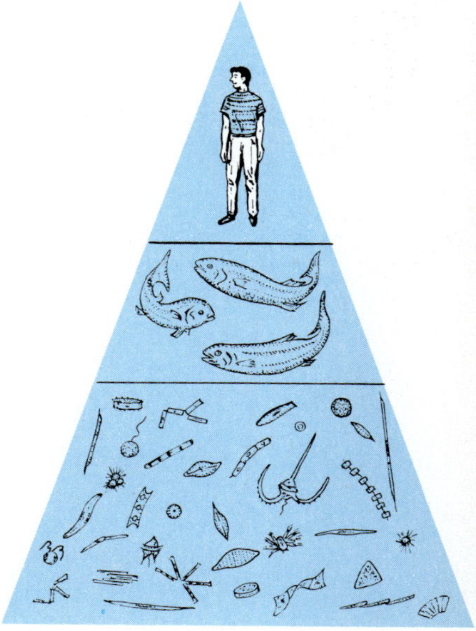

♦

What is at the beginning of the food chain? What uses the sun's energy to make food? Green plants or animals?

Green Plants in the Food Chain

The tiger captures grass-eating animals, mainly. The grass-eating animals get their energy from grass. Grass plants get their energy from the sun. Green plants, such as grass, make food.

Tigers, birds, whales, and fish all need food. Sparrows, elephants, you, and your friends all need a food supply.

Where do all animals get food for energy? From green plants.

Green plants are at the beginning of all food chains. They are the producers. Without green plants, animals would soon die off.

There is a saying that all flesh is grass. In a way, this is true. The saying means animals and green plants are tied to each other in a food chain.

A Try at Explanation

Look at the grasshoppers in the field near the stream.■ Each one is getting energy. What do grasshoppers use as food? Grass and the leaves of other green plants. Grasshoppers get energy from green plants.

This is the beginning of a food, or energy, chain. In the same picture you can find frogs and snakes. They need food for energy, too.

Now look for the food chain in the picture.■ Begin with grass and end with a snake. Then check your answer below.*

Are there more grasshoppers or more frogs in this field? Which do you think? Are there more frogs or more snakes in the field?

Snakes eat frogs. A small number of snakes need a large number of frogs. What about the frogs? This **population** (pop·yuh·LAY·shun) of frogs needs a *very* large number of grasshoppers for food. Population means the number of one kind of living thing (like the grasshoppers) in one place.

■ And the grasshoppers? What do they need? They need many fields of grass.

What would happen to the food chain if grasshoppers disappeared? There would be plenty of grass. But the frogs would die off. Frogs need the grasshoppers for food. The snakes would soon die, too. Why?

7. THE FOOD CHAIN IN WATER AND ON LAND

Near Alaska are some islands where many seals live. Some of these seals are trapped for their fur.● A few years ago the fur trappers were worried because the seals were dying off.

It had been a strange year. The ocean waters were colder than usual. When the waters had been warmer,

* *Grass* is eaten by *grasshoppers*. *Frogs* eat the *grasshoppers*. Where do snakes get their energy? *Snakes* eat the *frogs*.

they had carried many living things. These warmer waters usually brought seaweeds and the tiny plants and animals called plankton to the islands. The ocean near these islands had usually been like a big field full of food.

Large groups of fish used to follow the plankton. The fish fed on the plankton. The plankton were food for the fish.

These fish were food for the fur seals. The seals got their food for energy by catching fish.

But all this changed. The ocean waters changed their route. The warmer waters no longer came near these islands off Alaska. The water became cold. There were very few plankton in this cold water.

Now can you say why the seals were dying off? If enough plankton did not come near the islands, the fish did not come either. The fish follow the plankton. If enough fish did not come, the seals had no food.

Look how all these livings things are tied together. ▲ Take away one part of the chain. Suppose the floating plankton disappear. What happens? No fish and, soon, no seals either.

Which are the producers in this food chain? Which are the consumers?

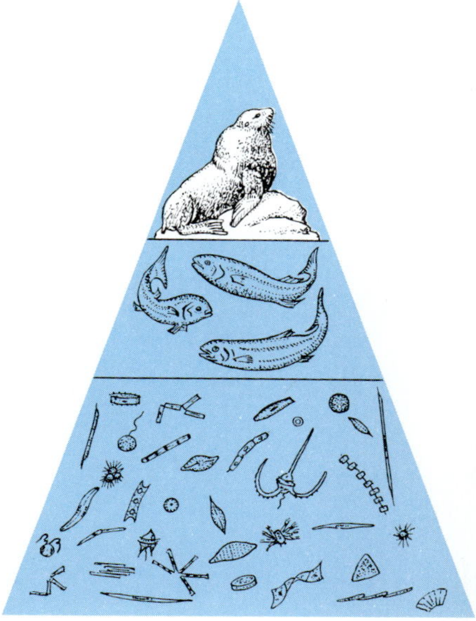

AN INVESTIGATION

Can Paramecia Live Alone?

Can tiny living animals like paramecia live alone? Do they really need green algae like chlorella? Let's find out if paramecia can live alone.

You will need three slides like those in the picture. ■ This kind of slide has a place in the middle for water. It is a kind of well in the glass. The well is deep enough for a drop or so of water.

Put a drop of water full of paramecia in the bottom of the well of one slide. Then put a drop of water with chlorella on the second slide.

On the third slide, put the *two together*. That is, put a drop of water with paramecia and a drop of water with green algae (chlorella) *together* on one slide.

Seal each slide so it is airtight. Do this by dipping a toothpick into Vaseline (VASS·uh·leen). Make a circle of Vaseline around the well on each slide. ■

Carefully put a coverslip on top of each of the slides. The cover slip sticks in the Vaseline. ● See how the water is trapped inside. It cannot dry out.

Look at the slides under the microscope. Try to remember about how many paramecia and algae are on each slide. Leave the slides in light, but not in strong sunlight.

Then look at your slides again after three days. Has anything changed? After five days, look again. What has happened to the cells on the slides?

■

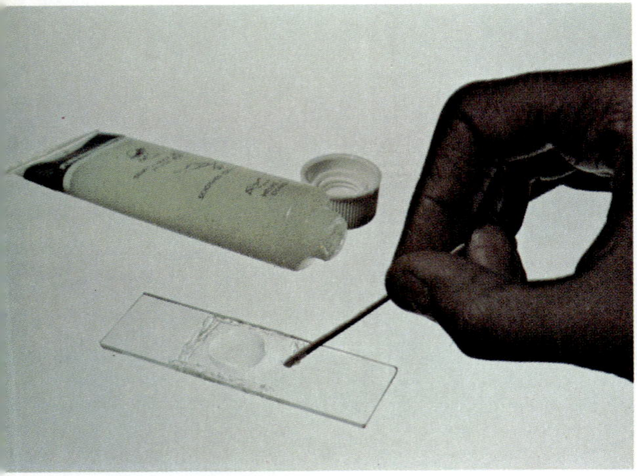

●

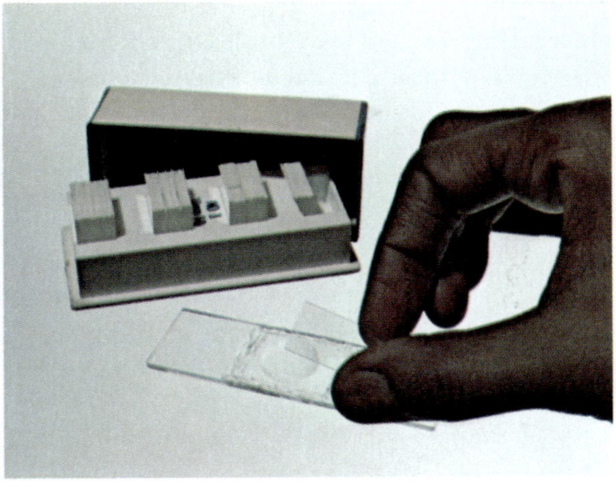

Do the paramecia seem to be dying out? On which slides? *Only on the one where the paramecia are alone,* without any green algae. ▲ Here the paramecia have no food for energy.

Notice how fast the paramecia swim in the drop that has green algae. ◆ They have food, plenty of food. They have energy. They can swim fast. They can reproduce. Do you find more paramecia than you had when you made the slide?

Can the green algae live alone? ★ Of course they can. Green algae can use the energy from sunlight. *They make their own food.*

The Food Chain on Land

You have probably seen pictures of hawks. This is a chicken hawk. ◇ You would not think of a chicken hawk as your enemy. But farmers know chicken hawks are enemies of chickens.

Chicken hawks attack chickens for food. Many farmers try to kill chicken hawks whenever they see one. These farmers do not know how useful a chicken hawk is. The hawk is useful even if it kills a few chickens.

In one place, farmers began to kill the chicken hawks. Then their chickens were not eaten. But pretty soon these farmers began to find something else wrong.

Their store of grain in the barns was eaten by rats. Soon the rats were overrunning the farms. Can you say where all the rats came from?

Again, the Food Chain—and a Natural Enemy

Think of it this way. We know that chicken hawks do kill *some* chickens. But what is the *main food* for the chicken hawks? Their main food is *rats.* The hawks eat more rats than chickens.

Hawks, rats, and grain fit together in a food chain. Each one is part of the chain. One living thing depends on the other.

It is true that chicken hawks feed on chickens. They also feed on frogs, rats, and other small animals. You cannot look at only one part of the food chain.

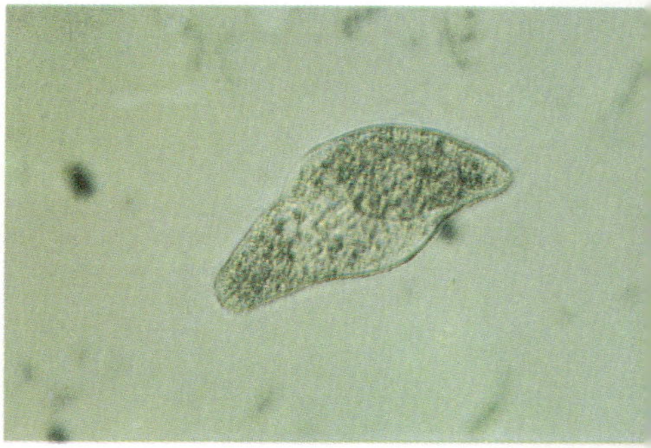
▲

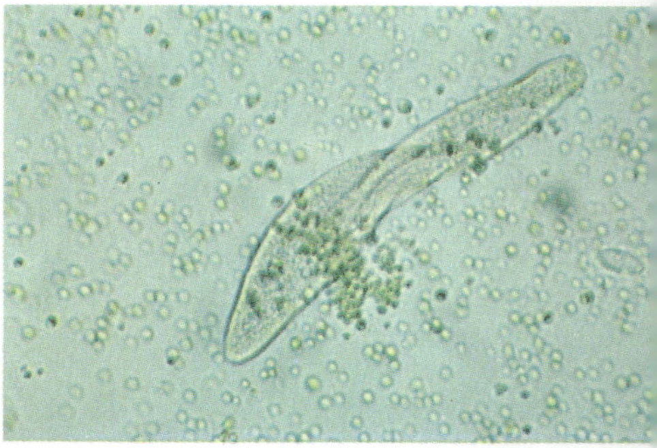
◆

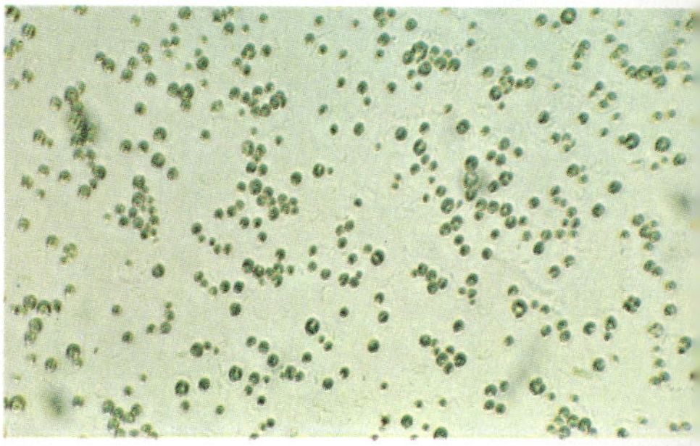
★

◇

In this case the farmers looked only at what seemed to bother them. Their mistake was seeing the chicken hawks in only one way. They counted only the number of chickens the hawks were taking.

No one seemed to have looked at the other parts of the food chain. The farmers forgot about the whole chain of living things in the place where they lived.

The hawks kept down the number of rats.■ With the hawks gone, *rats had fewer enemies*. Rats reproduce very fast. Soon, there were too many rats. They lived in the same barns full of stored grain. Before, there had been some rats there. They took very little food. Now, there were so many rats they could ruin the farmers.

The farmers had made a mistake. They had removed a **natural enemy** of the rats—the chicken hawk.● Now one of the main enemies of the rats was gone. Rats were the farmers' new pest.

A Try at Explanation

Is there a food chain that does not start with a green plant—a producer? Let's see.

Here is a picture of a pelican (PEL·ih·can).▲ A pelican is a bird that feeds on fish. These fish feed on smaller fish. What do the smaller fish use for energy? The very small fish feed on some kinds of baby shrimp. These tiny animals are part of the floating plankton.

Are there green plants in this food chain? Look closely at the floating plankton. What would you see if you looked at the plankton under a microscope?

Of course, there are green cells here. The tiny animals that are part of the plankton eat the algae. The algae are the start of the food chain that leads to the pelican.

Try to name the steps in this food chain. Begin with the producers and end with the pelican. Write down the consumers in the food chain step by step. Is your food chain like ours? Check below to see if you are right.*

* Green algae capture the sun's energy. Baby shrimp and other tiny animals live on the green algae. The consumers, in turn, are eaten by larger animals, the small fish. Still larger consumers, the bigger fish, eat the small fish. Pelicans swoop down into the ocean and eat the herring or larger fish. In a roundabout way, the pelicans are really living on the sun's energy.

A Try at Prediction

Look at the otter (OTT·er) and all the other animals and the plants in this environment.◆

Can you describe the food chain? Start with the living things that first capture the sun's energy. The otter eats fish. Would it matter if the plants died off? What is your "educated guess"? Scientists have a word for an educated guess. They call it a hypothesis (hy·POTH·ih·sis). They base an educated guess, or hypothesis, on what they have learned from what they have seen and done.

◆ 

Once Again—The Concept

In a small way you can do what scientists who study the environment do. Let's say someone shows you a *frog*. You know the frog did not come from the ocean. The frog would come from a place that had fresh water. You know the frog is adapted to a fresh-water environment.

Suppose you saw some blue crabs and oyster shells and seaweeds together in one place. You would know they came from a salt-water environment. You would not say they came from a desert. You would not say they came from a lake or a woods.

Here is another example of what we mean. Suppose you have a box of mosses and willow branches. There are also some cattails in the box.

In what kind of environment would these plants be growing? Would you say desert? open meadows? damp woods? Where?

You may not know all of these plants, but you might know moss. You would say that moss grows in damp places. Cattails also grow in wet swamps. Do these plants tell you what their environment is? Where do mosses and willows and cattails live? Would you expect them to live in a place like this?★

Plants are adapted, or fitted, to grow only in a certain environment. Animals, too, can live only in very special environments.

★ 

AN INDEPENDENT STUDY

A Study of Two Environments

You have been learning about the many different environments on earth. You have read about the kinds of plants and animals that live in different environments. What will happen if you put an organism that lives in one environment into a different one? You know now that it will die. That is, it will die unless we copy the right environment for it. Plants and animals are adapted to live in their own special environment.

Remember the two water environments? What did you learn about the plants and animals that live in the ocean? and about the ones that live in a pond or a lake?

You can *find out* for yourself in this independent study. Read the questions and put your answers on your answer sheet. Then look across to the next page to check your answer.

Are you ready? Be sure to check each answer before you go on the next question.

1

Look at the picture on the next page. ■ The boys are fishing in the ocean.

When the tide comes in, the water is high. This is called high tide. When the water goes out, it is low tide. At low tide you can see algae on the rocks.

Is the tide *low* or *high* in the picture? (How can you tell?)

(Write your answer on your answer sheet. Look across to box 1 to see if you are right.)

1

The tide is **low.**

You can see many animals and seaweeds on the rocks.

(Turn the page to question 2.)

2

Look at the picture on page 183 again. Have you ever seen seaweeds like those on the rocks under the pier? Or barnacles (BAR·na·kulz) growing on the rocks? Barnacles grow on rocks or boats. You can cut yourself on the sharp shells of barnacles. But inside the shells are small animals.

The seaweeds are marine. That is, they grow in salt water. At low tide the seaweeds are in the air. How can they live for hours out of water? They have thick leaves with air pockets. They are fitted for their environment.

The barnacles live in salt water, too. On the outside, each barnacle has shells that can open and close. How can the barnacles live at low tide? When the water goes out, the barnacles are in air. The shells close, and the animal is safe inside. They are fitted for living in air at low tide.

The seaweeds and barnacles are fitted for their e__________.

3

When the tide comes in, water covers the barnacles again. Then the animals in the barnacle shells can open their shells and take in food and oxygen. Barnacles must catch their food.

Under water, the *seaweeds* can take in gases and make their own food. The seaweeds are wet enough even in the air to make their food. They do not die out.

Which living things make their own food? Which are producers—the *seaweeds* or the *barnacles?*

(Write your answer. Check across to the next page to see if you are right.)

2

environment

They live in water, but they can live at low tide, too.

(*Go on to question 3.*)

3

the **seaweeds**

The seaweeds are producers.

(*Turn the page to question 4.*)

4

Why are seaweeds producers? Seaweeds make their own food.

Look back again at the picture on page 183. Flounders (FLOUN·derz) are the flatfish you can see feeding on the bottom. They get their food from the ocean floor. They may eat fast-swimming fish that die and fall to the bottom. Or they may eat small animals that live on the bottom.

A flounder must catch and consume its own food. This is why a flounder is called a ______________.

(Write your answer on your answer sheet. Then check across to the next page to box 4 to see if you are right.)

5

There are other consumers you can see in the picture on page 183. There are other fish—large, fast-swimming fish. There are starfish and barnacles, too.

Find one other consumer you see in the picture. This consumer is called a ______________.

6

What would happen if all the green seaweeds disappeared from the ocean? What would happen if they all died out?

Would it make any difference?

Yes or *no?*

4

consumer

(Go on to question 5.)

5

crab

The fish, the starfish, the barnacles, and the crabs are consumers. The boys are consumers, too.

(Go on to question 6.)

6

yes

Why? If the seaweeds died, the fish would die. The consumers, the animals, would die.

Green plants make the food supply.

(Turn the page to question 7.)

7

That's true. Now, what would happen if one of the boys in the picture tried to keep a small flounder alive? What would happen if he put it in a tank of *fresh* water?

Would the flounder *live?* or *die?* Which?

8

Why? Because flounders are fitted to live in salt water. That is their environment. The liquid in their cells is a lot like sea water. Sea water is salty. The flounders' cells would die in fresh water. Their cells swell up in fresh water.

If a flounder were put in fresh water, its cells would swell. It would be like putting a dry raisin in water.

What do you think would happen to the raisin? Would the raisin *shrink more?* or *swell up?* Which?

9

Why? Because there is almost no water in a dry raisin. If a dry raisin is put in water, there is more water on the outside than on the inside. Water goes into the raisin, and it swells up.

Would the same thing happen if the salt-water seaweeds were put in fresh water? *Yes* or *no?*

10

Seaweeds and other plants and animals living in salt water swell up when put in fresh water.

Cells of salt-water animals and plants are fitted to live in ____________ water.

7

die

The flounder would die in the fresh water.

(Go on to question 8.)

8

swell up

A dry raisin in water would swell.

(Go on to question 9.)

9

yes

Too much water would go into the cells of the seaweed.

Salt-water seaweed cells would swell up in fresh water.

(Go on to question 10.)

10

salt

In the ocean are salt-water plants and animals. Salt water is the right environment for these plants and animals.

(Turn the page to question 11.)

11

Look at the picture above. Notice the kinds of plants and animals in this water. You can see some tadpoles, some snails, and some crayfish (KRAY·fish). Crayfish are a kind of lobster-like animal that live in fresh water. There are many green plants in the picture, too.

What kind of environment is this? *Fresh water* or *salt water?*

12

How do you know? You can tell by the living things in the environment. Tadpoles are found in a lake or pond. They are found in fresh water. Tadpoles will grow up to be frogs. Frogs are found in fresh water. Crayfish are also found in fresh water.

Look again at the green tadpoles and crayfish. Can they make their own food? *Yes* or *no?*

13

Tadpoles may be green, but they are not green plants. They feed on green plants in the water.

What would happen to the cells in the tadpoles and in the plants if you put them in the ocean?

Would the cells *swell up?* Or would the cells *shrink?* What do you think?

11

fresh water

(Go on to question 12.)

12

no

Tadpoles and crayfish are animals. They are *consumers*. They cannot make their own food.

(Go on to question 13.)

13

shrink

The cells would shrink in the ocean.

The ocean is salt water. The fresh-water cells would lose water. They would shrink in salt water. The cells would die.

(Turn the page to question 14.)

14

You can say it many ways. It is like the example of the potato slices you put in salt water. Remember how they shrank and got soft?

Why do the fresh-water cells shrink in the ocean? Because water goes *into* or *out of* cells. Which?

15

Of course. Cells in fresh water live best in fresh water. They have about the same amount of water inside the cells as outside.

Plants and animals are fitted for where they live. Where they live is their ________________.

16

The food chain is the same. A food chain starts with a green plant that makes food.

The green plant traps energy from the sun. The sun's energy is locked up in the plant cells. Animals get this locked-up energy from plants. You can trace the energy. Green plants trap energy from the ________________.

14

out of the cells

The water goes out of the cells.

The fresh-water cells are in the wrong environment.

Fresh-water cells are fitted to live in fresh water.

15

environment

Is the food chain in the ocean environment different from the food chain in the lake environment?

No. The food chain starts with green plants.

16

sun

8. THE RISE AND FALL OF TWO POPULATIONS

For a living thing to grow, it must have a certain environment. Each living thing needs certain things to live. For example, there must be the right amount of shade and light. There must be the right amount of heat. There must be enough food and water. All that a plant or animal needs must be in the environment.

Let's take a close look at an environment at its best. This is a jar of yeast cells growing in sugar water. ■ The jar is kept in a warm place. Everything is just right for the yeast cells to grow. The temperature is right. The food is there, too. At the beginning, everything is at its best for these plant cells to grow.

What will happen to the population of yeast cells? It will grow. There will be many more yeast cells after one day.

There is nothing to hold the cells back. There is nothing to stop the cells from growing. If there were, you would say there was something wrong with the environment.

If the yeast plant cells grow every day, there will be more of them in the sugar water. Each day there will be more yeast cells—until they use up the sugar.

Let's add a few paramecia to the sugar water. ● Now there are paramecia and yeast cells together in the weak sugar water. What happens?

We only added a few paramecia at the start. But the paramecia feed on yeast cells. Then each of the paramecia divides in half. Each half soon becomes a full-sized animal. Now there are many *more* paramecia feeding on the yeast cells.

Large numbers of yeast cells are eaten by the many, many more paramecia now in the jar. What do you think will happen?

Pretty soon, the *few* paramecia will have become *hundreds* of paramecia. Now the space for living is crowded. There is hardly any room left. The paramecia population is large. This environment cannot hold any more living things.

When this happens, the paramecia population cannot grow so fast. Now the paramecia reproduce slowly. There are very few new paramecia.

What is holding down their population? What keeps the paramecia from reproducing? *Food.* The supply of yeast cells is low. Their food—in this environment—is yeast cells.

As the numbers of paramecia grow and feed, what happens to the yeast cells? The number of yeast cells goes down. There are too many paramecia. They eat the yeast cells faster than the yeast can reproduce.

Soon there are not enough yeast cells for all the paramecia. Then the paramecia stop reproducing. They do not have enough food. The paramecia population goes *down.*

What Happens Now?

The paramecia stop reproducing. Then the yeast cells get a chance to reproduce again. Yeast cells reproduce faster than paramecia. Soon there are *thousands of yeast* cells again.

This goes on and on. The population of paramecia goes up and down. The population gets larger and then gets smaller. That is, the number of living things in it gets larger and smaller. When the population of paramecia gets large, the yeast population gets small. When the paramecia population goes down, the yeast population goes up. Each population goes up and down like a seesaw. ▲

The Seesaw—Yeast and Paramecia

This seesaw keeps going up and down. The yeast population changes, just as the population of paramecia changes. When one population goes up, the other goes down.

The yeast cells get their energy from the sugar. The paramecia get their energy from the yeast cells. These two populations act with each other. Each has something to do with the other.

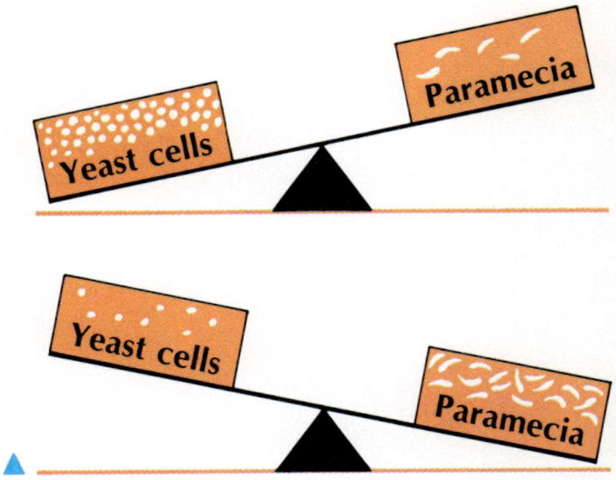

A Balance in the Food Chain

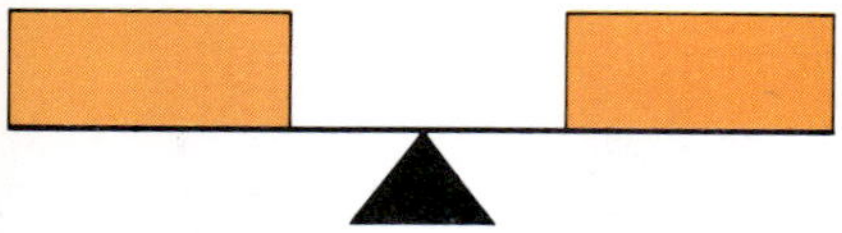

We say the yeast cells and the paramecia are like a seesaw. A seesaw goes up and down.

But a seesaw can stop. It can be even, like this. ■ This seesaw is *balanced.* It does not go up or down. It is steady.

In a food chain there is a *balance*. This means living things can live together in one place. They can live together in one environment. The number of living things will not go up and down. It will be steady.

We say *there is a balance in the number of living things that can live together in one place.*

In a balanced food chain there are consumers. ● Consumers are living things that feed on other things, living or non-living. For example, yeast cells live on sugar. Hydras are consumers that feed on water fleas.

There are also producers in a balanced food chain. ▲ Producers make their own food. They do not need other living things for food. These producers are green plants.

Green plants get energy from the sun. They get energy to make their own food.

Consumers cannot make their own food. We can say the chicken hawk is a consumer of energy. The chicken hawk eats rats and chickens. The frog is a consumer. So is the grasshopper a consumer of energy. It feeds on green plants and gets energy. You can say the energy of animals can be traced back to the sun.

Consumers Need Energy

You may be wondering about the paramecia and the yeast cells. Which are the consumers? The paramecia are consumers because they eat yeast cells.

But what about the yeast cells? They are also consumers. Yeast cells are plants. But they cannot make their own food.

They are not *green* plants. But what do they eat, or consume? Sugar. Yeast cells use sugar for their energy. But where does the sugar come from? Sugar is made by green plant cells.

Again, the food chain goes back to a green plant.

A Try at Prediction

Look at the pictures. In which of these environments would you expect to find a buffalo? In 1 or 2 or 3?

Would you expect to find a squirrel in 1 or 2 or 3?

And an alligator? Where? In 1 or 2 or 3?

How We Stay Alive

Now you know a great deal about how we all live on earth. You have studied some food chains.

You have found that all living things depend on green plants to live and grow.

You have learned about consumers and producers in a food chain. You have discovered how consumers and producers—all living things—depend on each other. Living things need each other to live.

There is also a balance of living things in one place. There must be enough food for all the populations in one place. There must be enough food for all the plants and animals in one place.

You have a special environment. An oak tree has a special environment. So do a chicken, a zebra, a hydra, and a polar bear.

Each living thing has its own special environment. In fact, when you see a plant or animal, you can almost always predict what its environment may be.

6

Man Remaking the Environment: Control of Disease

INTRODUCTION

Man Has Come a Long Way

Human beings have better brains than animals. A good brain is man's defense. He does not have eyes as good as a bird's. ■ Nor teeth as good as a lion's. ●

Human beings do not have claws as good as an eagle's. They do not have the keen sense of smell of a deer or an elephant. But they do have a better brain.

It's a good thing man has a better brain. With this brain man learns about the environment around him. He doesn't just look at his environment. No. Man learns to *use* the environment.

For example, man has learned much about sickness. Another word for sickness is **disease** (dih·ZEEZ). Man has learned slowly about the many causes of diseases. He has learned many ways to stay healthy. He has learned many ways to make a more healthy environment.

Man has learned how to live better—and how to have a longer lifetime. ▲

Now you can learn a little about man's control of some kinds of diseases.

▲
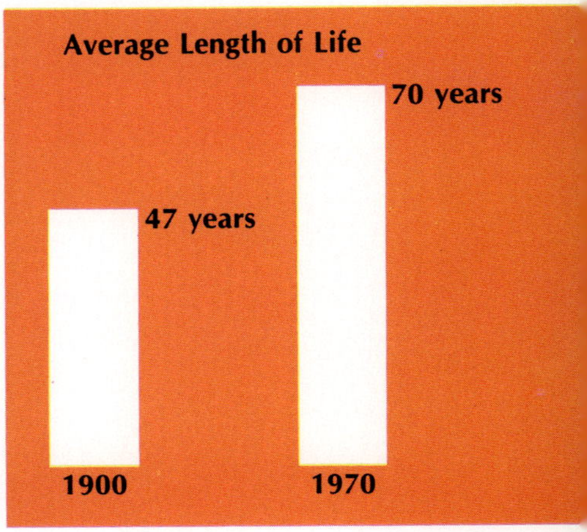

1. THE CONTROL OF THE ENVIRONMENT: DISEASE

Of course, man does not live alone on earth. As you know, he shares the earth with other animals and plants. Some of the plants and animals are not friendly to man. Some of the plants and animals cause diseases.

Over the years, diseases have wiped out thousands of plants and animals. Diseases have wiped out millions of people at a time. In the past, the world population was small. A disease could kill off a big part of this small population.

For example, in the 1300's a disease called bubonic plague (BYOO·bon·ick PLAYG) killed about 25 million people in Europe. This was about ¼ *of the whole population of Europe.*

A plant, such as a pine tree, can't do much about a disease. For example, the white pine tree can't do much about the disease called white pine blister rust. ■ A tiger can't do much about the worm diseases it has.

But man can do a great deal about the plants and animals that cause diseases. That is, man can destroy living things that would kill him. He can destroy many animals and plants that cause diseases. In other words, *man remakes his environment.*

An Example: The Plague

How did people get plague? Plague is caused by bacteria. Fleas carry these bacteria. Fleas are on rats. Ships carry rats and their fleas to all parts of the world.

A flea may bite a rat. This rat may have bacteria that cause plague. The bacteria grow in the flea. Later on, the flea may bite a person. Then the bacteria pass from the flea to the person. ●

How can man control plague? A never-ending war on rats is the best way to control plague. And many other diseases, too. Rats are a problem in big cities. The rats should be destroyed.

Control of disease is one way man remakes his environment. In one way or another, man has been trying to control disease for a long time.

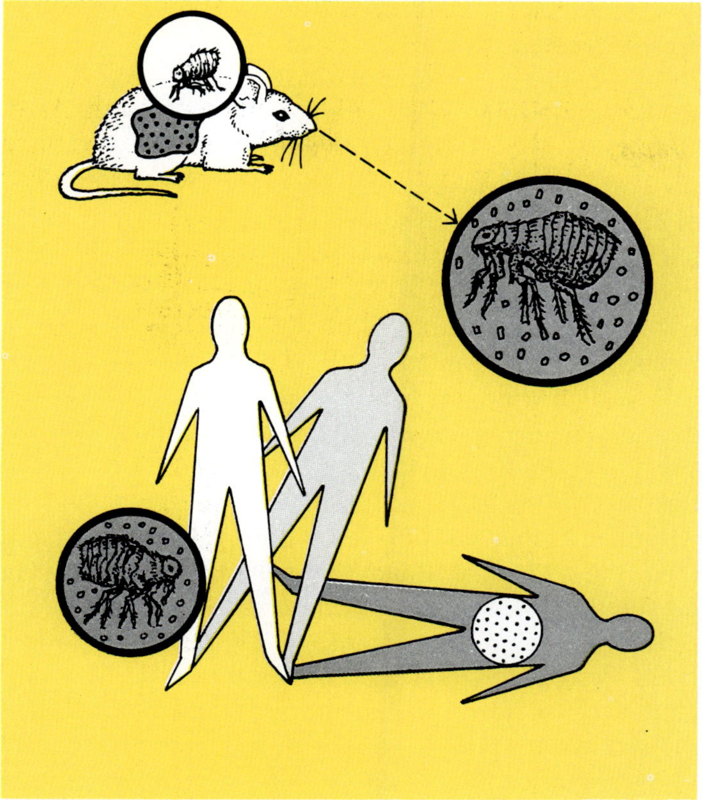

Some of the earliest writings tell about diseases. Even the Bible talks of diseases. The Bible gave directions about the kinds of pots to use to cook meat or milk. This was one way to control disease.

It was not easy to clean or wash pots when tribes of people traveled. It was not easy to keep clean. People had short lives. There were many diseases that killed them. Many children never lived to their teens.

Some Beginnings

In the early days, keeping clean was one of the few ways known to control disease. And destroying rats was another. But little by little, man began to learn better ways. One of the big steps forward was the invention of the microscope. This tool led to the discovery of a

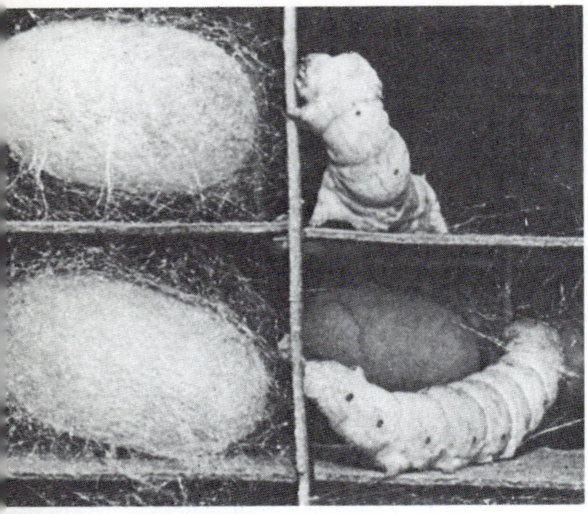

new world of living things. Were some of these tiny organisms the cause of diseases?

Let's begin with Louis Pasteur (LOO·ee pas·TEHR). He helped find the *cause* of a disease in silkworms.

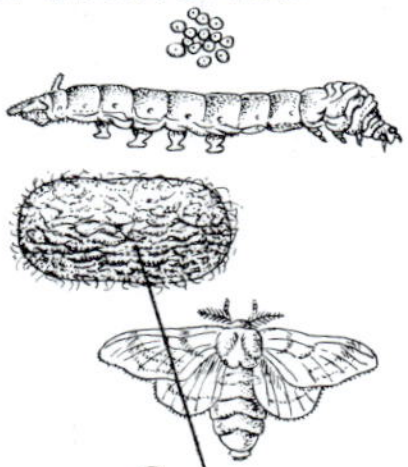

Cocoon

You may know that silk comes from silkworms. Silkworms are the young of certain moths. Each silkworm spins a long silk thread. It makes the thread into a *cocoon* (kuh·KOON). Inside the cocoon, the silkworm turns into a moth. It becomes an adult.

If you unwind the cocoon of a silkworm, the silk thread is ready as a raw material. It can be made into silk cloth. Raising silkworms was a big industry (IN·dus·tree) in the 1800's. But during the 1850's the silkworms in France got sick. Spots appeared on them, and then they died. Without the silkworms, there were no silk threads. Without silk threads, no silk cloth.

The French government (GUV·urn·ment) asked Louis Pasteur to study the case of the sick silkworms. Under the microscope Pasteur found little organisms in the sick worms. There were none in the healthy worms.

Pasteur also discovered something else. He put the sick and the healthy worms together. The healthy ones also became sick. So the disease of the silkworms was one that spreads. To control the disease, Pasteur said that healthy worms should be kept away from sick ones. The silk industry was saved.

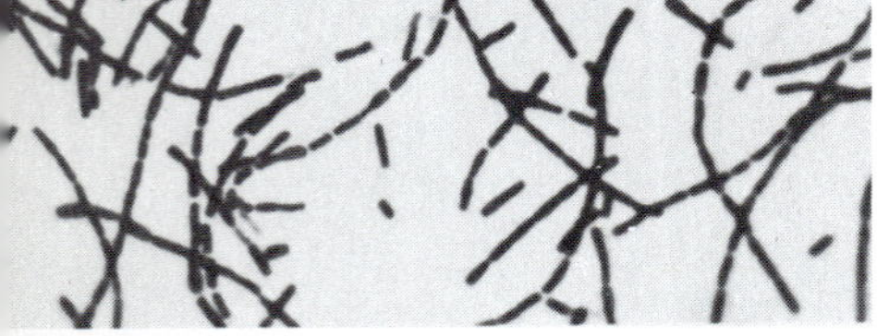

Pasteur said that *many* diseases are caused by organisms that can only be seen under the microscope. We call these organisms **micro-organisms.** Pasteur told doctors that if you find a sick animal you may find it teeming with micro-organisms.

The Battle Against Anthrax

The next part of our story is about another disease that was soon controlled. This is a deadly disease of cattle and sheep, called *anthrax* (AN·thrax). Anthrax killed off thousands of animals during the 1800's. Sick animals had rod-like micro-organisms in their blood.

Healthy animals did not. Pasteur studied these tiny rods from the blood of sick animals.

Then the scientist Robert Koch found out that the micro-organisms were bacteria. He took blood from the sheep and studied the bacteria. Then he thought out a way to grow these bacteria. He grew them *outside* the body of the sick animal. He grew the bacteria in special food.

Then Koch gave these bacteria to some healthy animals. These healthy animals got the deadly disease. They got anthrax. This was the real proof that these bacteria cause the disease.

Koch had made a great discovery. He had discovered the exact kind of bacteria that cause anthrax. He had found the *cause* of anthrax.

Koch also *stained* the bacteria. He could see the stained bacteria better. Koch knew exactly which bacteria cause anthrax.

Now scientists knew *how to separate the kinds of bacteria*. Scientists were now able to find the causes of many, many diseases.

Looking at Bacteria

Now you can look at bacteria. But you will not look at bacteria that cause disease. That would be too dangerous. You will look at some bacteria that are helpful. You will see the bacteria that give yogurt (YOH·girt) its taste.

You will need a carton of yogurt. You will also need slides, eyedroppers, and a microscope. And some blue dye to help you see the bacteria.

Put a drop of blue dye on a slide.▲ Let the blue dye dry. Then add water to a little yogurt to make it very thin.

Use the eyedropper to put a drop of yogurt on the spot of blue dye.◆ Add a cover slip. Watch how the blue dye spreads out on the slide.

Look at your slide under the microscope. Use low power, then high power. What do you see?

▲

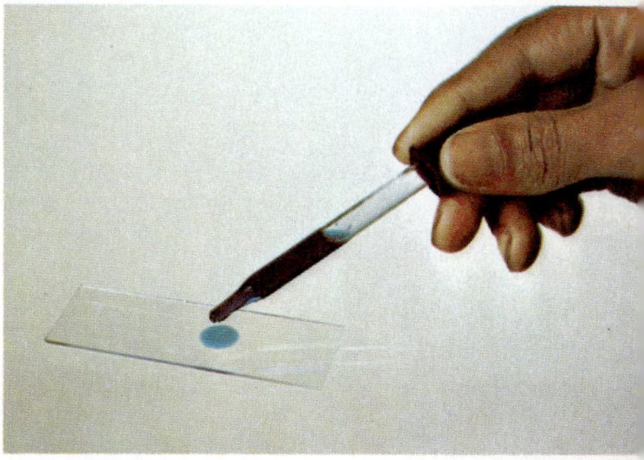

◆

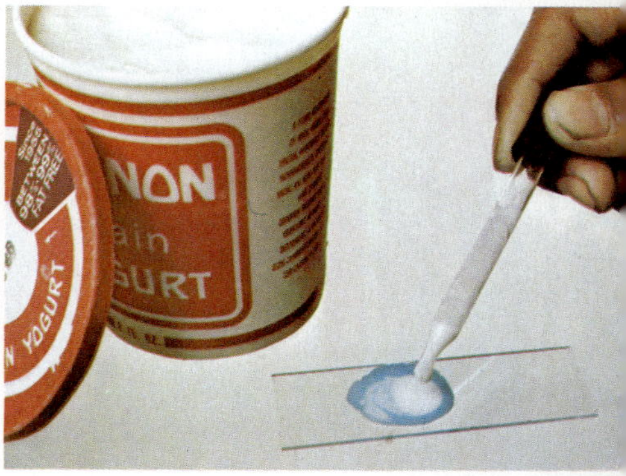

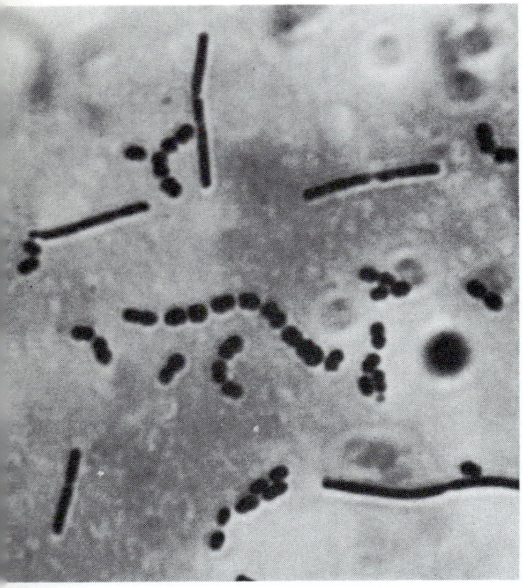

Do you see some rod-like shapes like the ones in this picture? These are bacteria. They look like the organisms Pasteur saw in the sick silkworms. And they look like the bacteria Koch found in the blood of sheep with anthrax. But your bacteria do not cause disease. They are *helpful* bacteria.

Perhaps you can see some round shapes like the ones in the picture. They may look like a string of beads. These are also helpful bacteria. But they look like some of the bacteria that cause disease.

Bacteria are very tiny. They are microscopic. Do you see why it took so long to discover them? It took a long time to learn which bacteria cause disease.

Jenner and Smallpox

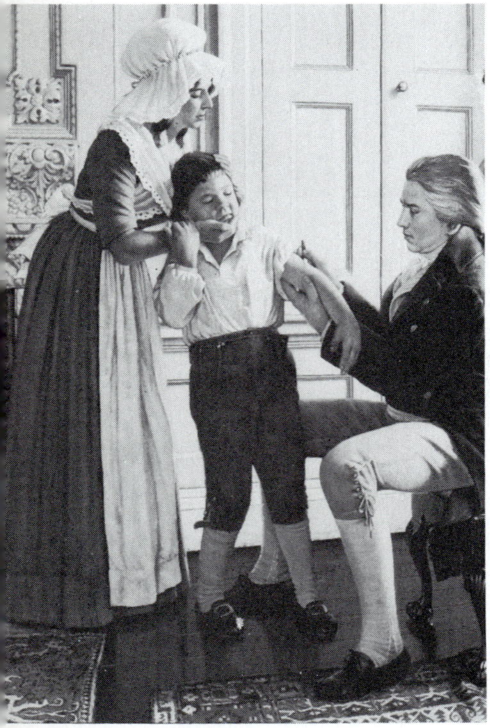

In some cases, man learned to control a disease even before the cause was known. This is the way it was with *smallpox.* Smallpox is an old disease. It was written about more than a thousand years ago. Many people died from the disease. People who did not die had very bad scars from the disease.

Dr. Edward Jenner (JEN·ur) did not know the cause of smallpox. But he found a way to protect man against the disease. This was in the 1790's.

Dr. Jenner saw that when smallpox broke out, some people did not get the disease. Strange. Then Dr. Jenner saw that the people who did not catch smallpox were those who took care of dairy cattle. Dr. Jenner knew that these people sometimes caught sores from cattle. These sores were caused by a disease of cattle. The disease is called *cowpox.* Dr. Jenner thought that maybe getting cowpox kept people from getting smallpox.

Dr. Jenner took some of the matter from the sores of cattle with cowpox. He scratched this matter on the arm of a young boy. The boy's arm got a little swollen. That was all. He did not get really sick with cowpox.

Later the boy was near people who had smallpox. Also, Dr. Jenner scratched the boy with a little of the matter from a smallpox sore. Did the boy catch smallpox? Not at all. Why not? The boy was protected against getting smallpox.

What was the matter Dr. Jenner first scratched the boy with? Matter from cowpox sores. What disease did the boy now have protection against? He was protected against cowpox and smallpox. It seems that his body must have made something that kept him from getting sick with cowpox. And what his body made also protected him against smallpox.

This is what happened. The boy's body made something that protected him from cowpox *and* smallpox. What is this protection that his body made?

Protectors in Your Body: Antibodies

We know now that cowpox is caused by a **virus** (VIE·rus). Smallpox is caused by another virus. Viruses are smaller than bacteria. They live inside cells.

When Dr. Jenner scratched some cowpox matter into the boy's arm, some viruses entered the boy's body. His body began to make special chemicals. These chemicals are called **antibodies** (AN·tih·bod·eez). Antibodies help destory micro-organisms that enter the body.

Here is how antibodies protect us from diseases. Let us imagine cowpox viruses look like this. ▲ The boy's body made antibodies that might look like this. ◆ Do you see that the antibodies might fit together with the viruses?

We think this is what happens. We think the antibodies fit over the viruses. Just the way a glove fits over a hand. ★ The antibodies cover the viruses. Then the viruses are harmless.

When viruses get into the body, antibodies are made. A different kind of antibody is made for each kind of virus. The boy's body made antibodies. They were the kind that should attack *only* cowpox viruses. But here is where luck comes into the story.

The body makes many more antibodies than it needs. This means that many extra antibodies float around in the blood. So when the boy was later scratched with

▲

◆

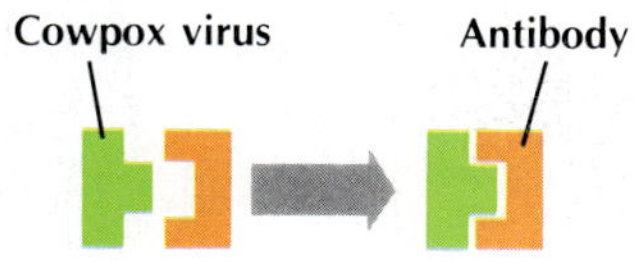

★ Antibodies fit over the viruses. The viruses are then harmless.

Extra antibodies made against cowpox viruses stay in a person's blood.

Maybe smallpox viruses look like this.

Smallpox viruses and cowpox antibodies match well. The cowpox antibodies cover the smallpox viruses and make them harmless. ■

matter from a smallpox sore, his body had antibodies. His antibodies kept him from getting sick with smallpox.

The extra antibodies against cowpox virus were in his blood. And here is the luck. Cowpox viruses have a shape like smallpox viruses. At least, it seems so. The antibodies against cowpox viruses also fit over the smallpox viruses. ■ The smallpox viruses then are harmless.

When Dr. Jenner scratched the boy's arm, he gave him a **vaccination** (vak·sih·NAY·shun). This means he gave the boy cowpox viruses. This was enough to make his body start making antibodies. Not enough to make the boy sick.

Preventing a Disease: Vaccination

When you are vaccinated, you get a **vaccine** (vak·SEEN). There are different kinds of vaccines. What is in a vaccine? Sometimes a vaccine has a small amount of weak bacteria or viruses. Sometimes the vaccine has dead micro-organisms. Sometimes it has wastes from the micro-organisms.

Once you have a vaccine, your body starts to make antibodies. Your body makes antibodies for each kind of disease micro-organism. Your body makes this protection. All you need is a small dose of the vaccine. *But you do not get the disease.*

This seems to be the way your body protects you from some diseases. Your body makes antibodies. It makes antibodies when you have a disease. And when you are vaccinated. The antibodies fit over viruses and some bacteria. Then the micro-organisms are harmless. It takes the right antibody to fit the right virus or bacteria. It's like the right key that fits in a lock. It's the right combination.

Sometimes the antibodies you make seem to give you lifetime protection. This means that your antibodies stay in your blood the rest of your life. But sometimes people get another vaccination if they are traveling outside the United States. Just to be safe. Just to be sure they have enough antibodies against diseases like

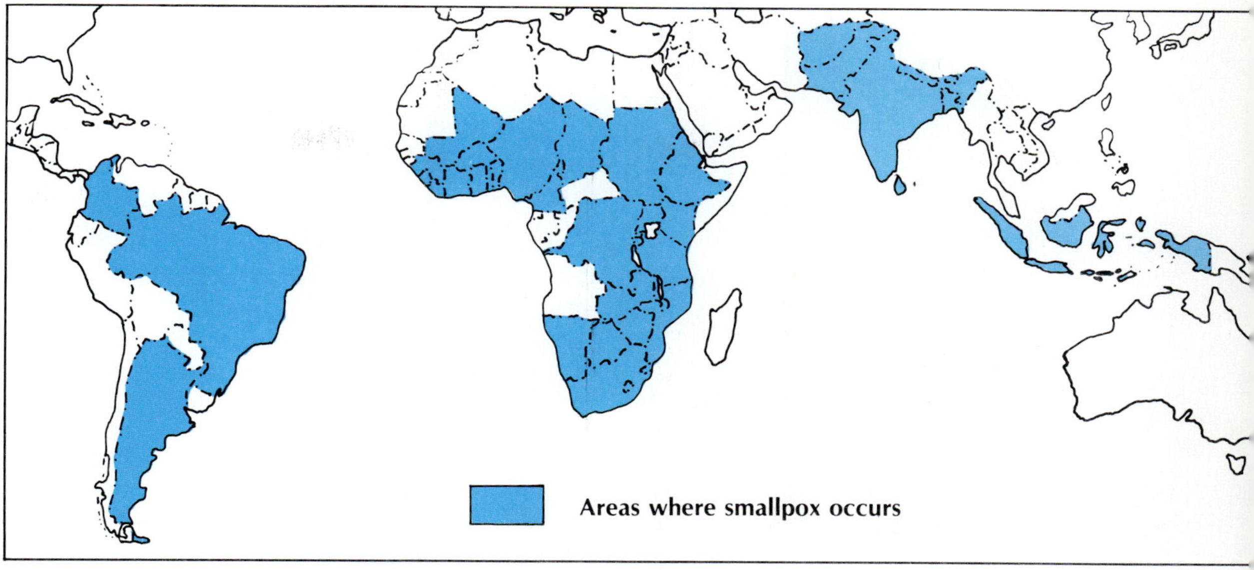

●

smallpox. The new vaccination builds up an extra supply of antibodies.

After all, 60,000 people somewhere in this world still die *each year* of smallpox. ● Not many in the United States. Especially not in places where children must get smallpox vaccinations before they enter school. Many states have laws about this.

Winning the Battle over Polio

Another disease we do not have to get today is *polio* (POH·lee·oh). We can make our own protection. This is one of the latest battles man has won. We have won the battle against polio.

Polio is a disease that cripples people. Polio viruses attack nerve cells. If nerve cells do not work, muscle cells cannot move.

Your body can make antibodies against polio. It can make this protection without getting the disease. There are two kinds of vaccines you can get against polio.

Dr. Jonas Salk made a dose of very weak *living* polio viruses. Small doses of these weak viruses can be injected (in·JEK·ted) into a person. ▲

▲

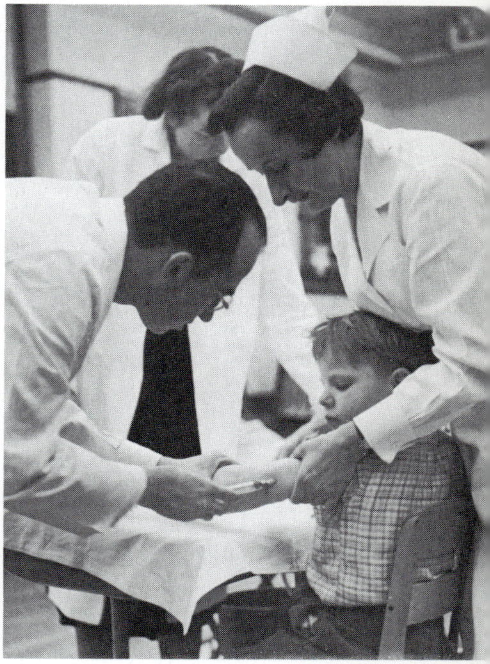

Dr. Albert Sabin made a dose of *dead* viruses of polio. This dose can be swallowed. It can be taken by mouth. ■

In each case, the body reacts the same way. Your body makes antibodies against the polio viruses. This means that you stay well, even if polio viruses get into you later on. You have some polio antibodies already made. These antibodies will float around in your blood for many years.

Giving Away Antibodies

Most people have had some diseases. Sometimes they know they have had the disease. Other times they may have had a touch of a disease, but didn't know it. They never got too sick. But their bodies made antibodies against the disease micro-organisms. And these antibodies stayed in their blood. There are antibodies against many kinds of diseases in the blood of people.

People often give blood to a blood bank. This blood in the blood bank has all kinds of antibodies. There is a gold mine of antibodies in the blood bank.

Children may be given these antibodies that have already been made by other people. Then they have a start in the making of antibodies. In case these children get infected, they already have some kinds of antibodies on hand. They might have the right antibodies to fit over the micro-organisms that cause the disease. These antibodies start to fight the infection. They give the children a headstart in fighting the infection. And then the children make their own antibodies too.

Antibodies from Horses

Sometimes people are given antibodies made by animals. For example, a healthy horse can be given bacteria that cause *diphtheria* (dif·THIR·ee·ah). The horse's body makes special antibodies against these bacteria. The horse is not harmed.

Now some of the horse's blood is collected. The antibodies are taken out of its blood. Then suppose a child gets sick with diphtheria. The child may be too weak to make antibodies fast enough on his own. He may be given the antibodies the horse made. These antibodies will destroy the poisons of the diphtheria bacteria. The child is helped by the horse's antibodies.

This is how antibodies that are made by a horse or another animal can save the life of a sick child.

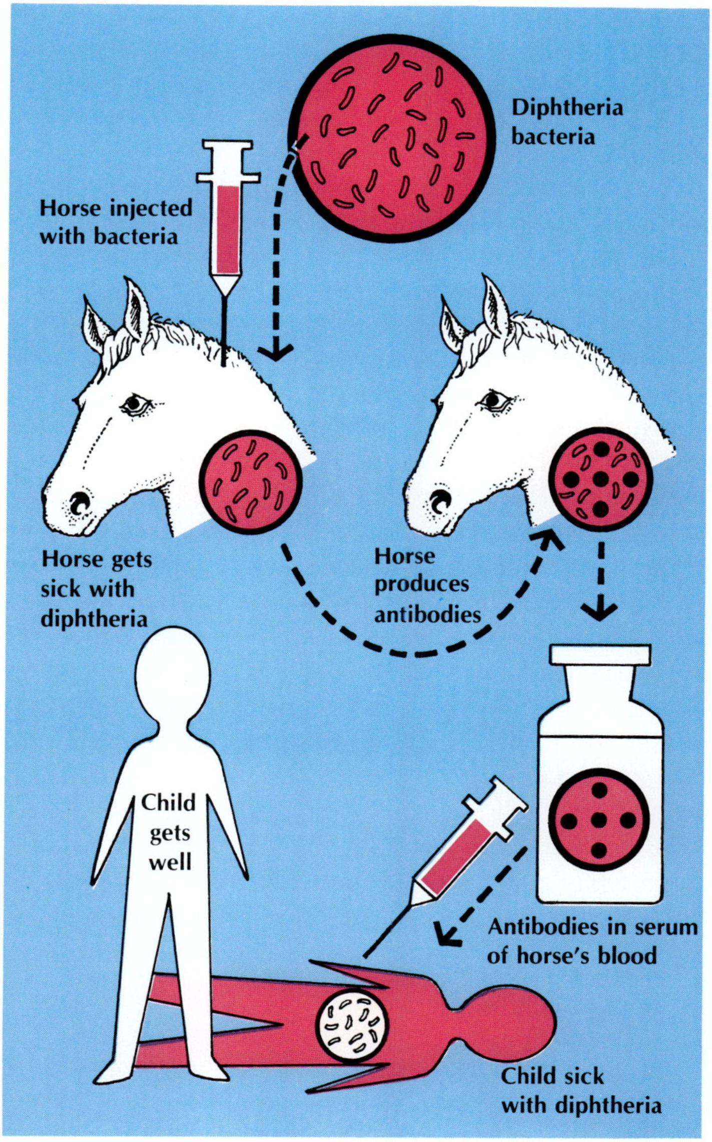

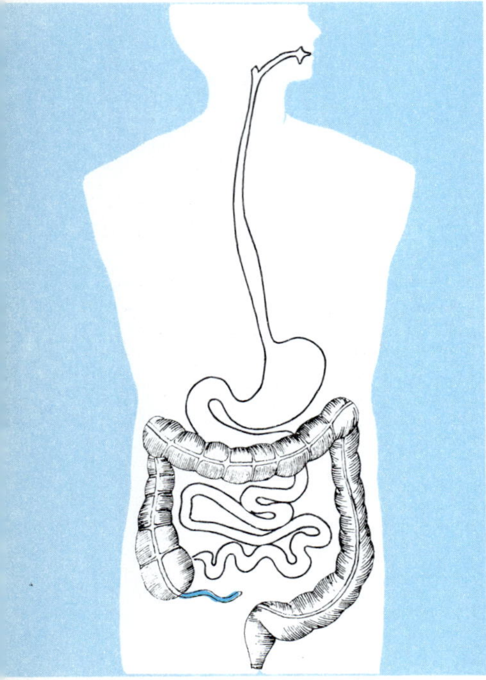

Other Protectors in Your Blood

Let us say a boy complains of pain in his lower right side. This is where the *appendix* (uh·PEN·dix) is. The appendix may get infected with bacteria. This is called *appendicitis* (uh·pen·dih·SY·tis). Could the boy have appendicitis?

The doctor can make a test. He can look at a drop of the boy's blood under the microscope. He can see the different kinds of cells in the blood. He may see that the boy's blood has a very large number of one kind of white blood cells. Many more than the normal number.

A very large number of these special white cells in the blood usually means there is an infection (in·FEC·shun). In other words, there are many micro-organisms in the body. The micro-organisms are collecting among the cells of the body. Perhaps they are collecting in the appendix. If they are, there is no time to lose. The appendix must be removed at once.

White blood cells are protectors in the blood. Some of these cells can move out of the blood stream. They move out to reach any place where there is an infection. Where there are many bacteria growing. Thousands of white blood cells move to the place of infection.

The white cells cover the bacteria. Then the bacteria are *inside* the white cells. The white cells carry away the dead bacteria. The battle is over. The infection is gone.

The blood always has these white cells. In a drop of blood smaller than a pinhead there are about 8,000 white blood cells.

When there is an infection, the body makes thousands more white cells. Many more white cells are needed to help destroy the mass of micro-organisms in the body.

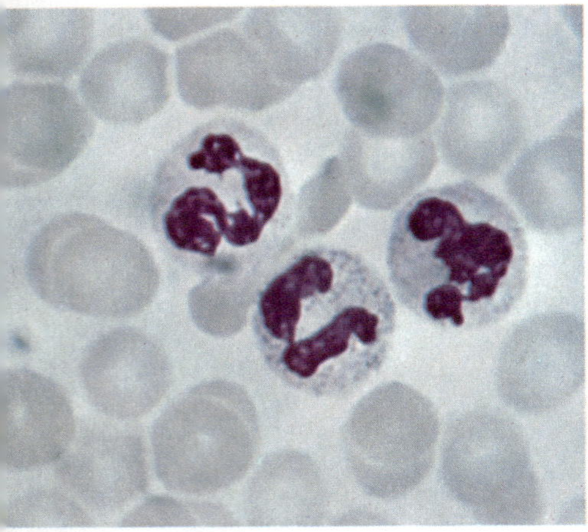

Paul Ehrlich's Idea

Strong chemicals can kill some micro-organisms. They can kill bacteria in an infected wound. Dr. Paul Ehrlich (AYR·lickh) knew this. This gave him an idea.

Would it be possible to put a chemical *inside* the body? Was there a chemical that would be like a chemical bullet? Could a chemical bullet pick out the right target? Was there a chemical bullet that could kill micro-organisms *inside* the body?

Of course, some chemicals are very strong. They might be too strong to put inside the body. The chemical must not kill the body's tissues. The chemical must only kill micro-organisms in the body.

Dr. Ehrlich tried. He tried out 100 chemicals on sick animals. Then 200 chemicals. Then the number went up to 605 chemicals. None of these chemicals worked. Sometimes they killed the micro-organisms, but they also killed the animals. Sometimes the chemicals did nothing at all.

But chemical number 606 did work. Chemical number 606 destroyed the micro-organisms that cause *syphilis* (SIF·ih·liss). It did not kill the animals.

Syphilis is a very serious disease if it is not treated. It harms the tissues of the body. It is caused by organisms bigger than most bacteria. They look like coils of hair under the microscope. ▲

Syphilis is a disease that spreads by sexual contact. It spreads from one person to another. ♦ Syphilis micro-organisms get into the blood. From the blood stream the micro-organisms can go to all parts of the body. They can harm the heart, the lungs, and the brain.

▲

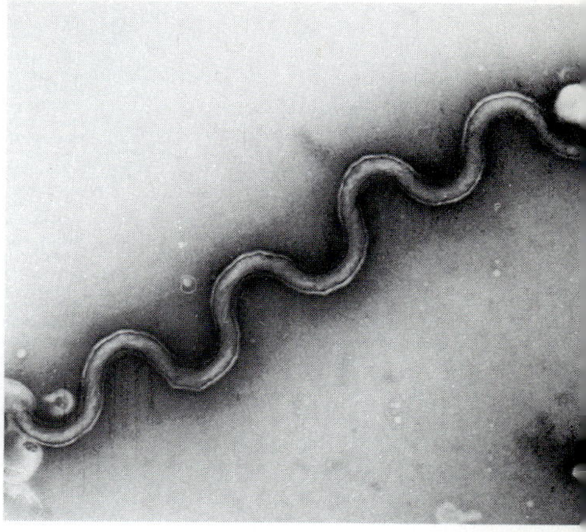

♦

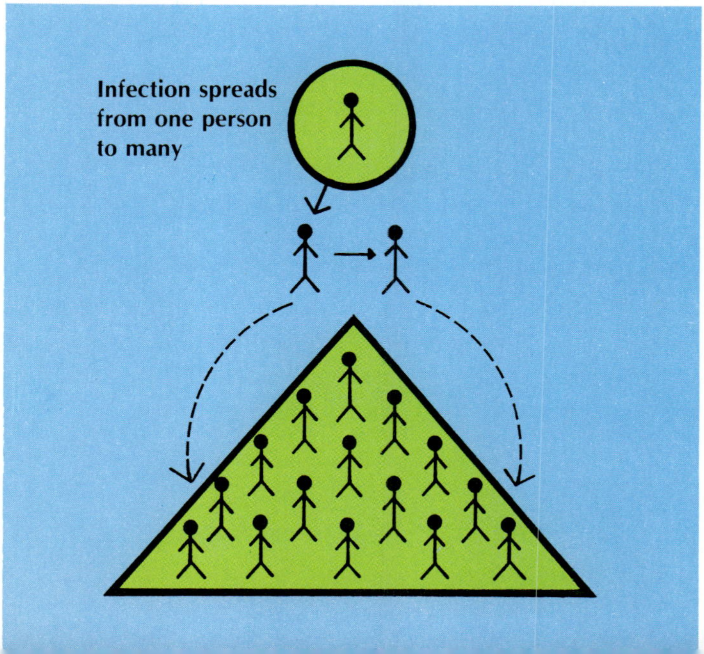

A baby can be born with syphilis micro-organisms in his blood. The micro-organisms go from the mother's blood into the baby's blood. The mother must be treated before the unborn baby is harmed. The micro-organisms must be killed so they cannot harm the growing baby's body.

Syphilis is a very old disease. Chemical 606 was discovered only in 1909. This was not so very long ago. Before then, there was no cure. Today some new kinds of drugs are used to treat syphilis.

But we are getting ahead of the story. Let's stop to see what man had learned about disease by the early 1900's.

What Man Had Learned About Diseases by 1900

So far, these were the great discoveries made by scientists. These were giant steps in man's control of part of his environment: diseases.

Proof that many diseases are caused by micro-organisms.

Proof that the body builds antibodies when it is given small doses of a vaccine—or when a person has some diseases.

Proof that white blood cells increase in great numbers when there is an infection in the body.

Proof that a chemical can be put into the body to kill disease micro-organisms, but not body cells.

The Next Slow Step: Penicillin

Paul Ehrlich's chemical 606 could kill one kind of micro-organism. But there are many kinds of organisms that can cause diseases. Chemical 606 did not kill all these organisms. Something more was needed.

In the late 1920's Dr. Alexander Fleming discovered something. He noticed some green mold plants growing on dishes of bacteria. ■ In a few days, the bacteria had all died off. Only the mold plants were left.

This is the explanation. For a long time scientists had known something interesting. Sometimes disease micro-organisms get into the soil. But then they seem to disappear. Some molds and bacteria that live in the soil seem to kill the disease bacteria. It seems some soil organisms kill bacteria.

The same kind of thing was happening in Dr. Fleming's dishes. The green mold plants were giving off a chemical. This chemical killed the bacteria that grew near them.

The chemical given off by green mold plants is penicillin (pen·ih·SILL·in). Pencillin kills off some kinds of bacteria. In the 1940's doctors began to use penicillin as a drug to kill many kinds of bacteria in the body.

●

Chemicals like penicillin are called **antibiotics** (an·tih·by·OT·iks). Antibiotics are chemicals made by some living micro-organisms. Some bacteria and molds give off these chemicals. These antibiotics destroy many kinds of bacteria.

Now we have these new drugs that can destroy disease micro-organisms. We have antibiotics. Antibiotics can destroy micro-organisms *inside* the body. They are made by some living things. They can kill *other* micro-organisms.

There are other antibiotics that are like penicillin. *Streptomycin* (strep·toe·MY·sin) and *Aureomycin* (or·ee·oh·MY·sin) are antibiotics. There are many more. All these chemicals are made by living organisms. They kill or slow down the growth of micro-organisms.

Short Stories About Some Diseases

What are some diseases caused by bacteria? There are many. We can only look at a few. Look how *tuberculosis* (too·burr·kyuh·LOW·sis) is passed from person to person. ●

Bacteria that cause tuberculosis can settle in the lungs. Or in the bones. You can see how easily the bacteria can pass from a cough or a sneeze to a healthy person.

How can tuberculosis be prevented? What is one way the disease is controlled today? ▲

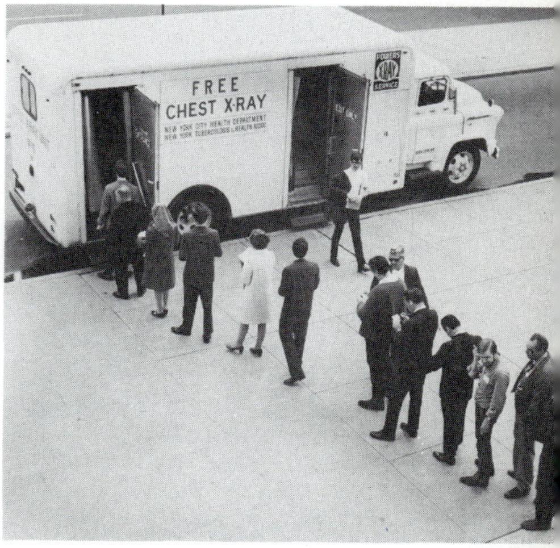

▲

How do people get *typhoid* (TIE·foid)? Typhoid bacteria live in the intestines of man. Water or food fouled with human wastes can pass on typhoid bacteria to a healthy person. ■

■

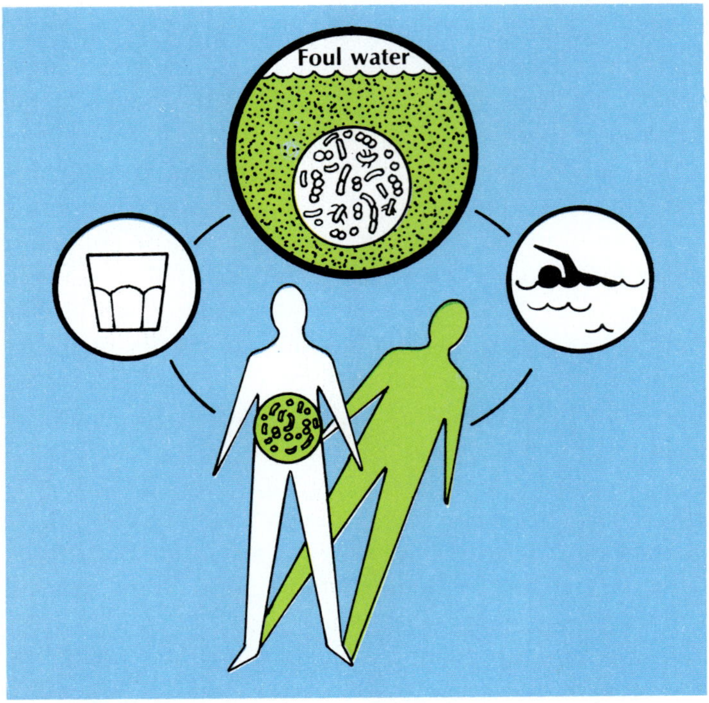

Sometimes a healthy person has typhoid bacteria in his body. He does not get sick. But he can pass along the bacteria to others. This kind of person is called a human carrier of typhoid. ●

●

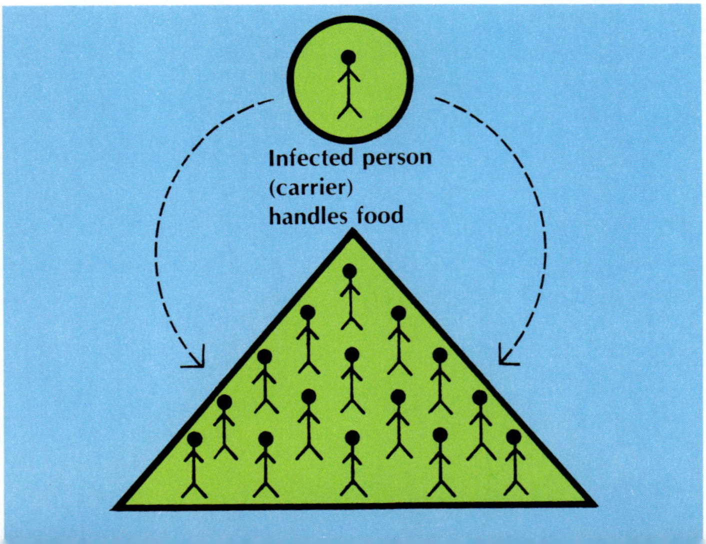

How can typhoid be prevented? How can it be cured today?▲

▲

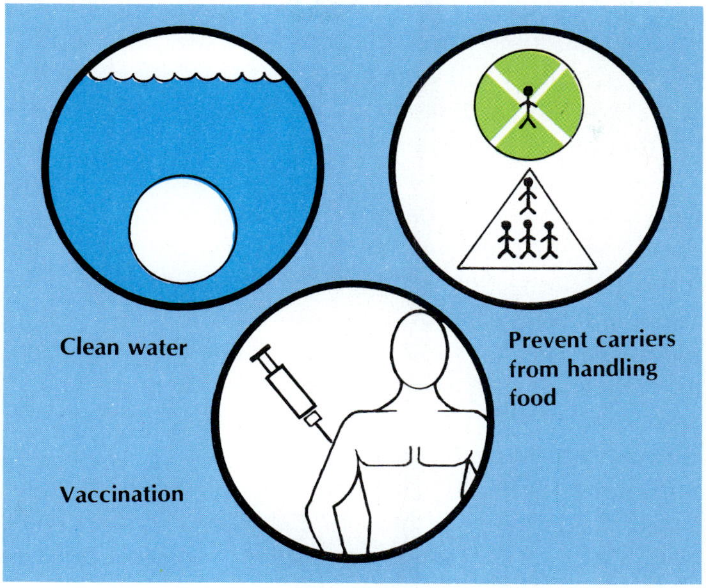

More About Virus Diseases

You know about some diseases caused by viruses. One is smallpox. There is a vaccination against smallpox. So we no longer need get the disease.

Another disease caused by viruses is polio. And, again, there is a vaccination against polio. This means that polio viruses will not get to nerve cells. They will not cause the crippling polio disease.

Many other diseases are caused by viruses. You surely have had some of them. You have had colds. Many kinds of viruses cause colds. *Flu* is also a virus disease. The virus spreads from person to person. Sometimes thousands of people in a city get flu at the same time.

People often have many virus diseases during childhood. *Measles* (MEE·zulz) used to be a common virus disease of children. Not many children get measles now. Why not? Since 1965 there has been a vaccine against measles. Children get this vaccine. Then they make their own antibodies. They make antibodies against measles viruses. Children don't have to get measles anymore.

There is another virus disease called *German measles*. If young children and teenagers have this disease, it is not very serious. It is a mild sickness that lasts for only a few days. But if a pregnant (PREG·nunt) woman gets German measles, it's a very different story.

These German measles viruses may harm the growing baby. They harm the baby if the mother gets the disease during the first few months the baby is growing in the mother's body. This is the time that the baby's brain, eyes, and ears are forming. And the heart is forming then, too. The baby may be born retarded (rih·TAR·did) or with heart disease. The baby may be born deaf or with damaged eyes. Sometimes if the mother has German measles, the growing baby may be so harmed it dies. This birth of a dead baby is a *stillbirth*.

Can the mother be helped? Can the baby be saved? Yes. She can get some antibodies from the pool of antibodies. Antibodies from the blood people give to the blood bank. The antibodies may stop the disease before the baby is harmed.

There is a better way to protect us against German measles. Today, young children and teenagers can get a vaccine. ■ They can get a vaccine against German measles. Why is this especially important for teenage girls? If they get the vaccine, they can build up antibodies. Then they have antibodies in the blood. The antibodies will stay in the blood. The girl will be protected against German measles.

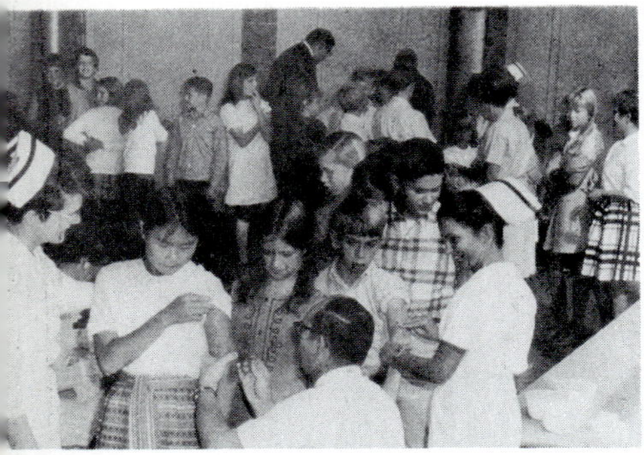

■

Another virus disease is *rabies* (RAY·beez). Dogs and many wild animals can get rabies. The viruses that cause rabies are in the saliva of a sick animal. If the animal then bites a person, the viruses can get into the person.

People can get rabies from the bite of a dog with rabies. The viruses get to nerve cells. There the rabies viruses grow in the nerve cells in the spinal cord. This happens in dogs and in human beings. Paralysis and death are the result. *There is no cure.*

But there is a treatment to *prevent* rabies. If a person is bitten by an animal with rabies, he must be given special shots. These shots will prevent the disease. The shots against rabies are called Pasteur shots. It was Louis Pasteur who first gave us this treatment.

Do you know what to do if a person is bitten by a dog, rat, or some wild animal? Yes, the animal must be tested to find out if it has the rabies viruses.

What if the animal cannot be found? Then the person who was bitten must get the special shots. No one wants to take a chance. The animal might have had rabies. After all, you cannot be cured if you get the disease.

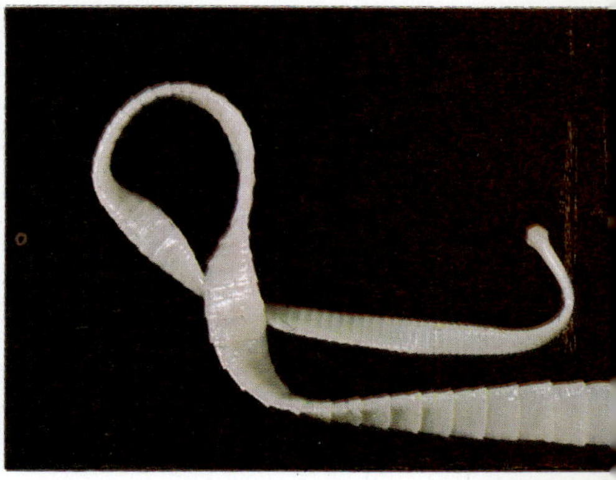

Some Worm Diseases

Worms can get into the body and cause diseases. Some kinds of worms may get into the body in the meat we eat. Maybe you have heard about *tapeworms*.

This tapeworm got into a person's intestine. The person may have eaten beef or pork from an animal infected with tapeworm. Then the meat was not well cooked before he ate it.

The tapeworm is several feet long now. But it was not this long at the time the meat was eaten. It was only about the size of a pinhead then.

Can you tell how this pig got the tapeworm?

You see that the pig ate food that had the eggs or the young of the tapeworms on it. Then the worms moved into the muscles of the pig. Now suppose a person eats meat from the pig. If this meat is not carefully cooked, the young of the worm will still be alive. Then it can grow in the human body.

Another worm that people can get from poorly cooked meat is the *trichina* (trih·KY·nuh) *worm*. This worm infects some animals, especially pigs. It lives in the muscles of the animal. If a person eats meat from the animal, the worms can live in his body. The infected person is said to have the disease *trichinosis* (trick·ih·NO·sis).

Is there a way to be safe from these worms? Yes. When the meat is cooked long enough, the worms are killed. This is the reason that pork must be fully cooked before it is eaten.

Other worms that cause disease can bore right through the skin. Like the *hookworm*. See if you can tell how the hookworm enters the body. ■ And where does it go in the body? ●

Diseases Caused by One-celled Animals

Over the world the number one killer is *malaria* (muh·LAYR·ee·uh). More than 2 million people die of malaria each year. ▲ What is the cause? A kind of one-celled animal that gets into the blood. These microorganisms live in red blood cells.

■

●

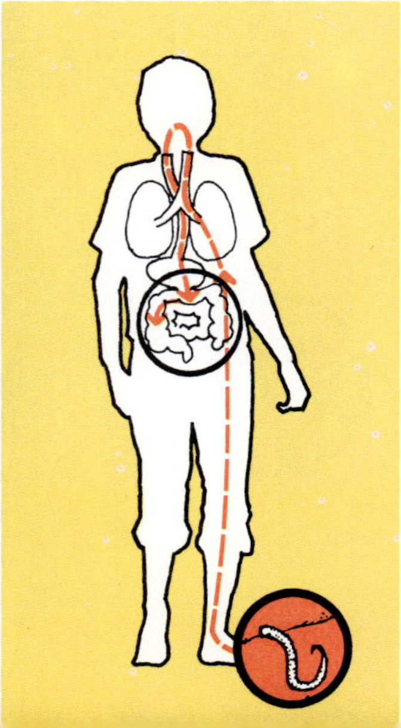

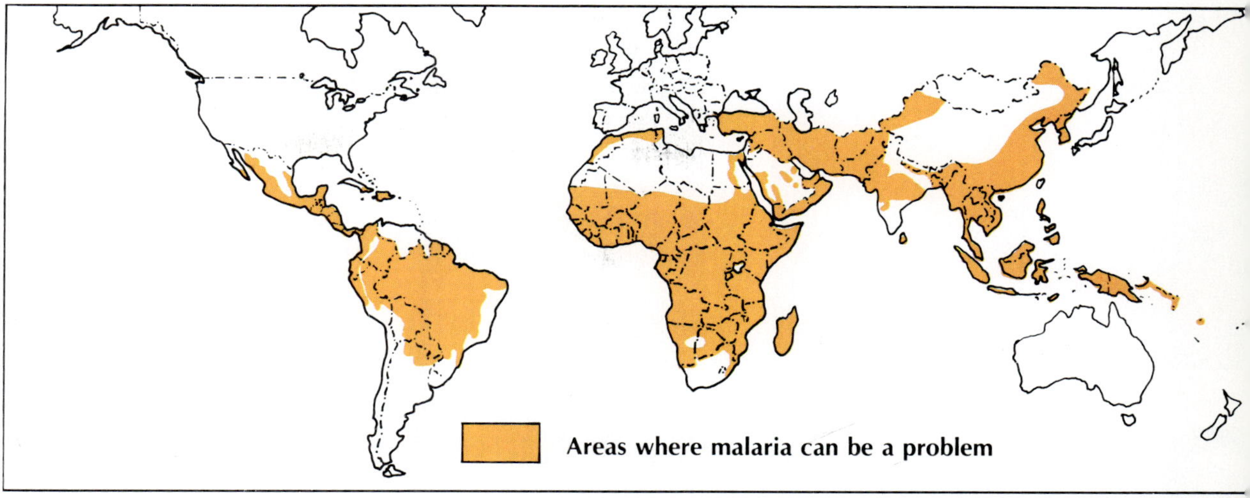

▲

In the picture start with the female mosquito. ♦ It is drawing blood from a person who has malaria.

What is the next step? How does a healthy person get malaria?

How can malaria be controlled? *Quinine* (KWY·nine) is a drug that was used for many years to treat malaria. Then an American scientist made a drug that seems even better than quinine. It is called *Atebrin* (AT·ih·brin).

♦

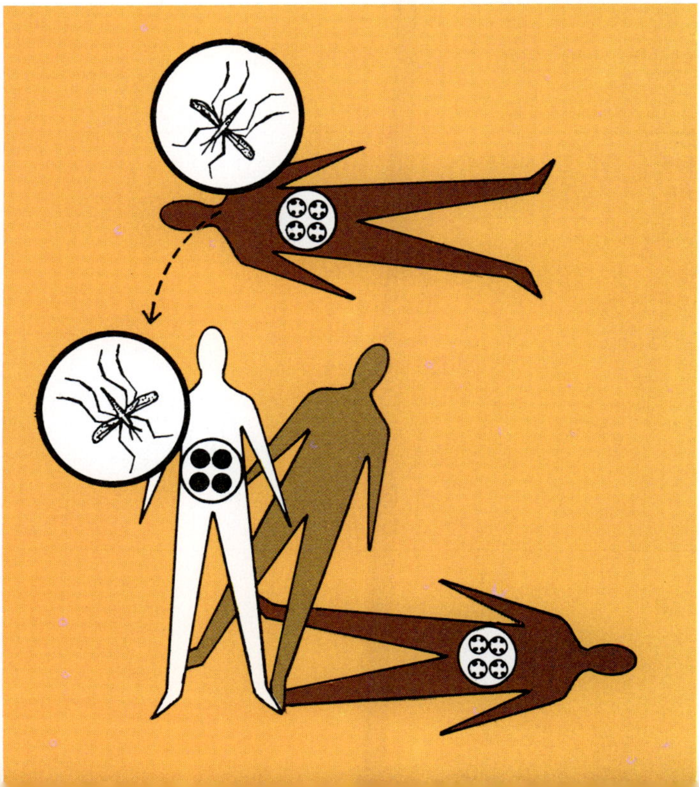

How can malaria be wiped out? Mosquitoes lay their eggs in swamps. What would happen if there were no more swamps? If there were no places for mosquitoes to lay eggs?

If we could dry up swamps, there would be no more malaria. It seems an endless job. As soon as some swamps are drained, heavy rains come and make more swamps. It seems to be a losing battle.

Disease Caused by Fungi

Many hundreds of diseases are caused by fungi. Molds are one kind of fungi. These fungus plants have no chlorophyll. Some fungi live in moist cracks in the skin. These fungi cause *athlete's foot*.

Another disease is *ringworm*. Ringworm is *not* caused by a worm. Fungi plant cells form round lumps under the skin. This is the reason for the name. The disease was named before people knew that it is a fungus disease.

Even plants get fungus diseases. Look at the fungi growing on corn. ■ And on wheat. ●

A Try at Explanation

Look at the chart. ▲ Which disease was the number one killer in the United States in 1900? Which were the second and third killers?

LEADING CAUSES OF DEATH IN THE UNITED STATES

1900	Today
1 Influenza and pneumonia	1 Heart diseases
2 Tuberculosis	2 Cancer
3 Intestinal infections	3 Stroke
4 Heart diseases	4 Accidents
5 Stroke	5 Influenza and pneumonia
6 Kidney diseases	6 Certain diseases of early infancy
7 Accidents	7 Diabetes
8 Cancer	8 Diseases of blood vessels
9 Old age	9 Respiratory diseases
10 Diphtheria	10 Liver diseases

Can you tell why these diseases have dropped down to a lower place today?

How is it different today? Which group of diseases is the number one killer today? And which is the next killer?

Do you see that the diseases we have control over are caused by micro-organisms? They have dropped down into lower places. We have antibodies. We can get vaccinations and make our own antibodies. Or we sometimes get antibodies already made for us. And there are new drugs today. There are many antibiotics that destroy bacteria in us.

People are living longer today. They are living long enough to suffer from some other kinds of diseases. If we live long enough, our organs show some wear and tear.

What kinds of wear and tear can the heart and blood vessels have? Look what one disease can do to blood vessels. ◆

Diseases like *cancer* are still with us. Cancer is a disease in which some cells grow fast. And they do not stop. They keep making more and more cancer cells. We do not know the cause of cancer. We know that cancer in some animals is caused by viruses. We do not know if this is also true about cancers in human beings.

2. CONTROL OF THE ENVIRONMENT: THE "CHEMICAL" DISEASES

You know now about many diseases caused by micro-organisms. And you know about some diseases that happen mostly in older people. Now you can learn about some diseases caused by chemicals.

Sometimes a disease is caused by a chemical that is missing from the body. For example, many diseases are caused by missing vitamins. Or by too few vitamins. Vitamins are needed for cells to do their work properly. They are needed for cells to keep healthy.

Vitamins, you remember, come from your environ-

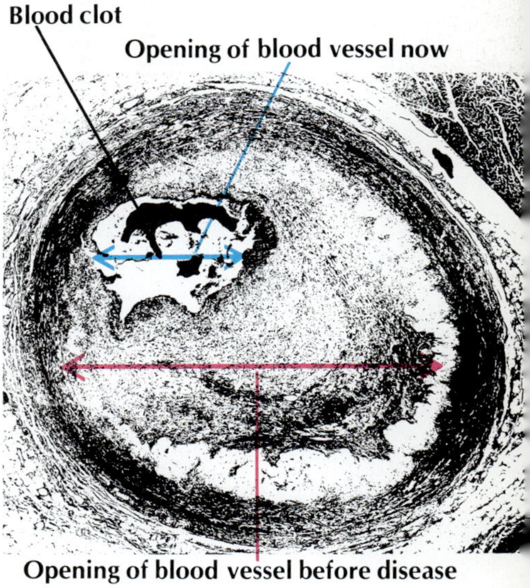

◆

■

●

▲

◆

ment. Mostly they come from the plants and animals that you use as food.

Vitamins in Foods

When a driver has trouble seeing the road at night he may have a disease. We call this disease *night blindness*. In this case, the nerve cells in the eyes do not have a chemical they need. They need *vitamin A* to make this chemical.

From the picture, can you tell what foods you need to eat to get vitamin A?■ You want to have normal night vision.

People can eat a lot of food. They can eat foods that cost a lot. But these foods can leave them hungry for vitamins. Just a little bit of vitamins means the difference between sickness and good health.

Another important vitamin is *vitamin C*. When vitamin C is missing a child may get *scurvy* (SKUR·vee). If a child gets scurvy, the tiny blood vessels in the skin break down. Can you say which foods you should eat to get vitamin C?●

A lack of another vitamin, *niacin* (NY·uh·sin), causes *pellagra* (puh·LAY·gruh). Pellagra is a disease that harms the skin, the nervous system, and the digestive system. It is a fairly common disease. But we know the foods we should eat to prevent it.▲ Niacin is one of many vitamins in the *vitamin B* family.

Vitamin K is another vitamin people need. This vitamin helps blood to clot. Do you see that there are some foods rich in vitamin K?◆

Look at these pigs.★ They are the same age. They are from the same litter. But look how much bigger one is than the other. The big pig has been fed a normal diet. The little pig has been fed a diet that has one important vitamin missing. This vitamin is *thiamin* (THY·uh·min). Thiamin is another of the vitamins in the vitamin B family.

Scientists found out about this vitamin from some sick chickens. The chickens could hardly stand up. They had a disease that harms nerve cells. Of course, this

★

means their muscles could not work as they should.

The chickens had been on a white rice diet. They had been eating only white rice. Then they were fed whole brown rice. After just a few feedings of brown rice the chickens got well. They could run around again. Why did they get better? Yes, the brown covering of the rice has thiamin in it.

People also get a disease like the one the chickens had. That is, they get a disease if they eat mostly white rice. They need to have some whole wheat or brown rice. What other foods have thiamin? Can you tell from the picture?◇ What foods protect us from getting the crippling disease *beriberi* (BER·ee·BER·ee)?

◇

Some foods can give us many vitamins all at one time. Is this a good group of foods?⊙ Are there many vitamins in these foods?

There are many, many vitamins in foods. Many more than the ones we have talked about. You need to eat red meat and green and yellow vegetables. You must eat some *fresh* vegetables and fruits. Then you are getting just about all the vitamins you need.

⊙

AN INVESTIGATION

A Test for Vitamin C

How much of a vitamin is in a food? How do scientists find out? They can do many kinds of tests to find out. Sometimes they use animals to test for a vitamin. Or they may use microscopic organisms. They see how fast these tiny organisms grow. Other times, scientists can use a chemical test.

You can see how a chemical is used. You know that orange juice has vitamin C. Two girls in Detroit wanted to find out *how much* vitamin C was in three brands of orange juice.

First they put an inch of brand 1 orange juice into beaker A. Then they put the same amount of brand 2 juice into beaker B. And they put the same amount of brand 3 juice into beaker C. Then they added an equal amount of water to each beaker. This water made the juices weaker.

Now watch how they did the test. The girls put 10 drops of a blue dye into a clean test tube. Then they added brand 1 juice a drop at a time. ■ They counted each drop. They kept adding the juice until the blue color disappeared. ● Vitamin C makes the blue color disappear. You can say vitamin C bleaches the color out of the blue dye. They wrote down the number of drops of orange juice that was needed.

■

Now they did the same thing with orange juice 2. They started with 10 drops of blue dye in a clean test tube. They counted the number of drops of brand 2 orange juice. They did the same with brand 3 orange juice.

You can do this yourself. You can see for yourself. Which juices have more vitamin C? How can you tell? The juice that bleaches the color *fast* has the *most* vitamin C.

Suppose it took 42 drops of brand 1 orange juice to bleach the blue color. And say it took 21 drops of brand 2. Which had more vitamin C? brand 1 or brand 2? Brand 2 had more vitamin C. It took only 21 drops to bleach the blue color.

This is one test that scientists use. This is one way that vitamin C in foods can be tested. This is one way that scientists know the foods we should eat.

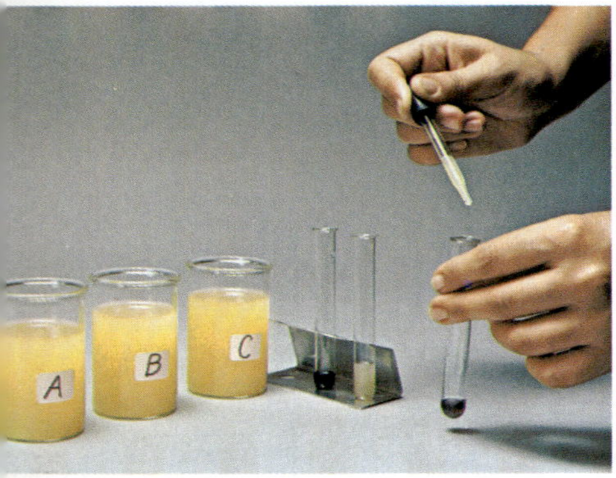

●

Chemical Messengers

Vitamins are not the only kind of chemicals that may be missing in the body. Other missing chemicals may also cause diseases. You may have heard of *diabetes* (die·uh·BEE·tis). A person with diabetes has too much sugar in the blood. In fact, extra sugar goes into the urine. This disease is caused by a chemical missing in the body. The missing chemical is **insulin** (IN·suh·lin).

Insulin is made by special cells in the *pancreas* (PAN·kree·us). The insulin goes into the blood stream and then to all cells. ▲ The insulin keeps the cells working properly. Then the blood does not have too much sugar.

We call chemicals like insulin special messengers. They are chemical messengers in the blood. The body has many kinds of these chemical messengers. They are made in special organs called **glands.** These chemical messengers go right into the blood. This is why they have a special name. They are called **hormones** (HOR·moanz). There are many chemical messengers, or hormones. They are made in many glands. ◆ The hormones get into the blood stream.

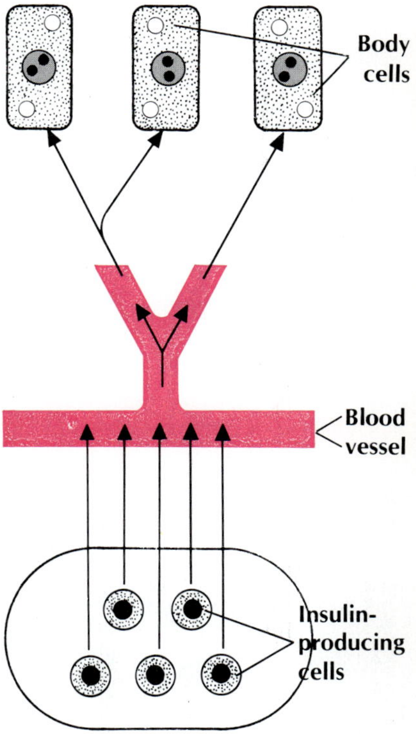

▲ Pancreas

◆

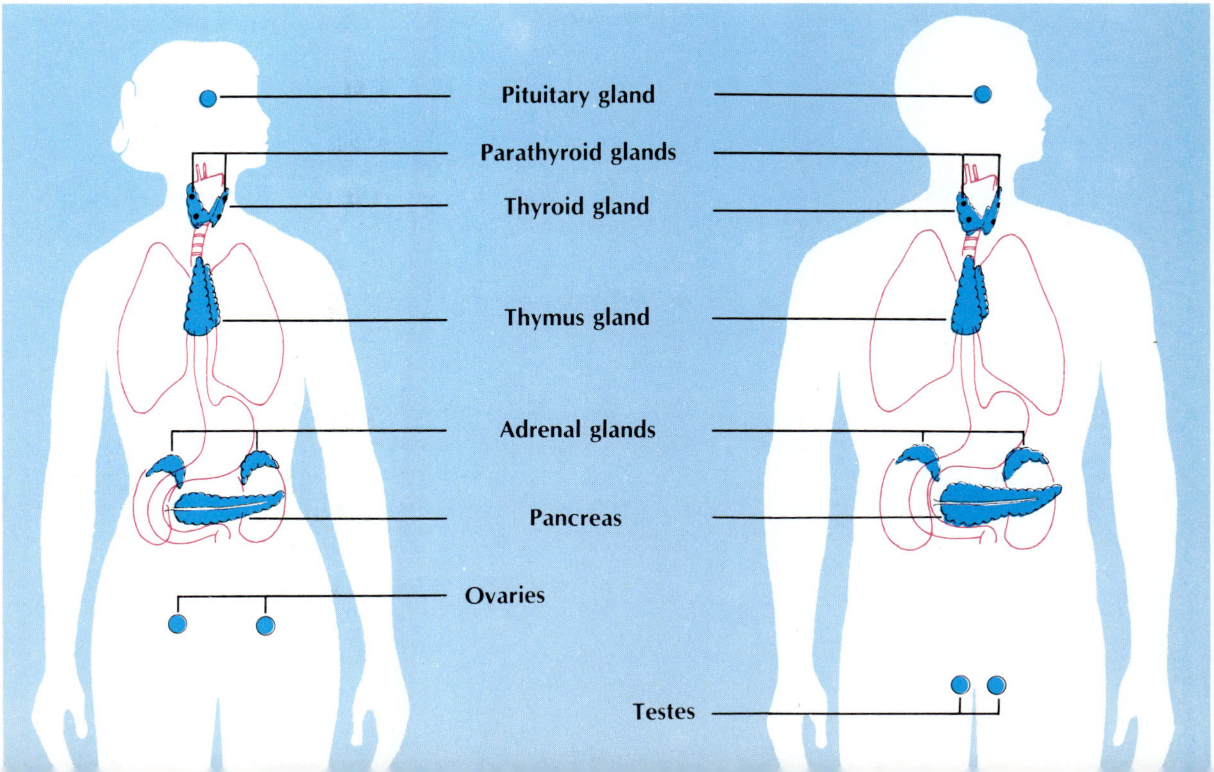

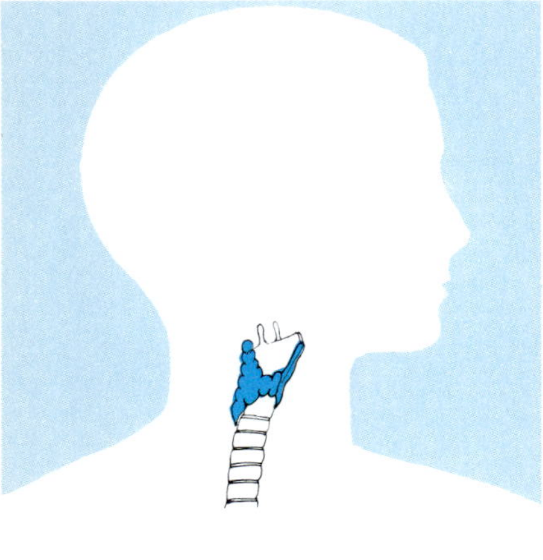

■

●

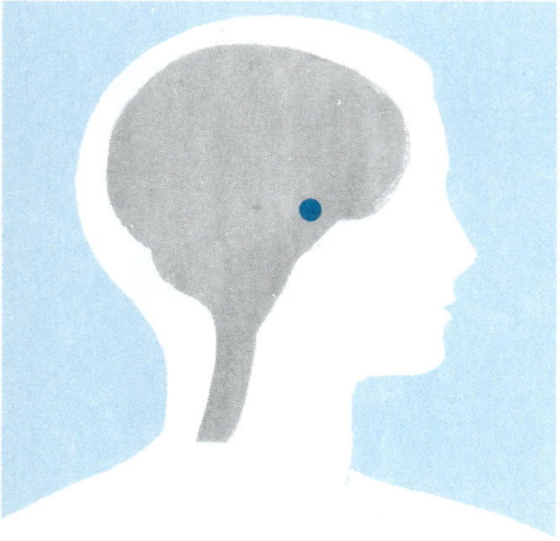

▲

The blood stream goes to all the cells. All the cells share the materials that are carried in the blood. All the cells are in touch with each other through the blood stream. Hormones keep all cells working together. Hormones keep the body in balance.

What happens if a person has a hormone missing? You may know someone who takes insulin. We get this insulin from the pancreas of cattle. The cattle's insulin works in us the same way our own insulin works. It keeps the right amount of sugar in the blood.

Hormones are powerful messengers in the blood stream. Look at this *thyroid* (THY·royd) *gland.* ■ It is in the front of the windpipe. It makes a hormone. *Thyroxin* (thy·ROCK·sin) is the name of the hormone. What does thyroxin do in the body? It sets the pace. It sets the pace for how fast or how slow you use energy from foods.

Sometimes the thyroid gland works overtime. Then it makes too much thyroxin. If there is too much thyroxin, a person may be very thin. He is using his store of energy too fast.

Sometimes a person may get very fat and move very slowly. He might say he doesn't have enough energy. Can you tell what may be wrong with him? This kind of a person is not making enough thyroxin.

See how tall one person in this picture is. ● And next to him is a midget (MIJ·it). What makes one person so short and the other so tall?

Just about in the center of the head is a small, powerful gland. It is the **pituitary** (pih·TOO·ih·ter·ee) **gland.** ▲ It makes many, many kinds of hormones. One of the hormones makes the bones grow the way they should. A very tall person may have too much of this growth hormone. And a midget? A very short person may not grow because he does not have enough growth hormone.

Of course, there is more to the story. It is not so simple. How tall or short a person is may be due to many reasons. It may be due to the amount of food he eats. It may be due to his health. It may be due to his hormones. And it may be because his parents are tall or short.

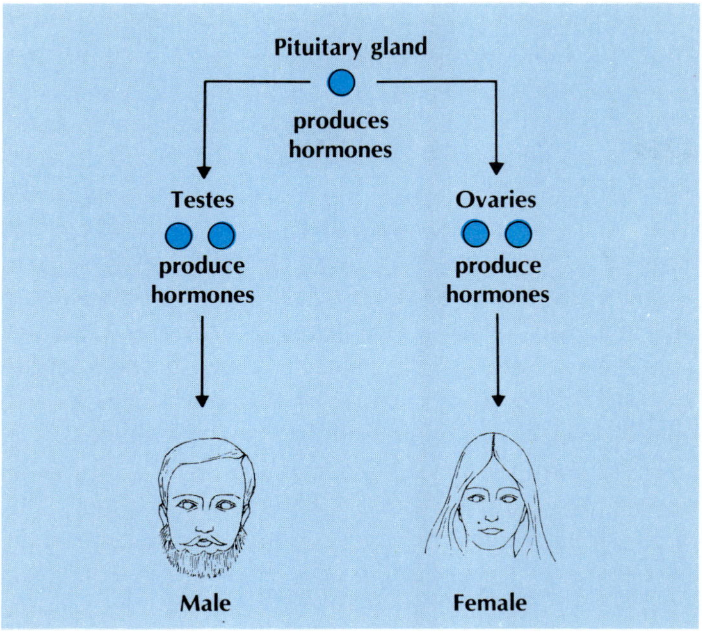

♦

Some other hormones made by the pituitary gland start the sex glands to work. Then the sex glands also make hormones. ♦

Diseases That Run in Families

Bleeder's disease, or *hemophilia* (hee·muh·FILL·ee·uh) is another kind of disease. It is not catching. It is not caused by a lack of a vitamin.

This is a disease that runs in families. This means it is **inherited** (in·HER·ih·ted). ★ It shows up more often in boys than in girls.

★

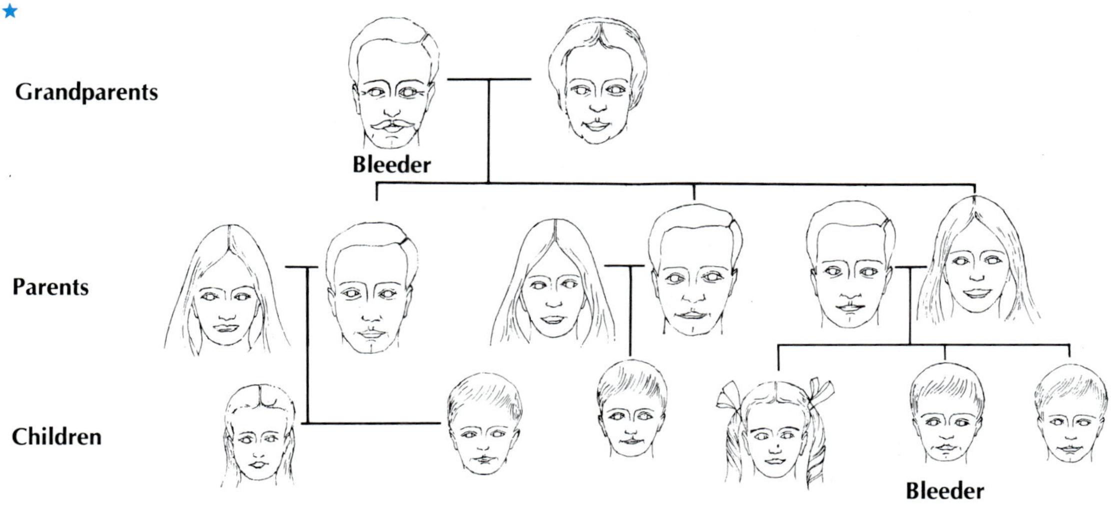

When a child is born with this disease, his blood does not clot the way it should. If he is cut, he may bleed to death.

There are many other inherited diseases. You will learn more later on about how chemicals may be missing from the body. When chemicals are missing in the body, cells do not work properly. That is why you can think of some diseases as chemical diseases.

There are some other diseases you may have heard of. Have you heard of *hay fever?* Or *asthma* (AZ·muh)? These are **allergies** (AL·ur·jeez). Many times allergies are caused by things in the environment. Things like dust or pollen. Or even some foods or medicines.

These things do not bother some people. But other people are extra sensitive (SEN·sih·tiv) to them. This means their bodies respond, or react, to the thing in the environment. These people may break out in a rash, get swellings, or have teary eyes. We say they are allergic (uh·LUR·jik) to dust, to fur, to egg white, or to milk. Or to some medicines.

About 10 percent of the population in the United States seems to be allergic to something. You can think of allergies as "chemical" diseases. The body reacts. The body gives off special chemicals into the blood stream. These chemicals cause the rash or swellings or teary eyes or sneezing. ■

A Healthy Body

There are many, many causes of diseases. And there are many kinds of diseases. It is hard to give a definition of disease. It is easier to tell what health means. And good health is what we want to think about most. Good health means we feel well and we have no pain.

A healthy body is a body that is working right. We need a body that can relax. We do not even think about the body when we are in good health.

There is more to keeping good health. Besides rest and relaxation, a healthy body needs exercise. It needs muscular action. The kind of action a person gets by playing in sports and by dancing. A healthy person also needs work—work that he enjoys. Work that he feels good about. Health also takes in the emotions (ih·MOH·shunz), a person's mental outlook. All these things make a balanced person—a healthy person.

A healthy person finds sports, dancing, friends, and work help to make him feel well and happy. ● A healthy person needs no harmful drugs. He needs no pills to pep him up. He needs no pills to calm him down. A healthy person lives well within his environment.

●

●

●

●

3. SPOILING THE ENVIRONMENT

There are some other diseases we can call "man-made." They are not really made by man. But the way that man lives makes some diseases worse. The way man lives makes these diseases a problem.

These man-made diseases may result because *man sometimes spoils his environment.* He sometimes spoils his *outer environment.* Sometimes he spoils his *inner environment.*

You may live in a city. There were many cities as far back as 5,000 years ago. But in those days the number of people in a city was small. The number of people—the population—in the whole world was small. But now we have big cities. We have big cities where very large numbers of people live close together. Today almost three quarters of all Americans live in big cities.

What is the environment man has made for himself in the cities? Some cities look beautiful. ■ Some do not. Look at the air the people in this city breathe. ●

AN INVESTIGATION

How Dirty Is the Air Around You?

How dirty is the air around you? You can find out. Suppose you put a thin layer of Vaseline or clear oil on two white dishes. ▲ Put one dish outside. Keep the second dish inside.

Or, at home, you can hang a white cloth outdoors. Also, keep a white cloth indoors. Why do you use two cloths? The one inside will show you what the other cloth looked like before you put it outdoors.

Look at the dishes or the cloths in a week. Is there a difference? Or is there a difference in 10 days or so? Do you see much black matter in the dishes or on the cloths? This black matter is called soot (SUHT). It is unburned matter thrown into the air from smokestacks.

Is the air just as dirty one day as it is the next? How could you find out? Is the air just as dirty at all hours of the day? Or is it worse at some hours than at others? Plan a way that you could find the answers to these questions.

▲

Other Wastes in the Air

Waste gases also make the air dirty. You cannot see all the gases that make the air dirty. But you can see signs that they are in the air. Look at this statue. ◆ And look at these plants. ★ The statue and the plants have been harmed by waste matter that dirties our air. Most of the waste matter that did the harm you see in these pictures was a gas.

Let us look into this matter of dirt and wastes in the air. Can they harm us? Can they make us ill?

◆

★

The Air We Breathe

The smoker puffs smoke into the air. Smoke from a burning steak on a grill goes into the air. Someone burns leaves in the fall. These are some of the ways waste matter is thrown into the air.

But much of the dirt and wastes in the air comes from factories. Some factories burn coal to make electricity. Some factories burn coal to make steel. The waste from what they burn goes into the air.

Factory smokestacks throw soot into the air. Soot is in the thick smoke that goes into the air. What else do factories throw into the air?

When cheap fuels are burned a gas called sulfur dioxide (SUL·fur die·OK·side) goes into the air. What does it do to your inner environment? You do not see this gas. But it burns like acid. It burns your eyes, nose, and throat. It is the gas that did the harm to the statue in the picture on page 231. And it harmed the plants in the other picture on the page.

Higher-priced fuels have much less sulfur than the cheaper fuels. Companies that bring us the gas and electricity we use need to burn better fuels. Then less sulfur dioxide will get into the air we breathe.

Cars are another cause of foul air. ■ Unburned matter from the exhaust of cars gets into air. Carbon monoxide (CAR·bun mun·OK·side) is a gas given off in exhaust matter.

Carbon monoxide has no smell. It is a deadly gas. This gas gets into the red blood cells. Then the cells cannot carry on their normal job. They cannot bring oxygen to all the other cells in the body. The cells die without oxygen.

■

This is the reason lots of air is needed while driving a car. It is also the reason that some streets are closed during heavy traffic. There is too much deadly carbon monoxide in the air.

Never drive a car with all the windows closed. Do not keep a motor going in a closed garage. Since there is no odor to carbon monoxide, a driver has to remember to be careful about it.

Smog

Sometimes the dirty air piles up over a city. When there is a layer of warm air over the city there is a chance the dirty air will pile up. The warm air forms a blanket over the city. It does not let the smoke and soot rise high up into the upper air. If the dirt could rise up, the winds could carry it away. The thick dirty air is called smog.

Where Does the Air Go?

Where does the air that you breathe in go? It goes into your lungs. It goes through the nose and down the windpipe.

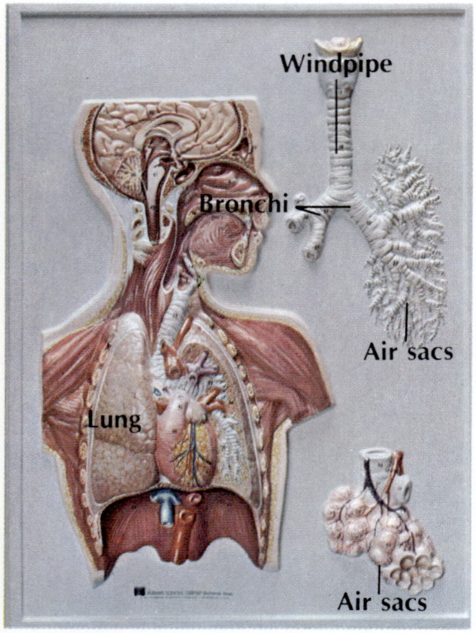

You can see that the windpipe divides into two large air tubes. ■ These large air tubes are *bronchi* (BRONG·kye). The bronchi branch off into many smaller tubes. At the end of these small tubes are *air sacs*. The air sacs are the parts of the lungs where oxygen goes into the blood. And carbon dioxide goes out of the blood. When you exhale the carbon dioxide goes out of the body.

What happens to people who smoke tobacco? What harm do they do to their breathing system? There are small bits of unburned matter in tobacco smoke. There are also tars. And there is a chemical called *nicotine* (NICK·ah·teen). All these parts of smoke dirty the thin wet membranes of the breathing system.

Some of the tars may change the cells. These cells may turn into cells that grow too fast. This is *cancer*.

The nicotine in tobacco smoke causes many changes. For example, it changes the size of blood vessels. Small blood vessels get much smaller when a smoker takes one puff. Nicotine also damages the heart. The heart beats faster. This puts extra strain on the heart muscles.

Each year the air around us gets worse. Our lungs take in more dirty air. Our inner environment is fouled. More people have lung diseases, like lung cancer.

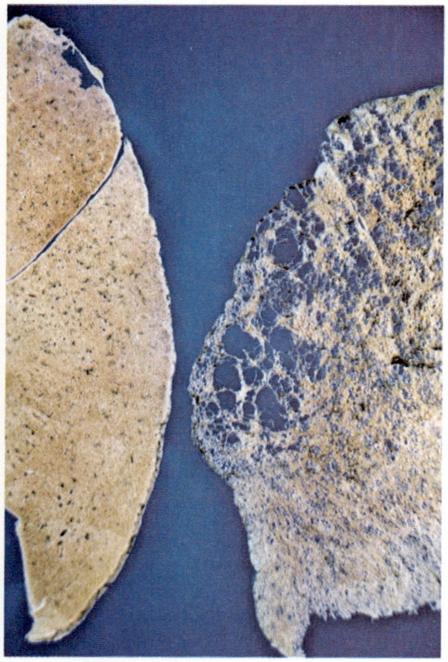

There are other lung diseases that may be caused by dirty air. Or that may be made worse by dirty air. One is called emphysema (em·fih·SEE·mah). The tiny air tubes and air sacs break down. Then the lung looks like the piece of lung you see in the right of the picture. ● See how different it is from the piece of normal lung in the left of the picture. People with emphysema have a hard time getting enough oxygen.

Another lung disease that is made worse by dirty air is bronchitis (brong·KYE·tis). People with bronchitis have a bad cough. And it may be hard to breathe.

Lung diseases seem much more common in smokers than in people who don't smoke. And they are made worse by heavy air pollution. There is extra work in breathing. This means extra work for the heart.

The time to do something about dirty air is now. We have to make some changes. We need to change this air we breathe. We all need a breath of fresh air.

A Try at Prediction

Look at these pictures. ▲ The same city is shown in each picture. But one picture was taken in 1967. The other was taken in 1960. Which do you think was taken in 1967? Why do you think so? Check your prediction at the bottom of page 236.

▲

▲

■

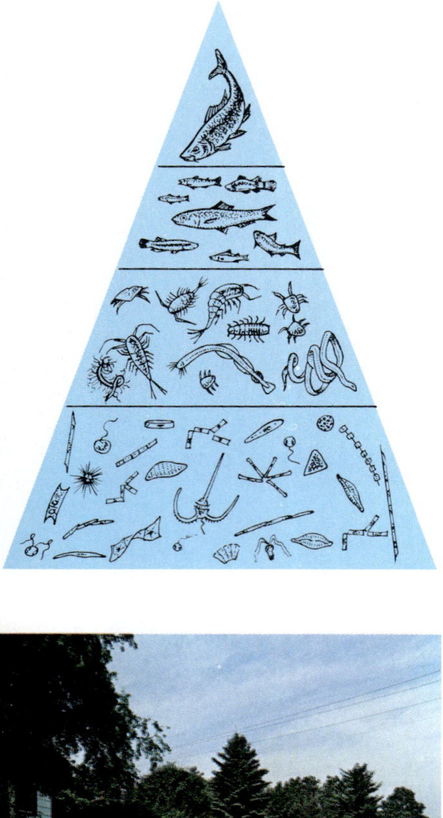

●

▲

The Water We Drink

Men have built machines. Men have built factories and tall buildings. And huge ships, trains, and jet planes. These help men in many ways. But they also do much harm. Plants and animals are harmed. And man himself is beginning to be hurt from what he has built.

Look at this clean river.■ There are communities of plants and animals living here.● There are balanced food chains. Big fish eat small fish. Small fish eat worms. Worms eat plants.

Here you can see white foam floating on the top of the river.▲ This foam is detergent (di·TER·junt). It comes from homes and factories.

The foam looks soapy. But it isn't. Bacteria can break down soap. But bacteria cannot break down this foam. It floats around and clogs up rivers. In recent years, detergents that make less foam have been made. We could not put up with all the foam we were getting from older detergents.

But there is more to this story. Detergents have chemicals called phosphates (FOSS·fates). These chemicals get into the rivers. They make small plant cells in the water grow better. Soon there are too many plants. Then they

The picture of the *clean* city was taken in 1967. Does this surprise you? We *can* clean up our air. But we must all care about it. We must all do our part.

die. Then bacteria that cause decay go to work on the dead plants. These bacteria use up a lot of oxygen. There isn't enough oxygen left for fish and other animals.

Food chains in the lakes and rivers and oceans are upset. They are upset by man's wastes. And by factory wastes. Some factories dump wastes containing mercury (MUR·ku·ree). This chemical gets into the rivers and oceans. Mercury gets into small fish. Then bigger fish in the food chain eat many small fish. This means the big fish get a lot of mercury.

Then man eats the big fish. You can see that a lot of mercury can get into man. Too much can cause illness. And the brains of children may be harmed.

Have you ever seen this kind of a sign on a beach? ◆ Why is no swimming allowed? Sometimes it is because *sewage* (SOO·ij) has gotten into the water. The water is not fit to swim in.

Scientists from the health department take samples of water. Sometimes they find large numbers of bacteria of a certain kind. The kind of bacteria that live in man's intestines. Then up goes this sign. They know that human sewage has gotten into the water. No swimming allowed!

Would you eat raw oysters or clams that live in this foul water? No. Think of the gallons of water that go through these animals each day. You can guess they have a large amount of bacteria in them. This is one way we can get diseases from the water we drink.

People can get typhoid fever and cholera (KOL·er·ah) from foul water. They can get other intestinal upsets. It depends on what harmful bacteria are in the water.

The earth is our spaceship. It has only a certain amount of clean air, water, and good soil. When man spoils this outer environment there will be nothing left.

Man's Inner Environment

You know that chemicals have been sprayed on most fruits and vegetables. You can wash the fruits and vegetables you eat. Then you will not eat these chemicals. But you may not know which chemicals are added to the other foods you eat.

◆ 

Chemicals Added to Food

Have you ever read the labels on the boxes or cans of food you eat? Try it. For example, dried fruits may have sulfur dioxide added. This chemical makes the color of the food lighter. Another chemical may be food coloring. Food colorings are added to many foods you eat.

Bread may have added chemicals, also. One prevents bread from getting moldy. Other chemicals are sprayed on potatoes and onions. These chemicals stop them from sprouting when stored.

There are many chemicals added to the meat, milk, and vegetables we eat. We do not know how much harm they do inside us. There are strict laws to protect us. One is the Pure Food and Drug Act. But the laws do not take care of all the foods we eat.

The Extras We Take In

What extras besides food, air, and water do we take into us? What extras do we take into *our inner environment?* Look around you. Look at the ads on television. And in newspapers. What are many, many of the ads for? Pills. Pills for headaches. Pills for upset stomachs. Pills to help us go to sleep. Pills to keep us awake. Pills to stop colds.

These pills you can buy anywhere. ■ Or doctors give prescriptions (pre·SKRIP·shunz) for some drugs. But there are not ads for many kinds of drugs. You can buy these drugs only if a doctor says you should have them. A doctor has to make sure you need the drug. And he has to make sure you do not take a harmful amount.

All drugs have effects on the body. They change the body's inner environment. A doctor prescribes (prih·SCRIBES) a drug because he *wants* it to change your inner environment in a way that will help you. Then the drug is useful. But it may be very harmful if you take it without your doctor's orders.

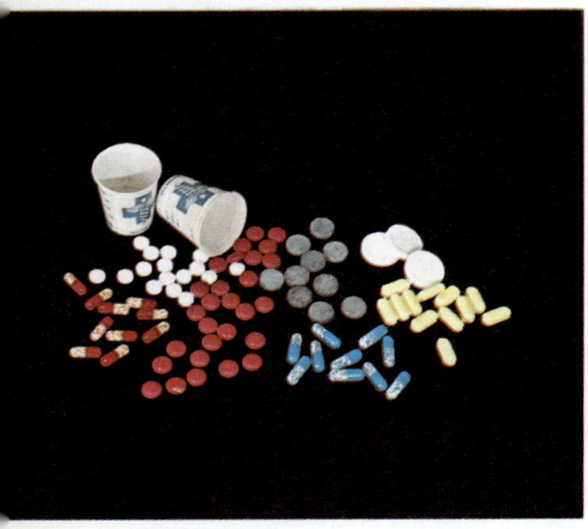

■

Maybe 10 million people take some kinds of drugs legally. That means they take the drugs under their doctor's care. But close to a million people may take these same drugs illegally. They take drugs on their own. They do not use the drugs the way they should be used. To get drugs illegally, they have to pay a lot of money. They may have to pay much more for illegal drugs.

What are some of the drugs that people take on their own? Some young people begin drug-taking by sniffing airplane glue, gasoline, or lighter fluid. This is the start of a dangerous habit. You know that these drugs have dangerous and often deadly effects. A person who takes these drugs gets a ringing sound in the ears, and he sees double. He goes into a kind of fog. He may pass out or go into a deep sleep.

Beginners also sometimes drink cough syrups that have codeine (KOH·deen) in them. Codeine is made from opium (OH·pea·um). Codeine is used in cough medicines to stop coughing. It is a useful drug. But if a person takes a lot of codeine, it may be dangerous. He may go into a deep sleep.

You may know someone who takes *tranquilizers* (TRANG·kwul·eye·zerz). These are useful drugs in the hands of a doctor. They help to quiet people down. But they do not put a person to sleep. These drugs can be dangerous when taken on one's own.

There are the drugs that put people to sleep. They are drugs called *narcotics* (nar·KOT·icks). Narcotics cause the nervous system to slow down. Doctors give narcotics to deaden pain. But a person who takes a narcotic on his own is in danger. He uses the drug as a way to escape from this world.

Opium is a narcotic. So are codeine, morphine (MOR·feen), and heroin (HER·oh·in). Opium is made from poppies grown mostly in Turkey, Mexico, and South Asia. ● It is brought illegally into other countries to be made into heroin.

People who push heroin usually cut the bags with other white powders. They may add milk sugar to the heroin. Then they can make much more money on the same amount of heroin. The danger is that sometimes they don't cut the bag as much as at other times. So the heroin a person buys may be stronger than he is used to. Then the person can die from an overdose.

People who are tense or unhappy may think they need narcotics to slow them down. What a mistake they make! They become addicted (uh·DIK·ted) to the drug. This means they cannot get along without it. They are hooked. Their bodies can't function without the drug. In fact, if they can't get the drug, they get very sick. They get withdrawal (with·DRAW·ul) sickness. So their lives become centered around drugs. They forget about school, their work, their families. They forget about their hopes to get ahead.

There are strict laws about buying and selling narcotics. A person caught pushing narcotics may get a big fine and be sent to prison for many years. A person who is caught with narcotics may also get a fine and be sent to prison.

Narcotics and other kinds of drugs that quiet down a person and make him sleepy are *depressants* (dih·PRESS·unts). You know now that codeine and heroin are depressants. So is airplane glue. So are goofballs.

Doctors call goofballs barbiturates (bar·BICH·er·its). Barbiturates may make a person stumble and fall and have slurred speech. They may make him go to sleep.

Again, barbiturates have many good uses in the hands of doctors. But abusers of barbiturates get to depend on them. They can't get along without the drug. In fact, it is more dangerous to try to withdraw from barbiturates than it is to stop taking most narcotics. A person really needs a doctor's help if he wants to withdraw from barbiturates.

Some kinds of drugs make a person feel high, or excited. This kind of drug is a *stimulant* (STIM·yuh·lunt). Drugs like amphetamines (am·FET·uh·means), or speed, are stimulants. Speed makes the pupils of the eyes get larger, or dilated (DYE·lay·ted). It makes people talkative and excited. Another stimulant is cocaine (ko·KANE), or coke. It comes from the cocabush in South America. ■ It stimulates the whole nervous system. People who take it get very excited. Their hands may tremble or shake.

Which drugs have you heard about now? Narcotics from opium. Also other drugs that are depressants. Drugs that are tranquilizers. Drugs that are stimulants.

There are also some other drugs that put a person into a dream-like state. These drugs are called *hallucinogens* (huh·LOO·sin·uh·jinz). The dream-like state they cause is called a hallucination (huh·loo·suh·NAY·shun). In this state a person has an out-of-this-world feeling. Some people seem to get bored with things around them. Or they think life is too tough for them. They take these drugs as a way to escape for a while.

Pot or grass is a hallucinogen. Another name for this drug is marijuana (mar·uh·WAH·nuh). When people smoke marijuana they first talk fast and laugh loudly. The pupils of their eyes get larger than normal. Then they fall into a dream-like state.

Some people think it is all right to smoke marijuana. They say people don't get addicted to it. But many people seem to depend on it. They think they need it. There is another danger, too. Some people don't stop with marijuana. They go on and try harder drugs. They may want to try heroin for kicks. That is one of the reasons that marijuana is treated like a narcotic. It is not legal to have marijuana or to sell it.

Marijuana comes from a hemp plant. ● Another drug made from this plant is hashish (HAHSH·eesh). It is like marijuana but stronger.

Acid, or LSD (ELL·ESS·DEE), is another hallucinogen. No one can tell what effect LSD will have on a person. Some people go into a pleasant dream-like state. Other people have feelings of terror. The effect may be pleasant one time and terrifying another time.

Drops of LSD may be added to a drink. Or put on crackers or on a sugar cube. Wads of cloth or paper may also have drops of LSD. It is easy to pass around. The danger is that you may not know when someone in your group has put LSD on a cracker or in a drink.

There are many kinds of dangerous drugs taken illegally. And many reasons why people take them. But there is no *good* reason why a person goes on drugs. Most addicts are sorry they are hooked. You may have heard people who have kicked the habit tell how glad they are. They are glad they are living in this world again. Even if it's not a perfect world.

AN INDEPENDENT STUDY

Fouling the Inner Environment

You cannot *see* what goes on inside your body. But you know about some things that happen inside you. You know your cells are in a watery environment. An environment that is the same for all your cells.

How do our cells get oxygen and food? From the blood stream. The blood stream brings oxygen and food to all cells.

Here you can find out for yourself what you know. Do the following pages on your own. Read the question and put the answer on your answer sheet. You can check the answer by looking across to the next page.

1

Your cells cannot take oxygen and food from outside your body. They need the blood stream to bring oxygen and food.

The blood stream is part of your inner environment. It brings oxygen and food to all your ______.

(Write your answer and check on the next page.)

2

We have talked about foul air. This is the air we breathe. This air goes into our lungs. Oxygen from this air goes into the blood stream. But dirt is kept out of the blood. The nose and tubes to the lungs hold back the dirt.

The gas that gets to your blood stream is ______.

1

cells

Cells get oxygen and food from the blood stream.

(Go on to question 2.)

2

oxygen

When we breathe air, only oxygen goes into the blood stream.

(Turn the page to question 3.)

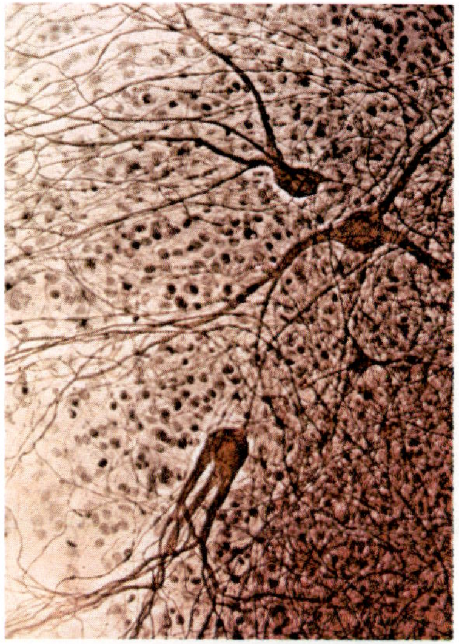

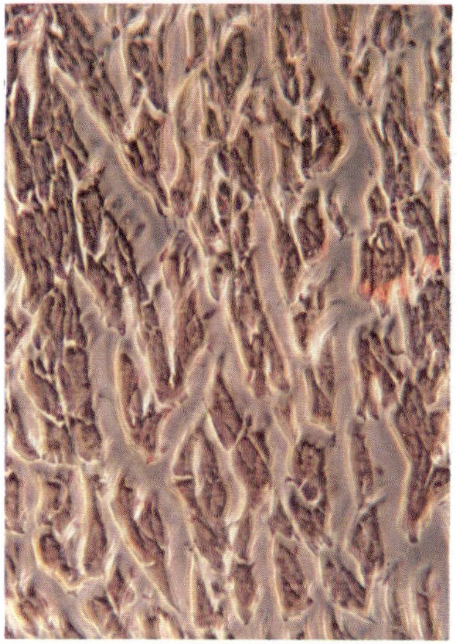

3

There are many causes of dirty air. Smokestacks and cars foul our air. We breathe in soot and a gas called sulfur dioxide. People who live in the country do not breathe so much dirty air.

The air in the country is cleaner than the air in the cities. People who live in some cities breathe in soot and the gas sulfur d------.

4

This gas can harm the lungs. More people who live in cities with heavy smog have lung cancer than do people who live in the country.

Not as many cases of lung cancer are found in people who live in the ________.

5

Another cause of foul air is cigarettes. Cigarette smokers breathe in tars that may lead to cancer.

Tars and nicotine from smoking cigarettes are harmful. They are bad because they can cause diseases of the heart and blood vessels. They can cause lung c-----.

6

We wash fresh fruit and vegetables before we eat them. We wash off the insect sprays. This is the safest way to keep our inside environment clean. Then these chemicals will not get into our cells.

Chemicals sprayed on fruit can harm the work of our c----.

3

dioxide

What harm does sulfur dioxide do to us?

(Go on to question 4.)

4

country

There are not as many smokestacks and cars in the country as there are in the cities.

(Go on to question 5.)

5

cancer

(Go on to question 6.)

6

cells or **our bodies**

Either answer is right.

(Turn the page to question 7.)

7

You are right. Our bodies are made of cells. Some kinds of chemicals harm nerve cells. Some harm skin cells. Or bone cells. Many drugs are harmful. Pep pills and sleeping pills are a few.

Many drugs like sleeping pills are *harmless* or *harmful*. Which?

8

There is fouling of more than our food and air. There is also fouling of the waters of our earth.

When you go swimming in a river or jump off a pier into the ocean, the water may be foul.

Which disease might you get from swimming in foul water—*malaria* or *typhoid*? Which?

9

A chemical that may be in our water is mercury. Factories may dump mercury in the water. Then the mercury can get into fish. Man may eat the fish. The mercury can get into man.

As the mercury passes along the food chain, larger and larger amounts pile up in the larger organisms in the food chain.

Mercury is most dangerous to man, because he is at the end of the f--- c----.

7

harmful

Many drugs harm the inner environment of cells.

(Go on to question 8.)

8

typhoid

Typhoid is caused by bacteria that live in the intestines of people. When human sewage gets into water, typhoid bacteria may get into water. If you swallow this water, you might get typhoid.

9

food chain

More and more mercury piles up in organisms as you go up the food chain.

(Turn the page to question 10.)

10

Sometimes chemicals are added to foods we eat. We don't think about this. But some people do think about drugs. Some people take drugs to pep them up. Drugs like amphetamines.

Which are stimulants—*amphetamines* or *barbiturates?* Which?

11

Some drugs slow down the nervous system. These drugs are known as depressants. Which of these are depressants—*codeine, goofballs, speed, morphine?* Which ones?

12

Some drugs put people into a trance or hallucination. Acid, or LSD, is a hallucinogen. What do you call grass, or marijuana? Is it a hallucinogen? *Yes* or *no?*

13

The environment inside us keeps all our cells in touch. All cells get food and oxygen. Wastes are removed.

The liquid that keeps this inside environment in balance for all our cells is the ______ stream.

10

amphetamines

These are pep pills, or stimulants.

(Go on to question 11.)

11

codeine, goofballs, and **morphine**

Speed is a stimulant.

(Go on to question 12.)

12

yes

Marijuana is called a hallucinogen.

(Go on to question 13.)

13

blood stream

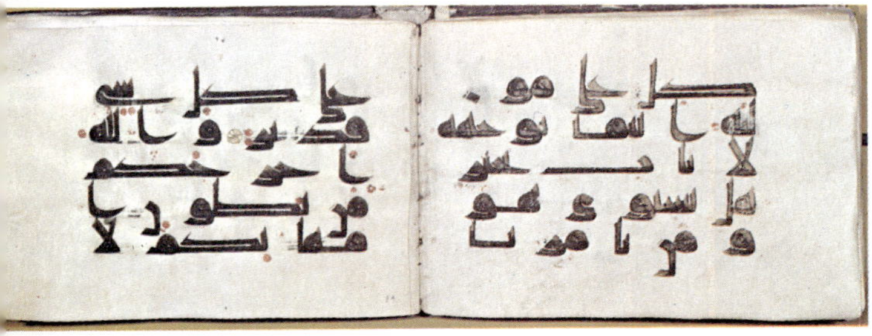

Man Remaking the Environment: Providing Food

7

INTRODUCTION

The Works of Man

Man is the one living thing who has learned to *change* his surroundings. Look at some of the works of man.

Man has learned slowly. But he has remembered his experiences. He has learned from his experiences. He has learned from other people's experiences, too. He has learned from the many people who lived before him.

People have lived together in groups, or societies (suh·SIE·ih·teez), for a few thousand years. They have kept records. ■ They have passed along what they have learned. Records are kept today. ● People who live in the future will learn from our records. This is the way culture (KUL·cher) and history are passed along. This is the way man has learned from all the people who lived before him.

Can you tell some ways man has changed his environment? In what ways has man spoiled his environment? Name some ways he has made it better. Has he made this environment better? ▲ How? How not?

You know man remakes his environment by controlling diseases. Now let's look into some other ways man has learned to *remake* his environment.

▲

1. THE CONTROL OF THE ENVIRONMENT: FAMINE

How many people would you say live on the earth? There are more than *3½ billion.* ■ Way back in the year A.D. 1 there were only about 250 million people. That is only ¼ billion.

Slowly, the number of people grew. By the 1600's there were 500 million. That's ½ billion people.

It took about 1,600 years for the number of people in the world to double. ● In the early days, many people died in childhood from diseases and hunger. People who lived through childhood did not have as long lives as we have. Many died during **epidemics** (ep·ih·DEM·icks) of disease. An epidemic means that many, many people get a disease at one time. Hundreds, even thousands, might die during an epidemic. Also, hundreds or thousands might die at one time from hunger. When many people die at one time from hunger, we say there is a **famine** (FAM·in).

Hunger and disease kept the world population low for a long time. You know that man has learned to control many diseases. Now let us see how he has tried to control hunger.

Food-gatherer to Food-grower

Early man was a hunter. He hunted animals and fish. He had to be a hunter. He knew nothing about growing crops. People ate berries and seeds. But they did not know that new plants grew from seeds.

A hunter probably needed about two square miles of land for hunting. In this space, he could hunt enough food for himself and his family.

Sometimes hunters grouped together. Then they could chase larger animals over a cliff. ▲

Early man probably spent all his time in search of food or in escape from animals. Life was hard. Life was also short.

▲

Growing Crops

At some time in history man learned to grow crops. Perhaps only about 10,000 years ago. The people who lived in warmer climates became the early farmers. They learned to pick out healthy, large seeds. Then they saved these seeds for the next year's crops.

In some parts of the world, ancient farmers grew barley. In other places, they grew rice or wheat. They grew what they could with the land, the climate, and the water supply in the places they lived.

Raising Animals

We know that early man learned to raise animals. The ancestors of dogs, sheep, pigs, goats, and cattle lived in the camps of early people. ◆

When men moved into new places, their animals may have followed them. Maybe the young animals were cared for by early people.

In Asia, Europe, and Africa, people lived like this over 10,000 years ago. This we know from the remains of plants and animals that are found buried in rocks.

◆

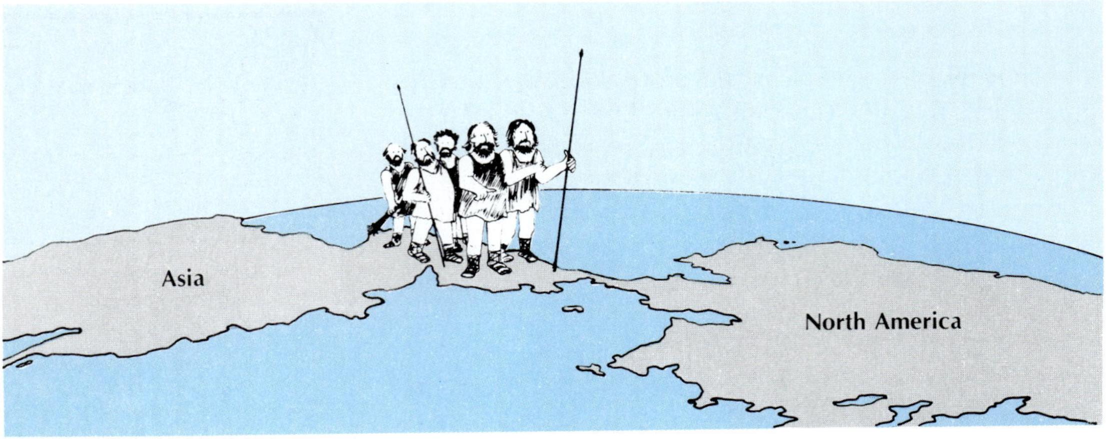

In the New World

Several times Mongoloid (MONG·guh·loyd) peoples had come across the Bering Strait (BER·ing STRATE) into the New World. ■ These people who came to the New World did not bring animals with them. They did not bring any cereal plants either.

But they found new plants in the New World. For example, they found wild corn plants. They learned to raise corn plants. Corn was easy for the Indians to raise.

Many years later the Europeans came to the New World. They found life different from life in the Old World. For example, in the Old World people had used more animals for food and work. They had used milk and meat and wool from animals.

Many different crops were grown in the New World. There were more plants. There were white potatoes, sweet potatoes, peanuts, and beans. Also crops of tobacco, tomatoes, pumpkins, and vanilla beans.

These are the same crops that we grow today. Even the making of wine and beer is not new. The Egyptians knew how to make beer a few thousand years ago. ● They used mash from barley. American Indians made a beer from the mash of corn.

No, modern man has not added any new crops. The aim today is to grow *more* crops—and to grow *better* crops.

The Year 2000

Today we are still growing crops. We are growing more and more food each year. But the number of people is growing much faster than it was. Scientists are worried. They say the world population will double to 6 billion. How soon? Within 30 years. Within 30 years the world population may be *double* the number now. That means there will be 6 billion human beings on earth.

There has never before been anything like this. Will there be enough food for so many people on earth? Have we land enough for farming? Will there be standing room or living space for so many people?

In some countries, like Japan, there is not much land to farm. In Japan, people are beginning to farm the oceans. They breed lobsters and fish. Then they put millions of young lobsters and fish back into the ocean. The young grow. This means more food for the fishermen to catch. More food for the population of Japan.

In America today, only 7 out of every 100 people live on farms. Only 7 percent of the people supply food for the other 93 percent of people—for the rest of the population. ▲

▲

A Thin Layer of Soil

We and all living things live in a thin layer of air. And *on a thin layer of soil*. We grow our crops in good soil only about 6–12 inches deep. ◆ Can soil wear out?

We know that gases go in and out of living things. Oxygen and carbon dioxide go back into the air. And water goes in and out.

But we also know some things can be used up. Then they are gone. For example, oil and petroleum (puh·TROH·lee·um) can be used up. They cannot be used again and again like water. We burn oil in furnaces and gasoline in cars. Petroleum is oil that took millions of years to form. We also burn coal. Coal also was formed millions of years ago. It was formed from wood. Our supplies of petroleum and coal are being used up. And other things are, too.

◆

■

Plants take minerals from the soil. If the plants are eaten, the minerals are not all put back into the soil. Farmers ship crops to the cities. Then the minerals the crops took from the soil are gone.

Crops need minerals. Farmers put minerals back into the soil. They do this by adding a little fertilizer (FUR·tul·eye·zer) to the soil. Fertilizers are special minerals that plants need in order to grow well. ■

Farming – Good and Bad

We need to have good soil for our crops. We have to save our soil.

You do not have to be a farmer to see what is wrong in this picture. ● And in this picture. ▲

See how two crops are growing on the *same* land here? ♦ What is good about farming this way? Why is this a good way to stop soil from slipping away? ★ How do trees help to hold down the soil?

The trees do more than hold the soil down. There are birds in the trees. What good can they do? Can the birds eat the insects that might harm the crops?

As the early settlers went west across the United States, there were giant green forests everywhere they turned. It must have seemed as if there would be enough trees forever. Men cut down the forests. They built wagons, houses, furniture. They burned wood to keep warm and to cook meals. They did not think about the future. Or so it seems.

●

▲

◆

★

Today we know there are not enough trees for all the people's needs. We use almost all of each tree that is cut down. Now new trees are planted to take the places of trees we cut down. We treat trees and forests as if they were a crop. They *are* a crop. They are our wood crop.

When trees are cut down, the animals and birds of the forest disappear. Thousands of tons of soil are carried away each year. We need trees and plants to hold the soil. We need to have good soil to grow crops.

The early settlers brought sheep and cattle with them. They let sheep and cattle eat up the countryside. Sheep eat grass so far down it cannot grow again. They eat roots, too. The soil becomes bare without grass or other plants.

Plants are needed to hold the soil. No roots, nothing to hold the soil. When it rains, the water carries the soil away.

Floods come every spring and carry more soil away. What causes the floods? The snows of winter melt and there is nothing to hold the water. Nothing to soak up the water. Just hard soil.

Men have been spoiling the land. This flood is man's fault. ■ So is this hard land. ● So is this dust storm. ▲

■

●

▲

♦

★

◊

A Try at Explanation

1. Study this picture. ♦ This might be a road in your neighborhood. Why is the soil slipping into the road? What should be done to correct it? Could plants be planted to hold the soil?

2. This pretty water plant is stopping up the rivers. ★ How has this happened?

3. Why is this a good way to farm? ◊ What is the good in having trees to stop the winds around crop land?

■

Farmers' Helpers

Look at the tractor in the picture. ■ It is pulling a large machine through a field of growing crops.

Tractors take the place of many men on a farm. With these machine workers, one farmer can take care of a large farm.

The machine in the picture is spraying *insecticides* (in·SECK·tih·sides) on the crops. ■ Insecticides are chemicals that kill insects. They kill the insects that try to eat the crops before man can get to them.

Animals like cattle and chickens are also sprayed with chemicals. This is one way farmers keep down diseases in crops and animal herds.

Insecticides and Food Chains

But there is more to this story. There is another side of the picture. This is another example of how man doesn't think ahead. We see only potato crops and the pests—potato beetles. ● Corn crops and the pests—corn borers. ▲ We see cotton crops and the pests—cotton weevils (WEE·vuls). ♦ Apple trees and codling (KOD·ling) moths. ★

●

▲

♦

★

For many years, one insecticide called DDT was thought to be very useful. But what happened when DDT was sprayed over acres of land? It killed useful bees and other *good* insects. It didn't kill only the insect pests. Even birds were found dead on the ground.

What killed the birds? Many times, the DDT in their food. The insects they ate were sprayed with insecticide. The birds ate many insects. The DDT added up in the birds. There was enough DDT in birds to kill them.

DDT was washed out of the air by rain. Rain ran off into rivers and seas. In fact, DDT was found in penguins in the Antarctic, not far from the South Pole. This insecticide had not been sprayed in the Antarctic. Where did the DDT come from?

Do farmers need to get rid of insect pests? Of course they do. They need to hold down the number of insect pests. The insect pests destroy our food crops. But we need to look around us. We need to predict what may happen. We need to know what can go wrong. What happens when the new insecticide gets into food chains? We must not foul the environment.

Can we find a way to kill only the insects that do us harm? There are new insecticides that don't stay in the soil or water. They don't add up to dangerous amounts. There are some other ways too.

There are other ways besides using insecticides. We are learning how to catch insect pests in traps. This is done with chemicals that have a special smell. Insect pests come from miles around. They fly to this special chemical. Then they are caught in the traps. ◆

Unwelcome Visitors and Natural Enemies

The United States ships many goods to other countries. Other countries ship goods to us. Sometimes we send unwelcome visitors to other countries. But we also get unwelcome visitors.

Here is only one example. Some flower bulbs were shipped from overseas to New Jersey. In the soil the young of some insects were hidden. No one thought of looking for anything in the soil with the bulbs.

■

●

▲

Only a few years later people were talking about the new insect around. It was the Japanese beetle. In only a few more years the beetles were eating up leaves of shade trees. ■ A dozen beetles might be found on one leaf. These insects spread into New York, Pennsylvania, Connecticut. Through New England and down the East Coast. All this happened fast.

The beetles were not found in such large numbers in Japan. Why did they become such a pest here? Because the beetles had no natural enemies here.

What keeps the beetles under control in Japan? We found out. Small wasps like this one are the natural enemy of the Japanese beetle. ●

So you see, insecticides are not always needed. Sometimes we can find a *natural enemy* to keep the pests in check.

We cannot, however, let strange insects come into this country. We may not have the natural enemy of the insect. The insects might grow in large numbers.

We know now the dangers in breaking the food chains. We know that when we remove one part of the chain we may pay a high price. We may have more insect pests.

Sometimes we can only see the need to do what we want right away. We cut down forests to make land for crops. But we are removing birds as well as trees. Birds live in trees. Without birds, insects may increase. Then there are more insect pests to harm the crops. This field of corn has been badly hurt by insects. ▲

Plant Good Seeds—Get Good Crops

Farmers can have good soil and a good water supply. There can be good climate. Farmers can have big tractors. What else do farmers need to get good crops? What do they need to get a large crop to feed large numbers of people?

They must use *good seeds* to start with. Good seeds come from parent plants that have good *heredity* (huh·RED·ih·tee). Heredity means the *traits* that parents pass on to their young.

What do we mean by traits? What do you say about the apple the girl is eating? ◆ It is big. Juicy? Oh yes! And sweet. These are traits we like in apples.

These are crab apples. ★ They are wild apples. They grow in the country. They are small and sour. These are traits we do not want in apples.

How can we get apples with the traits we like? By starting new apple trees from trees that make apples with the good traits. Man picks the right parents to get the sweet, juicy apples.

More Meat, More Milk, More Eggs

What traits do we want in cattle? We want tender meat like the meat this kind of cattle has. ◇ We want more meat on each animal, too.

What traits does a dairy farmer look for? Milk with cream—and a lot of milk. This is a prize dairy cow. ⊙ It comes from families of animals that have traits for good health and good milk. And animals that make lots of milk. These good traits are passed from the parents to the young, or offspring.

Chickens hatch from the eggs of hens. What traits do farmers want in their chickens? They want chickens that make a lot of eggs. A farmer picks parents that come from families of good egg layers.

Good soil and climate are not enough. A good environment alone is not enough.

Crops can only be as good as the heredity packed inside the seeds. Cattle and chickens can only be as good as the heredity packed inside the eggs.

Let's look into this.

◆

★

◇

⊙

■

●

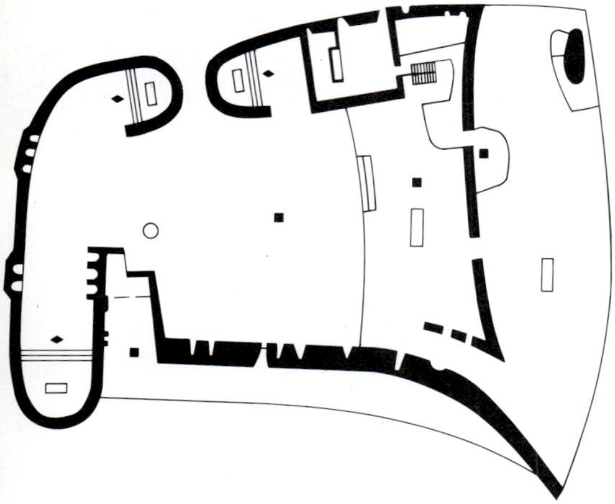

■

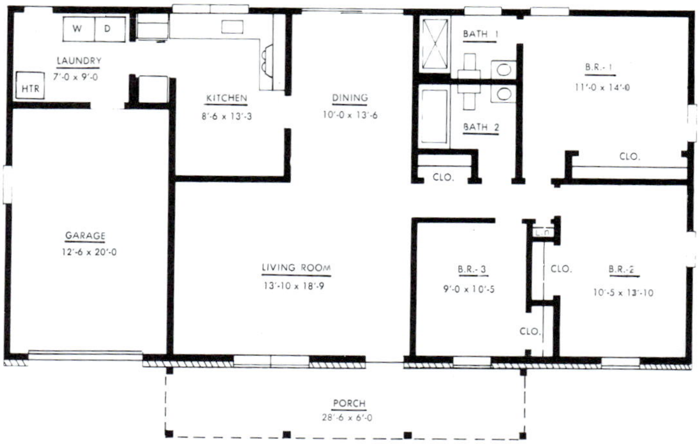

●

2. THE CONTROL OF THE ENVIRONMENT: THE SEARCH FOR PRIZE GENES

▲

This building is made from a pattern. ■ The building is a copy of the pattern.

A different kind of building comes from this pattern. ● This building is an exact copy of the pattern.

This baby hippo is a copy of its parents. ▲ The pattern the baby hippo gets is *not* the lion pattern. It is the hippo pattern. The pattern that is passed on to the offspring is its heredity.

One church cannot make another church. One house cannot make another house. Only living things make copies of themselves. This is *reproduction*.

◆

★

★

◈

◈

Two Parents—Mixed Heredity

This calf is a Jersey calf. ◆ Did it come from this set of parents? ★ or from this set? ◈

How do you know? You expect the offspring to look a lot like its parents.

Not exactly like the father. Not exactly like the mother. The calf has *half* its heredity from its mother. It has *half* its heredity from its father.

All these puppies are from the same litter. ⊙ You know their parents are Dalmatians (dal·MAY·shunz). They have many likenesses. But they have different kinds of spots. They have *half* their heredity from their father. They have *half* from their mother.

Offspring get their heredity from *both* parents. Half from their mother. And half from their father.

From the Beginning

Most living things begin the same way. They begin as an egg. They begin as a **fertilized** (FUR·tul·eyezd) **egg.** This means a sperm cell unites with an egg cell. Then the egg cell grows into the new organism.

⊙

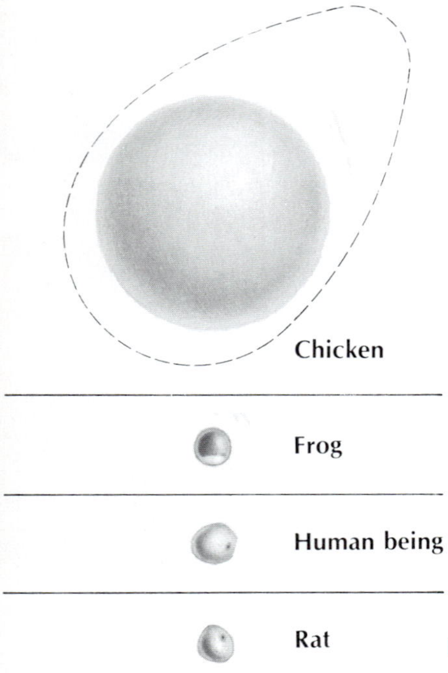

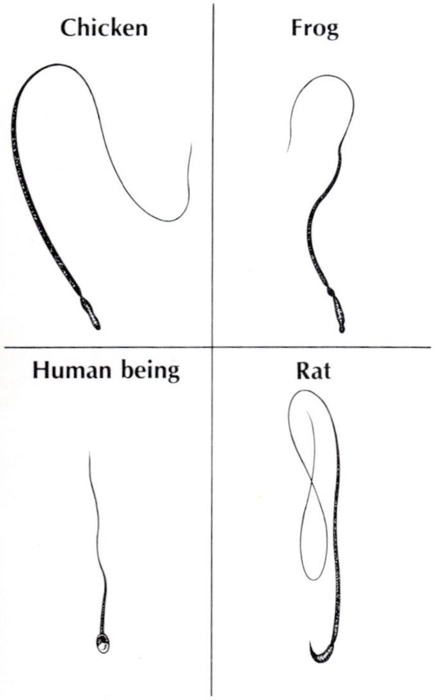

There are egg cells in different sizes. Large eggs and small ones. Egg cells come from the mother. They have the heredity from the mother. They have the heredity from the mother chicken. Or from the mother frog. Or from the mother human being.

An egg cell cannot grow into a new organism until a sperm cell unites with it. **Fertilization** is the union of the sperm cell with the egg cell.

Sperm cells are all very small. So small you can only see them with a microscope. Sperm cells come from the father. They may have different shapes. Almost all of them have a tail. They swim to the eggs.

It makes no difference if the sperms are from a whale, a frog, or a fish. Or if they are from a bird or from a human being. Sperm cells are microscopic. They swim to the eggs.

Where is the heredity? In the *nucleus* of the egg. And in the nucleus of the sperm. The nucleus of the sperm cell unites with the nucleus of the egg cell. This is fertilization.

The pattern, the heredity, is in the nucleus. No matter what size the egg is, the heredity from the mother is in the nucleus of the egg. The head of the sperm cell is mostly nucleus. The heredity from the father is in the nucleus of the sperm cell.

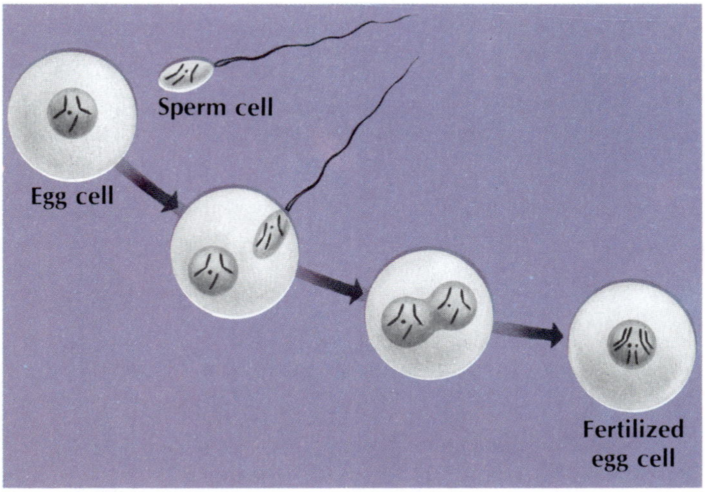

Inside the Nucleus

What is inside the nucleus of the egg cell and of the sperm cell?

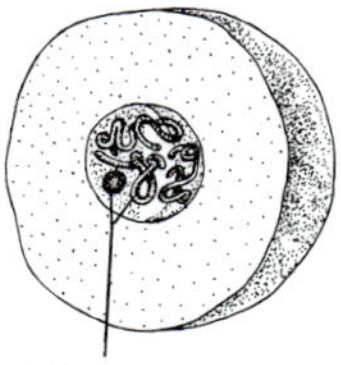

Chromosomes

Inside the nucleus are small parts called **chromosomes** (KRO·muh·sohmz). Chromosomes are made of chemicals. These chromosomes have the pattern, or heredity, that passes from parents to children. These are the chromosomes of a human being. ◆

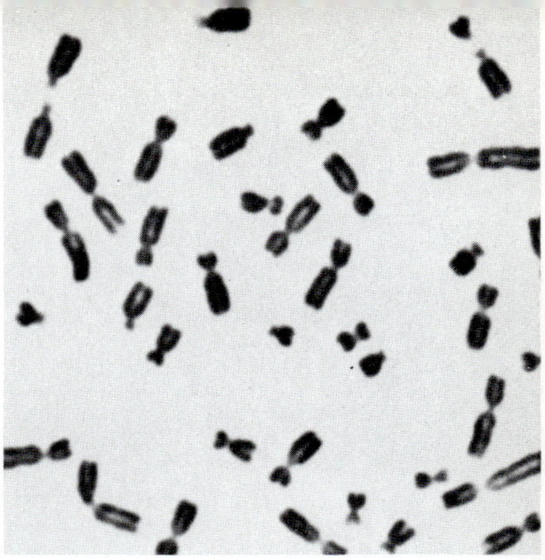

◆

Each kind of plant or animal has a right number of chromosomes. Look at the chart. ★ How many chromosomes are in the cells of human beings?

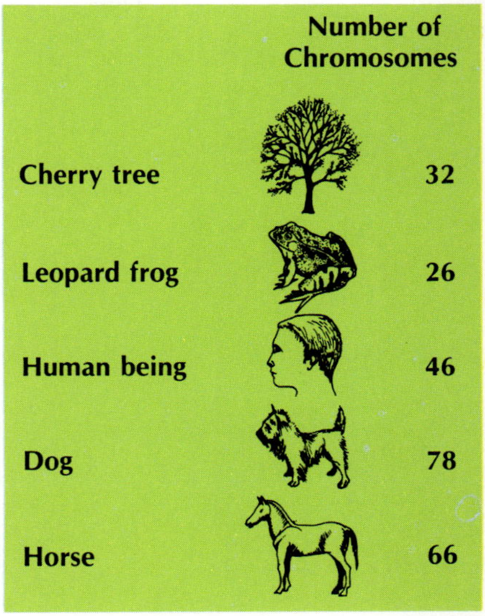

★

How many chromosomes are in the nucleus of a cell of a horse? in the nucleus of a cell of a cherry tree? And how many chromosomes are in the nucleus of a cell of a frog?

A Try at Explanation

A fertilized egg cell that will grow into a human being has 46 chromosomes. The child gets *half* its heredity from the father and *half* from the mother. This means each egg or sperm has half of 46 chromosomes.

Can you say how many chromosomes are in the egg cell? Yes, 23. And the sperm cell also has 23 chromosomes.

When the sperm cell fertilizes an egg cell, the number will be 46 chromosomes. ◊ This is the right number for human beings. In the cells of each human being there are 23 chromosomes from the father. And also 23 chromosomes from the mother.

◊

This is what you have learned:

The heredity from the parents passes to the child.

The heredity is in the chromosomes.

Chromosomes are in the egg cell nucleus and in the nucleus of the sperm cell.

Half the chromosomes come from the sperm cell. Half the chromosomes come from the egg cell.

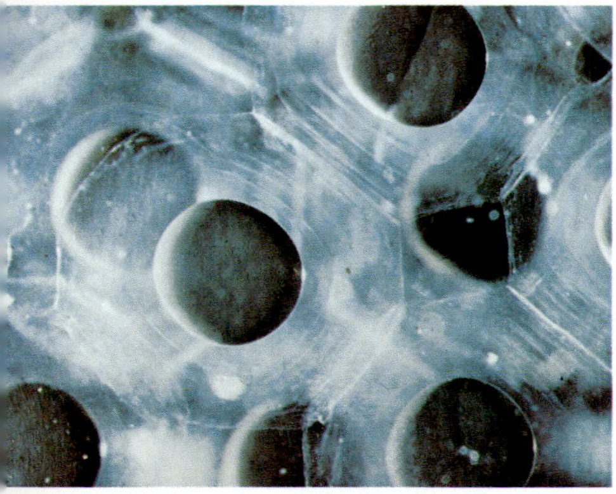

Carolina Biological Supply Co.

A Try at Prediction

This fertilized egg of a frog will grow into a tadpole. ■ Can you say how many chromosomes are in the cells of the tadpole? (*Hint:* Look again at the chart on page 267. Find the chromosome number for a frog.)

Can you tell what number of chromosomes must have been in the sperm cell? and in the egg cell? You know each parent gives half the heredity to the child.

Horses—Different Sizes

This horse has 66 chromosomes in its cells. ● So have these horses. ▲ What a difference there is in the size of the horses! Yes. Both kinds have the right chromosomes to be horses.

They have the *same number* of chromosomes. But the chromosomes are different. The *pattern of chemicals in the chromosomes* of each kind of horse is different.

A Shetland pony has its own special pattern of heredity in the 66 chromosomes. The large work horses have another special pattern of heredity in their 66 chromosomes.

As you see, horses are alike in many ways. But they are not exactly alike. They have some different patterns of chemicals for their heredity.

Inside the Chromosomes

The difference in the horses lies in the chromosomes. In the chromosomes are thousands of chemicals. The chemicals are lined up in patterns. The different patterns of the chemicals make the differences in the horses.

♦

These patterns of chemicals are called **genes** (jeenz). Two horses are not alike. Two puppies are not alike. Two human beings are not alike. They all have different genes.

All human beings have 46 chromosomes. But the genes in the chromosomes are not all alike. ♦ Some people have brown eyes. Some have green eyes. Some have curly hair. Some have straight hair. The patterns called genes are in charge of traits like these.

Genes in the chromosomes are passed from parents to their children.

The Improvement of Animals

Suppose a farmer wants to get better animals. He wants better sheep and cattle for better food and wool.

How would a farmer go about getting sheep with good wool? ★ He must pick parents with good wool. He picks sheep that have the heredity for good wool. They have the right genes in the chromosomes of the egg cells and the sperm cells.

★

A dairy farmer does the same thing. And so does a beef cattleman. They all pick the *right genes.* This means they pick the parents that have the right genes to pass to their offspring.

They pick genes for good beef. Or genes for creamy milk. Or genes for large animals with a lot of meat.

This is how farmers try to get better crops. This is how a horse breeder gets better horses. He picks two good horses to be parents. He hopes that the colt will have some of the good genes from each parent.

Sexual Reproduction

A child comes from a fertilized egg. This means a sperm cell fertilized an egg cell. This kind of reproduction is called **sexual reproduction.**

The child has a different heredity from its parents. It has half of its genes from each parent. This is how heredity is mixed in the child.

▲

Sexual Reproduction in Plants

This farmer is breeding a new kind of corn plant. ▲ Perhaps a plant with bigger ears of corn. He is taking pollen from one kind of corn plant. He will put the pollen on the flower of another kind of corn plant.

Inside the flowers of plants are egg cells. And pollen grains. Inside a pollen grain is a sperm nucleus. The sperm nucleus fertilizes the nucleus of an egg cell. ◆ Then the fertilized egg cell grows into a seed.

This is sexual reproduction in plants. Look at the tulip flower. ★ A sperm nucleus in the pollen grain will fertilize an egg nucleus. The egg nucleus is in the bottom part of the flower. Then a seed will grow.

Farmers choose their seeds. They want to get more food and better food for more people.

◆ Corn Plant

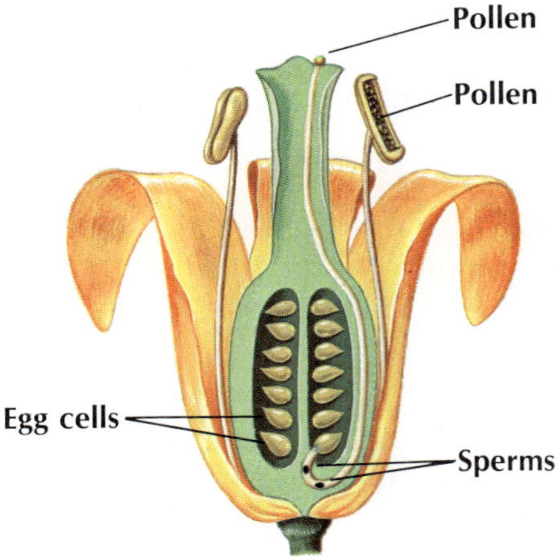

★ Tulip Flower

AN INDEPENDENT STUDY

About Eggs, Sperms, and Heredity

The story of man is only a million years old. What has man learned about heredity? What has he learned about growing a food supply? How has man made prize plants and animals?

Read each question and put your answer on your answer sheet. Check across to the next page to see if you are right. See how much you have learned.

Be sure to check each answer before you go on to the next question.

1

We have found that early men had no crops. They were food gatherers. They gathered fruits and seeds and roots of plants. They hunted animals.

Early man did not plant crops.

Early man was a food g-------.

2

What does a plant come from? It grows from a seed. Early man had to find this out.

Early man had to learn that plants come from ____________.

(Write your answer on your answer sheet. Check across to the next page to see if you are right.)

3

What is in a seed? Inside a seed is a tiny plant. A baby plant. This tiny plant comes from a fertilized egg. A sperm unites with an egg. The fertilized egg has the nucleus from an egg and the nucleus from a sperm.

In fertilization, a sperm nucleus unites with an ____________ nucleus.

1

food **gatherer**

Early man gathered and hunted food. He did not know much about how plants and animals grew.

(Go on to question 2.)

2

seeds

Man learned that plants grow from seeds.

(Go on to question 3.)

3

egg

(Turn the page to question 4.)

4

Before early people could become food growers, they had to find out that plants come from seeds. Man changed when he discovered this. *He changed from a food gatherer to a food grower.*

He learned that plants come from ____________.

5

Look at the corn plants below. Where are the sperm nuclei? Where are the eggs? Both the egg and sperm must unite. *Pollen grains* contain *sperm* nuclei. Powdery pollen grains are in the tops of the corn plants.

Where are the pollen grains? They are in the tops of the corn plants.

What is in a pollen grain—an *egg nucleus* or a *sperm nucleus*?

Which?

6

What happens in fertilization? The sperm nucleus in the pollen grain moves down to the egg nucleus in the kernel (KUR·nul) on the ear of corn.

What is in each kernel on the ear of corn—an egg *nucleus* or a *sperm nucleus?* Which?

4

seeds

Yes. Man could then grow his own crops.

(Go on to question 5.)

5

a sperm nucleus

A sperm nucleus is in a pollen grain. The tops of the corn plant make pollen grains.

The *egg cells* are in the kernels on the ears of corn. Look back at the picture in question 5. You can see the ears that are growing on the stems of the corn plants. Each kernel on an ear has a seed inside. Seeds come from fertilized eggs.

(Go on to question 6.)

6

an egg nucleus

Yes. After the egg is fertilized, the seeds grow. A seed is inside each kernel on the ear of corn.

(Turn the page to question 7.)

7

A farmer decides he wants better corn plants. He can pick the corn parents to get the right *seeds*. How would he do this? He would pick the biggest and best corn plants. He would bring the eggs and sperms together.

After eggs are fertilized, s__________ are formed.

8

Is it the same with animals? Yes. We know that animals come from fertilized eggs.

Man learned that egg cells have to unite with __________ cells. Then a new animal is formed.

9

A fertilized chicken egg grows into a chicken. This is sexual reproduction.

A rabbit or a fox also comes from a f__________ e___.

10

Why doesn't a chicken hatch from a rabbit's egg? Because in the rabbit's egg are the chromosomes for rabbit traits, or heredity. These chromosomes would not make a chicken.

In the chicken's egg are c__________ for chicken heredity.

7

seeds

These seeds may grow into more good corn plants.

(Go on to question 8.)

8

sperm cells

It is the same for both plants and animals.

(Go on to question 9.)

9

fertilized egg

(Go on to question 10.)

10

chromosomes

The chicken's egg has chromosomes for only chicken traits.

(Turn the page to question 11.)

11

When a farmer chooses parents to get better corn or chickens, he is picking out good chromosomes.

The good traits show up if the parent plants or animals have the good c_______________.

12

But having the right chromosomes is not enough. Man has learned that plants and animals do not grow as well as they should without the right environment.

A farmer wants to get good crops. He needs good soil as well as good ch_______________ in his corn plants.

13

What do chromosomes do in the fertilized egg? Chromosomes are in charge of what grows out of the fertilized egg. The egg on the left will be a chicken. The one on the right will be a duck.

Why? The duck egg has heredity only for a duck. The chromosomes in the duck egg will only make a _______________.

11

**chromo-
somes**

By choosing the parents, farmers get better corn, better cattle, better horses, better wheat, and better dogs.

12

chromosomes

The farmer needs plants and animals with good chromosomes. He needs the right environment for these good plants and animals.

(Go on to question 13.)

13

duck

Of course. Chromosomes are in charge of what comes out of the egg.

(Turn the page to question 14.)

14

Here is a picture of a Great Dane and a poodle. A dog is born a Great Dane or a poodle because of its c________________.

15

It is the genes in the chromosomes that make a child look like his parents or grandparents. Genes are patterns of chemicals in the chromosomes. All plant and animal chromosomes contain genes.

What animals and plants look like is due to patterns of chemicals called g____________ in chromosomes.

16

Half of the genes in a fertilized egg come from the mother. And half come from the father.

This is sexual reproduction. Two parents are needed. A child gets half its heredity from its father. And half from its ____________.

17

The genes from each parent can be different. One parent may have brown eyes. The other parent may have green eyes.

The kind of eye color in a child is due to the g____________ it inherits.

14

chromosomes

It is the chromosomes that make an egg grow into a man or a horse or a Great Dane or a poodle.

(Go on to question 15.)

15

genes

Genes are patterns of chemicals in chromosomes that make the plant or animal look the way it does. The chromosomes are in the nucleus of a cell.

(Go on to question 16.)

16

mother

The fertilized egg has genes from both mother and father. The heredity is in the genes.

(Go on to question 17.)

17

genes

8

Man — His Heredity

INTRODUCTION

What Is in an Egg Cell and a Sperm Cell?

Embryo

This growing **embryo** (EM·bree·oh) will be a human baby.■ It is in an early stage of development. It could become any one of these children.● How did the embryo start? The embryo grows from a fertilized egg.

Here are a few kinds of birds.▲ How did they start? You know they each began as a fertilized egg. How did they become so different?

There are many kinds of cats.♦ They may be the common cats you see in the street. They may be Siamese (Sy·uh·MEEZ) cats. Or Persian (PUR·shun) cats. They all started out the same way. They came from an egg cell that united with a sperm cell.

What is in the eggs and sperms of animals? What is passed along from the parents in an egg cell? and in a sperm cell?

Turn the page now. You need to start at the beginning. You need to learn how fertilized eggs grow into new animals. And what makes children like their parents. And what makes them different, too.

■

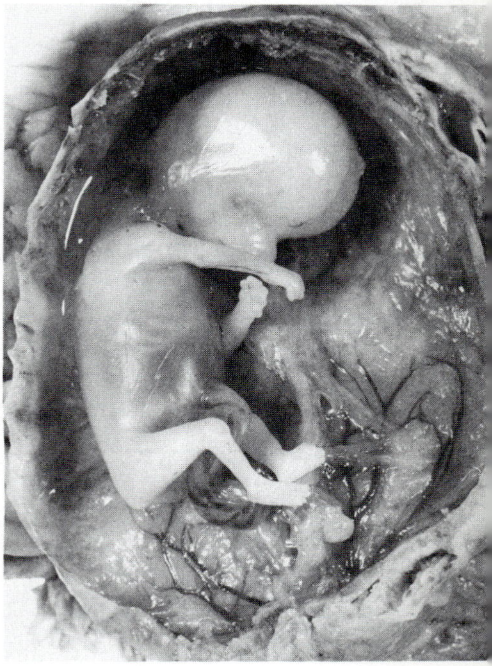

1. FROM PARENTS TO CHILD

■

●

These chicks have hatched out of their eggs. ■ They just hatched an hour or so ago. As you can see, their feathers are still wet.

It takes 21 days for a chick to grow from this spot on the yolk of the egg cell. ● This spot is the nucleus of the egg. The whole yolk and the nucleus make up *one* egg cell.

The yolk in the egg is the food supply for the growing embryo. It is the food supply for the growing chick. There is enough food for 21 days packed inside this large egg.

From an Egg

A chicken's egg starts in the *ovary* of the hen. In the ovary are many eggs. ▲ You see the eggs are different sizes. The big ones with a lot of yolk are almost ready to leave the ovary. Then one egg breaks away from the ovary. In the summer months this happens every day.

The egg moves into the **oviduct** (OH·vih·dukt). The oviduct is the long tube the eggs travel down from the ovary.

Before an egg is laid it gets a shell. Parts of the oviduct add the shell. Without a shell, the egg yolk and egg white would dry up.

A hen lays eggs regularly—even when the eggs are not fertilized. Indeed, most of the eggs you eat are not fertilized eggs.

▲

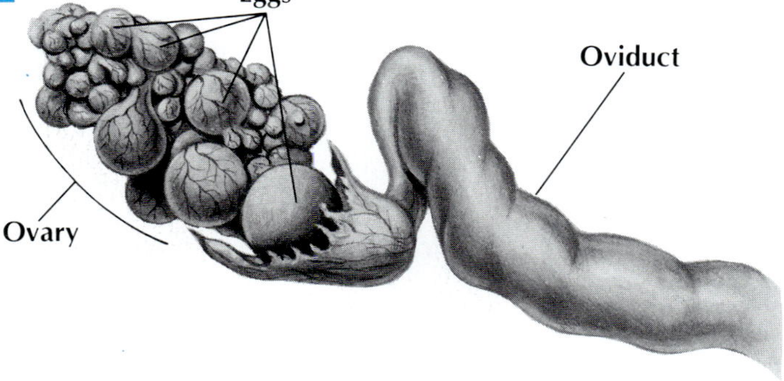

Egg and Sperm = Chick

When a rooster mates with a hen, her eggs may be *fertilized.* Sperm cells of the rooster swim up the oviduct.♦ A sperm cell may meet an egg cell. Only one sperm cell is needed to fertilize one egg cell.

This fertilized egg is moved down the oviduct. A shell is added. Then the egg is laid. This fertilized egg is protected with a shell. It can now grow into a chick.

In the fertilized egg, there is an *egg nucleus.* There is also a *sperm nucleus.* This means that half the chick's *heredity* comes from the hen. And half of the heredity comes from the rooster.

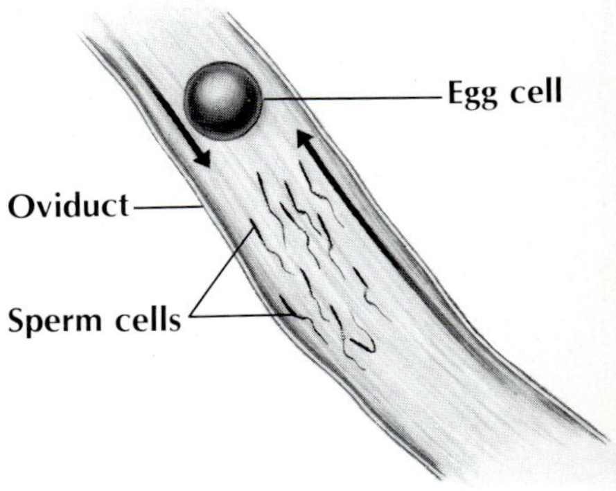

♦

A Fertilized Egg – A Child

It is the same for a human egg. Half the heredity comes from the mother. Half comes from the father.

A hen's egg must be fertilized before it can grow into a chick. A human egg must be fertilized before it can grow into a child.

This human egg is smaller than the spot on the yolk of the chicken's egg.★

A human egg doesn't have a shell. It doesn't need a shell. When the human egg is fertilized, it doesn't leave the body of the mother. The fertilized egg of a human stays in the body of the mother. It doesn't leave the body of the mother like a hen's egg. It grows in the body of the mother.

A human egg grows on nutrients and oxygen from the mother's bloodstream. The human egg does not have a food supply of its own. It does not have a food supply like a hen's egg.

★

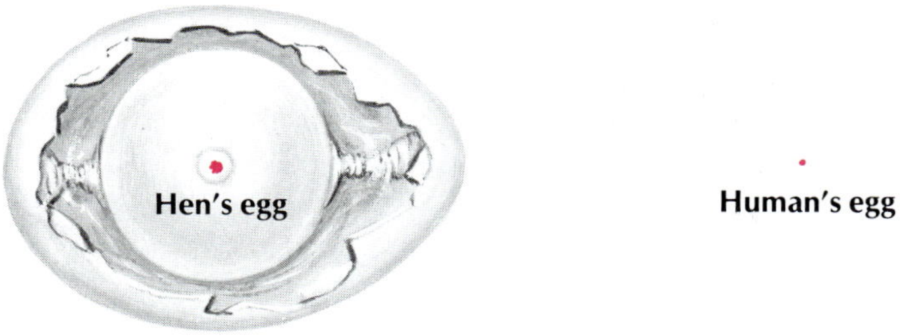

Let's start at the beginning. First, eggs are made in the ovaries of a human female. ■ Most of the time, only one egg leaves an ovary each month. This egg moves into the oviduct.

The egg cell is fertilized if sperm cells swim up the oviduct to the egg.

The fertilized egg gets half of its heredity from the father. And the other half from the mother.

Now the fertilized egg moves into the **uterus** (YOO·ter·us). ■ The uterus is the organ in the female where the fertilized egg will stay and grow.

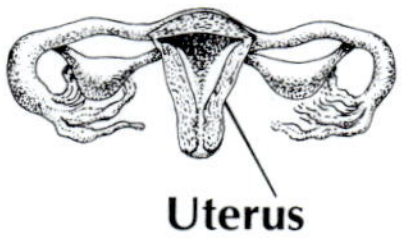
Uterus

What happens if there are no sperm cells in the oviduct? Then, the egg is not fertilized. What happens if there are sperm cells in the oviduct but no egg? Then, too, the egg is not fertilized. The sperm and egg must come together for an egg to be fertilized.

Human eggs and sperms that do not meet die. If the egg is not fertilized, it dies. It leaves the body. This happens every month.

Sometime each month the uterus is ready to hold a fertilized egg. The body gets the uterus ready to hold an egg sometime each month. But if the egg is not fertilized, the egg leaves the body.

If an egg is fertilized, it does not leave the body. It grows inside the uterus. ● In the uterus the fertilized egg becomes an embryo. ● Over a few months the embryo grows larger. It becomes a baby. It takes up more room. The uterus can stretch to hold the baby.

Look at the **umbilical** (um·BILL·ih·kul) **cord.** ▲ It is a life line. The embryo or baby gets nutrients and oxygen through this cord. And it carries wastes away.

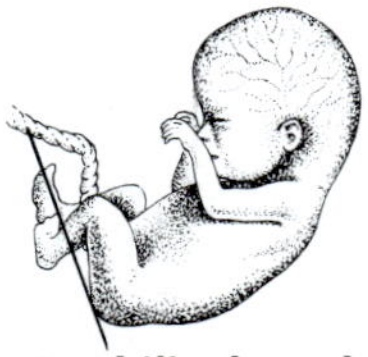
Umbilical cord

One end of the cord goes to the growing embryo or baby. The other end goes to a place on the inside lining of the uterus. This place is called the **placenta** (pla·SEN·tah). ▲ The placenta has a lot of blood vessels.

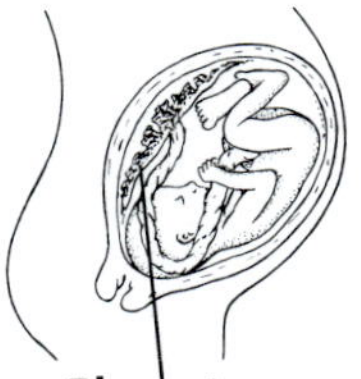
Placenta

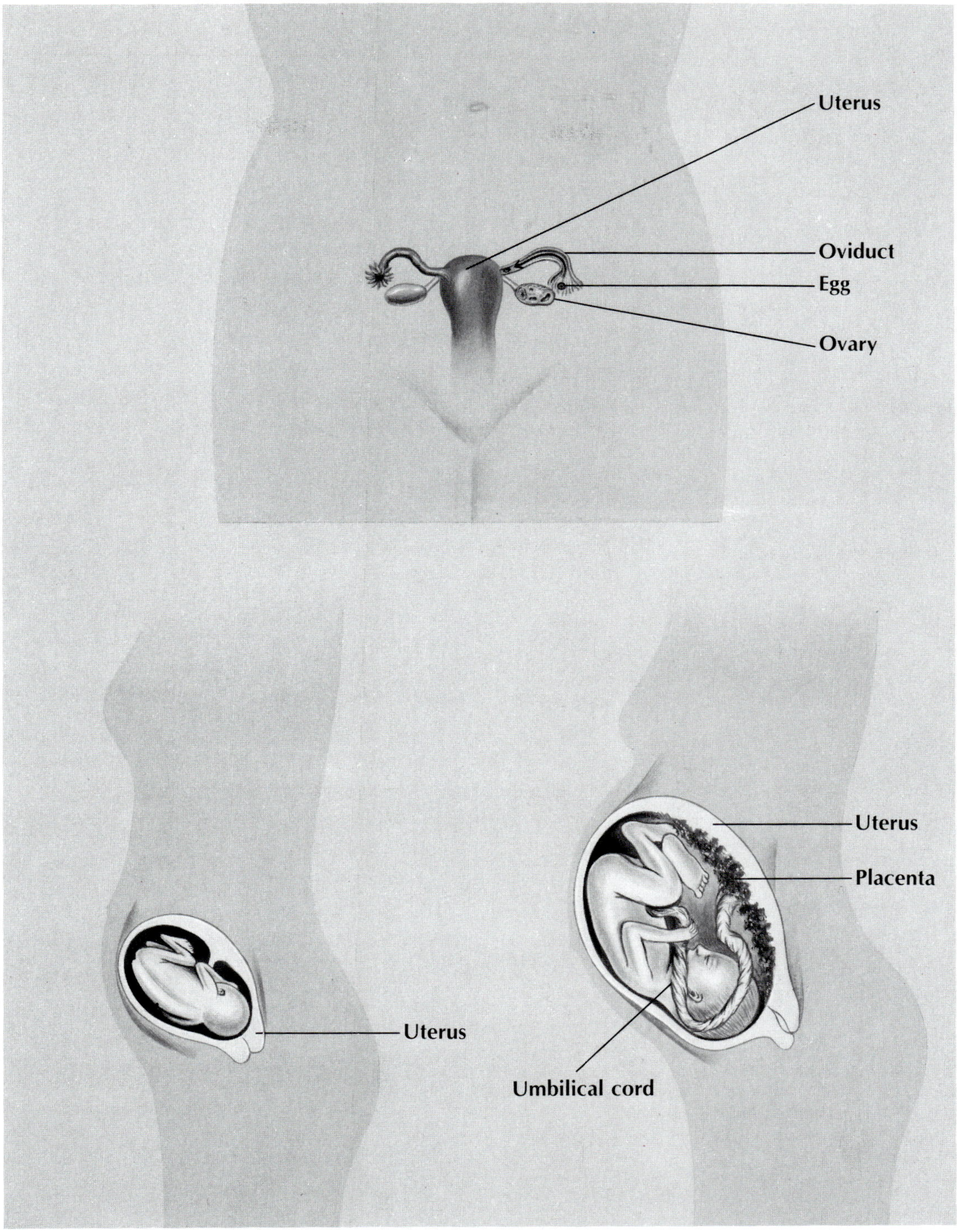
Uterus
Oviduct
Egg
Ovary
Uterus
Uterus
Placenta
Umbilical cord

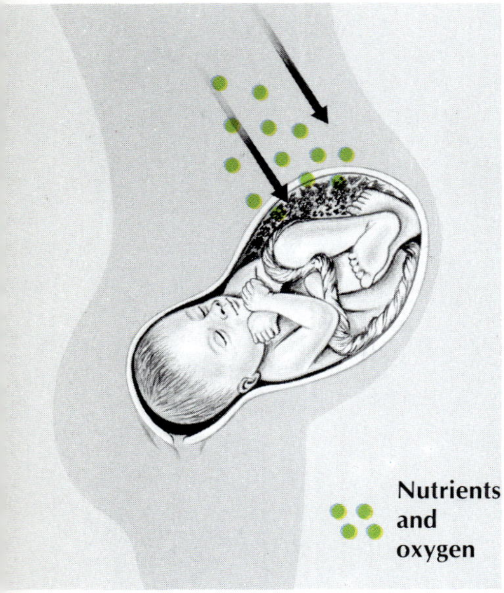

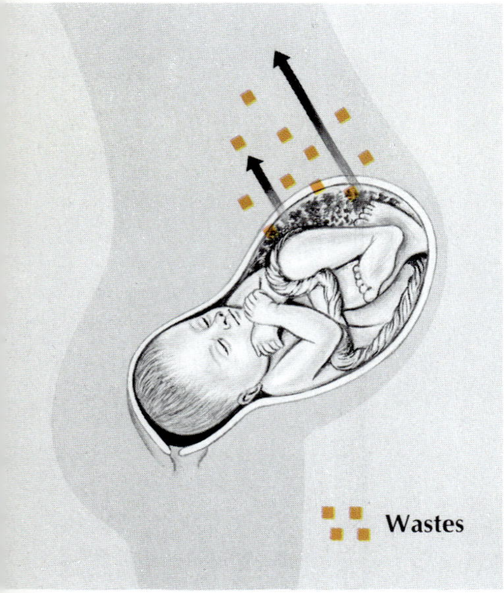

Molecules Move Across the Placenta

The mother's blood does not flow into the baby. The baby's blood does not mix with the mother's blood. Thin membranes of the blood vessels of the placenta keep the blood separate. The mother's blood does not mix with the blood of the baby.

We can picture it this way. ■ Molecules of nutrients and oxygen are in the mother's bloodstream. These molecules pass through the thin membranes in the placenta.

Where do all these molecules from the mother's blood go? Where do these molecules go when they pass through the placenta? The nutrients and oxygen go into blood vessels in the umbilical cord. These blood vessels bring nutrients and oxygen to the baby.

What is this nutrient supply? The baby needs all the nutrients you do. There must be simple sugars and digested proteins. There are also fats, water, minerals, and vitamins.

Sugars are needed for energy. Proteins are used for growth. Minerals like calcium (KAL·cee·um) make strong bones. Iron is needed to make red blood cells.

There are many more minerals and vitamins the baby needs. And all these move across the membranes in the placenta. All these nutrients move from the mother's blood across the placenta.

The umbilical cord is a two-way passage. ● The growing cells of the baby give off molecules of some wastes, like carbon dioxide. These go back to the mother's blood stream. The mother's body gets rid of the baby's wastes. These wastes move across the placenta to the mother's blood.

What might happen if a mother is on drugs? Some drugs go through the thin membranes of the placenta. Some drugs get into the baby. The growing baby can become a drug addict before it is born.

You may know that when a drug addict is taken off drugs, he suffers pain. He has withdrawal symptoms (SIM·tomz). The addict has great pain when he no longer gets drugs.

A baby can also suffer from withdrawal symptoms. The baby can have withdrawal symptoms after it is born. The baby has withdrawal symptoms when it no longer gets drugs from its mother's blood stream.

Sometimes the growing baby in the uterus can get microbes from the mother. There may be small breaks in the thin membranes of the placenta. This is the way the micro-organisms that cause syphilis can get into a baby's blood stream. Then the baby is born with syphilis.

A Baby Is Born

From the time the egg is fertilized, it takes about 9 months for the baby to grow. It takes about 9 months for the baby to be able to live outside the mother.

How does a fertilized egg grow into a baby?

Here is a fertilized egg. ▲ It divides into 2 cells. Then into 4 cells. ◆ Next into 8. The cells keep dividing to become thousands of cells.

Then the cells begin to be different. They are changed into special kinds of cells. Some cells grow into nerve cells. Some become muscles of the heart. Blood cells are made. Blood vessels grow. Then the embryo's heart begins to pump blood through the body. Blood cells are sent through the blood vessels.

Bone cells form the skeleton. Other cells form organs like the liver, lungs, kidneys. The food tube is made. Many of these organs do not work until the baby is born. The baby doesn't need most of these organs while it grows. It gets its supplies from the mother. The mother gets rid of its wastes, too.

▲

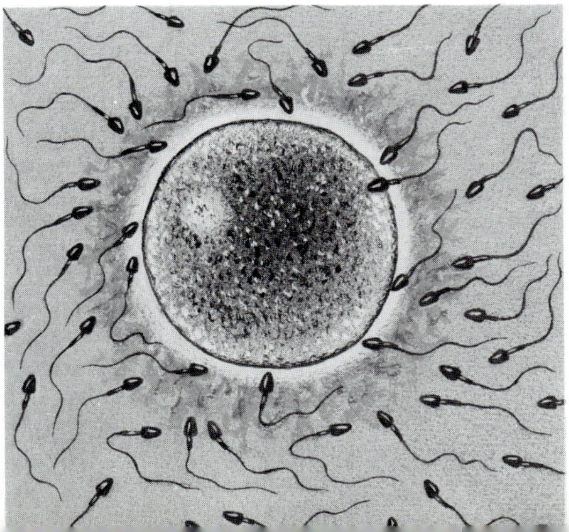

◆

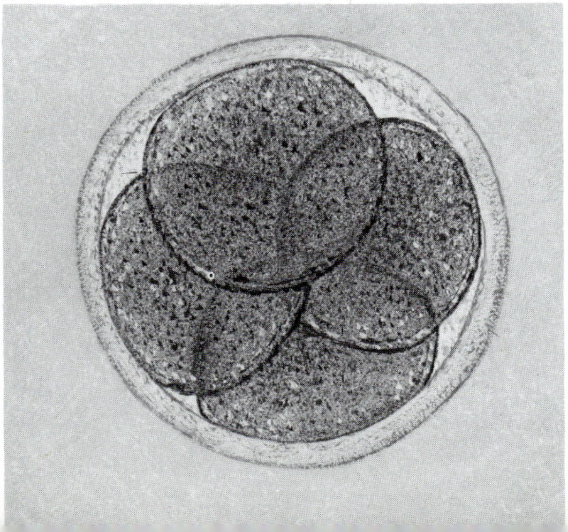

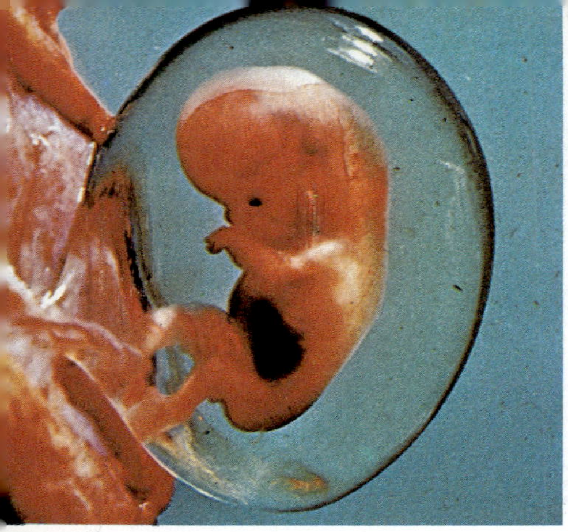

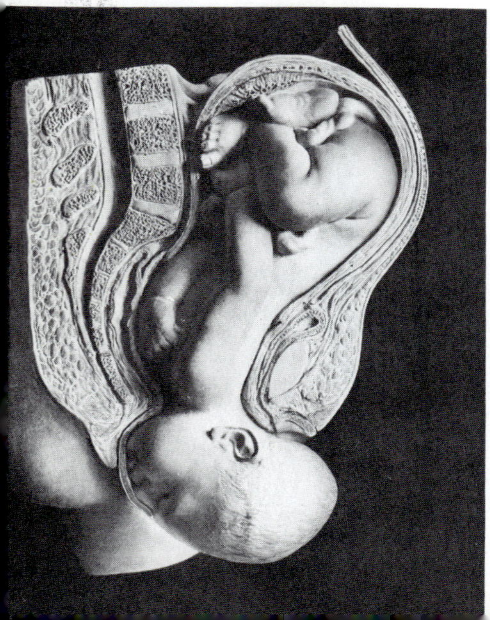

Most of this happens by the time the baby is 2 months old. The baby is *very* small. But it has ears, nose, and a mouth. And there are tiny legs and tiny arms.

After $3\frac{1}{2}$ to 4 months, you can hear the baby's heart beat. The baby is also much larger. It is *about* 8 inches long. The baby is growing very fast now. It has fingers and toes. And many of its bones have begun to harden. The muscles of the arms and legs can move.

By 5 months, the mother can feel the baby move its arms and legs. From this time until it is born, the baby can move around. It has fingernails. And it has hair on its head.

For 9 months the baby grows in a water bag. This bag is like a cushion inside the uterus. It protects the baby. The baby rocks gently in the water bag for 9 months.

The baby is a living, growing organism inside the mother's uterus. After 9 months the baby is ready to be born. It is ready to live outside the mother's body.

There are special chemical messengers that start the birth of the baby. These special chemicals pass into the mother's blood stream. They make the muscles of the uterus push together. These chemicals are called hormones. They cause the muscles of the uterus to push the baby out.

Then the water bag breaks. The baby is born sometime later. The baby is usually born head first. The baby comes head first into the world.

The umbilical cord is cut. Now the baby must breathe for itself. It has to use its own lungs. It has to take in food. It has to digest its own food. It must get rid of its own wastes. The baby is now on its own.

There is no more use for the placenta. The placenta is pushed out of the uterus. It is pushed out in the same way the baby was. The placenta comes out after the birth of the baby. So does what is left of the water bag. This is why they are now called the *afterbirth*.

Now the uterus is empty again. The uterus was the home of the baby for 9 months.

What sex will the baby be? Will it be a girl or a boy? Can we predict?

◆

★

2. WHAT WILL IT BE—BOY OR GIRL?

You can say it's a 50-50 chance. A girl or a boy. In some families it does not seem to be a 50-50 chance. This family has all boys.◆ Or some parents have all girls.★

It *is* chance though. Suppose you flip a penny 5 times. You know that sometimes you can get 5 heads in a row. Or sometimes 5 tails in a row. It can happen. But usually it doesn't happen that way.

The sex of a child is inherited. This is like the other traits a child is born with. What's in the fertilized egg that makes a girl? Or a boy baby? The sex of a child has to do with it's chromosomes.

The Pair of Chromosomes That Does Not Match

Chromosomes, you remember, come in pairs. There are 46 chromosomes in cells of humans. Or we can say there are 23 pairs.

You know each sperm cell has only half this number of chromosomes. Each sperm nucleus has 23 chromosomes.

Each egg nucleus also has 23 chromosomes. After the sperm fertilizes the egg, there are 46 chromosomes. There are 23 pairs of chromosomes in the fertilized egg. The fertilized human egg then has the full number of chromosomes. How many is that? 46.◇

When does the egg get the full number? When the sperm unites with the egg. Then the fertilized egg has the full number of chromosomes.

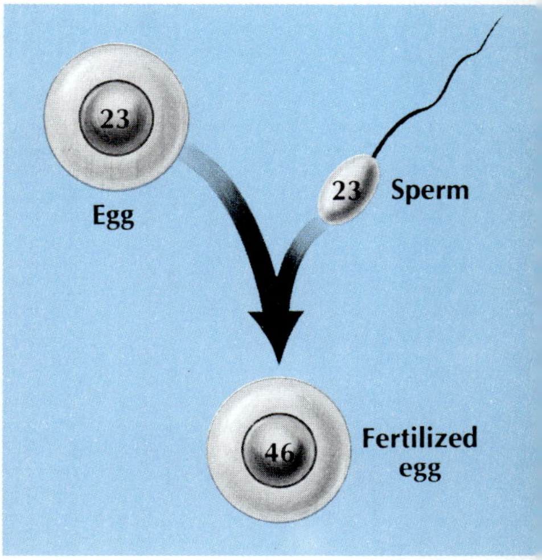

◇

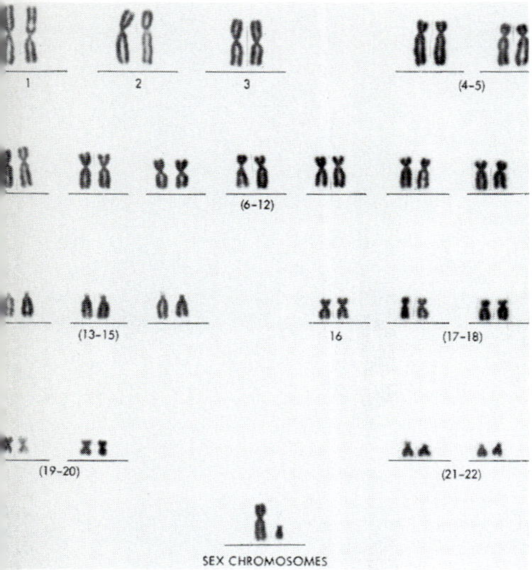

■

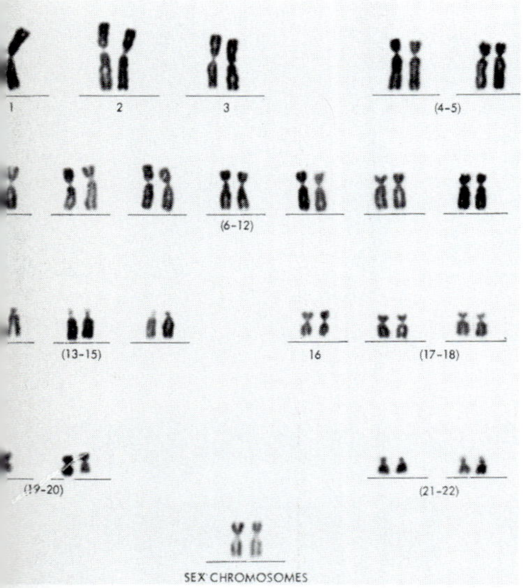

●

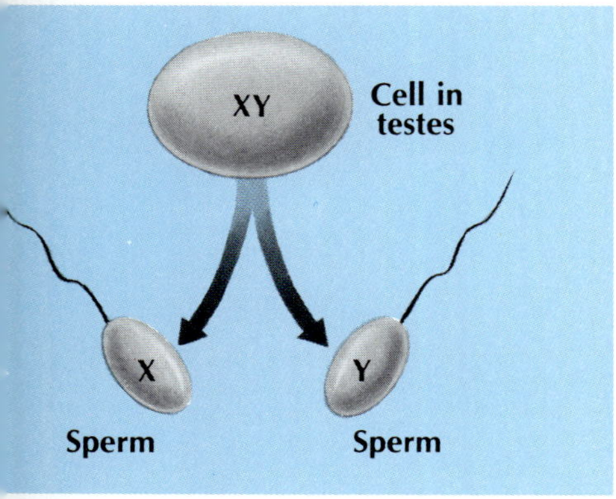

▲

In the cells of a human male there is one pair of chromosomes that does not match. ■ We call them X and Y chromosomes. The XY pair. The XY pair is found in males.

These are the **sex chromosomes.** They are one pair of the 23 pairs of chromosomes in the cells.

This means the father's cells have 22 pairs that match. And one pair does not match. You can see that 22 pairs and the XY pair add up. They add up to 23 pairs of chromosomes.

Males have this kind of chromosomes.

The Pair That Matches

Look at the pair of *sex chromosomes* in the cells of a female. ● The pair is XX. They match. Cells in the female have 22 pairs and the XX pair. That is the matched pair of sex chromosomes. This makes 23 pairs of matched chromosomes all together. Females have 23 pairs of matched chromosomes.

What Sex Chromosome Is in an Egg and Sperm?

What sex will the baby be?—a boy or a girl? It has to do with the pair of sex chromosomes. What pair of chromosomes is in the fertilized egg? Is it XX? Or is it XY?

This is what happens.

Sperm cells are made in the *testes* of the male. Cells in the testes divide to make hundreds of thousands of sperm cells.

Let us see what happens to the sex chromosomes. Let us only look at the sex chromosomes. The other 22 pairs of chromosomes do not have to do with the sex of the fertilized egg. Not in the beginning.

The cells in the testes divide. The pair of sex chromosomes separates. ▲ The sex chromosomes in the cells separate. One sperm cell gets the X chromosome. Another sperm cell gets the Y sex chromosome. A sperm cell gets *either* an X or a Y.

Suppose a male makes 100,000 sperm cells. What

amount of these sperms will have the Y chromosome? Probably about half the sperm cells. The other half of the sperm cells should have the X chromosome.

A sperm cell has *only one* of the pair of sex chromosomes. This means a sperm cell may have an X chromosome. Or a sperm cell may have a Y chromosome.

What is in the nucleus of an egg cell? When an egg is made, it has only half the XX pair. Each egg has one X chromosome. ♦

This is the way it is with one egg cell, or with a hundred egg cells. Each egg has only one X chromosome.

♦

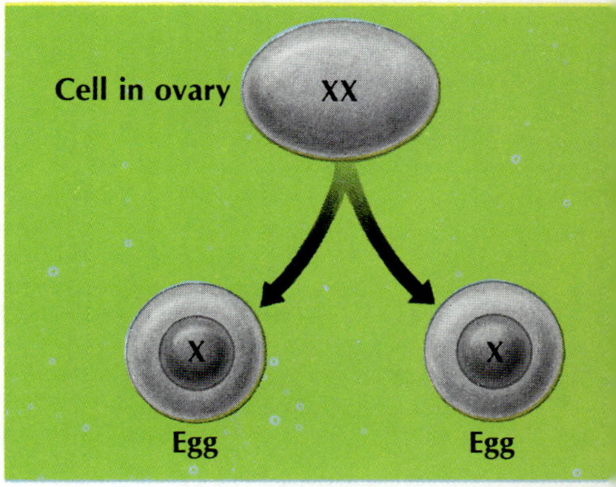

XX = A Girl

Which sperm cell will fertilize an egg cell? As you see, an egg cell has only an X sex chromosome. ♦

What kinds of sperms are there? Some sperms have an X chromosome. Some sperms have a Y chromosome. ▲

It's a case of chance which sperm will fertilize an egg.

What do the letters XX mean? This is the pair of sex chromosomes that a girl has. ★

How does a fertilized egg get this pair? A sperm cell with an X chromosome fertilized an egg with an X chromosome. ★ A fertilized egg that has an XX pair will grow into a girl.

★

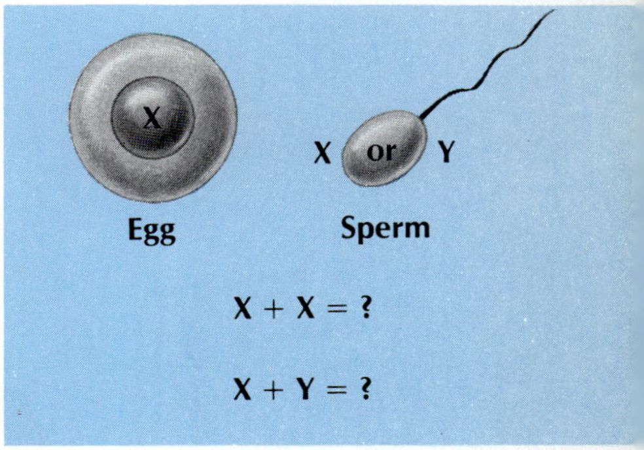

XY = Boy

A baby boy grows from a fertilized egg with the XY pair. ★

From which parent does the boy get his Y chromosome? From the father. The male has two kinds of sex chromosomes. This means that half his sperm cells will have the X chromosome. And the other half will have the Y chromosome.

So there is a 50-50 chance. Half the time an egg is fertilized by an X sperm. There will be an XX pair. This will be a girl. The other half of the time, a Y sperm will fertilize an egg. Then there is an XY pair. This will be a boy.

AN INDEPENDENT STUDY

The Beginning of Life—A Child Is Born

A child is born. It begins when a sperm fertilizes an egg cell. The baby could be a boy or a girl. There could be two babies, or three, or more.

How long does the baby grow inside the mother? Start with the egg and sperm. See what you have learned. First read each question. Write your answer on your answer sheet. Then check across the page to see if you are right. Check each answer before you go on to the next question.

1

Each of us begins as a fertilized egg. We begin with a sperm from our father and an egg from our mother's ovaries. A sperm cell swims to an egg cell. It fertilizes the egg cell. Egg cells are made in the female.

The eggs are made in the organs called o------.

(Write your answer on your answer sheet. Check across the page to see if you are right.)

2

Sperm cells are made in the testes. What do sperm cells look like? ■ They have a head and a tail. Sperm cells are made in the male. They are made in the t----- of the male.

3

Usually one egg cell leaves an ovary every month. An egg can be fertilized if a sperm cell swims and meets an egg cell.

An egg is fertilized when it unites with a ----- cell.

1

ovaries

Egg cells are made in the ovaries of females.

(Go on to question 2.)

2

testes

Sperm cells are made in the testes of males.

(Go on to question 3.)

3

sperm

Sperm cells swim to the egg in the oviduct.

(Turn the page to question 4.)

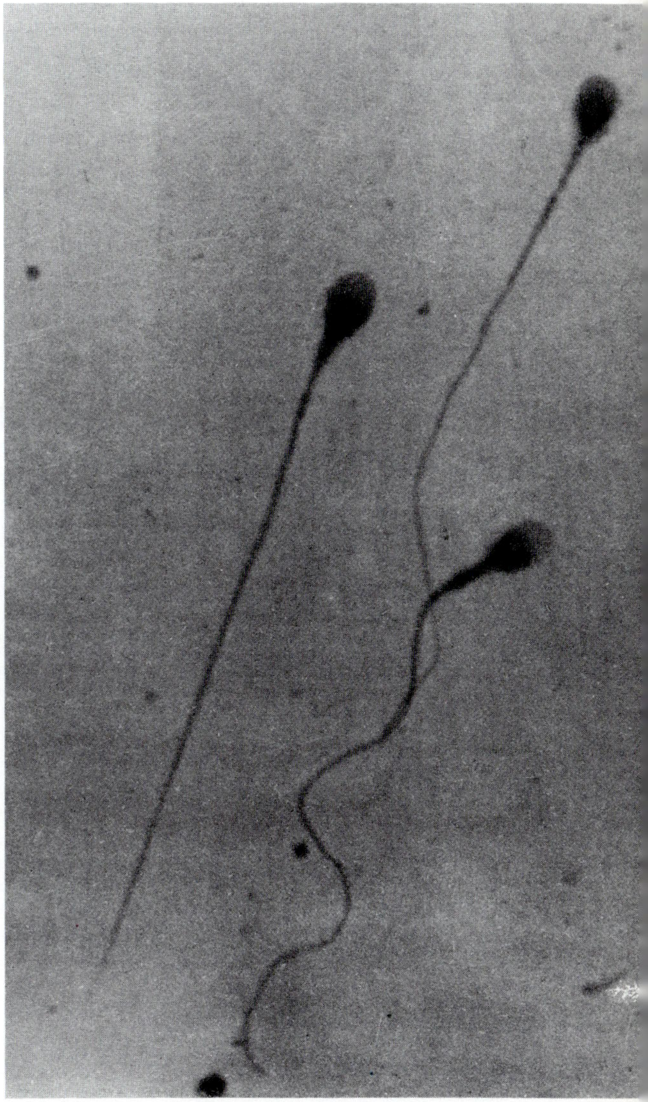

4

What happens when the egg and sperm meet? The nucleus in the head of the sperm unites with the nucleus of the egg. Where is the nucleus of the sperm cell? *in the head* or *tail*? Where?

5

Each human sperm has 23 chromosomes. These are in the nucleus of the sperm. Chromosomes pass heredity from parents to child.

There are the same number of chromosomes in the nucleus of the egg cell. This means the nucleus of an egg has ____ chromosomes.

6

Twenty-three is half of the full (or right) number of chromosomes found in humans. The number of chromosomes found in the body cells is 46. Humans have 46 chromosomes in their body cells.

The full number of chromosomes in your skin cells is 46. The right number in your lung cells is ____.

7

An egg cell has 23 chromosomes. The fertilized egg gets 23 chromosomes from the sperm. This means the fertilized egg then has 46 chromosomes.

This is adding 23 from the egg and ____ from the sperm.

4

in the head

The nucleus is in the head of the sperm cell. The tail is used for swimming.

(*Go on to question 5.*)

5

23

There are 23 chromosomes in an egg, and 23 chromosomes in a sperm.

(*Go on to question 6.*)

6

46

Yes. The number is 46 chromosomes for body cells found in humans. It's a different number for egg and sperm cells.

(*Go on to question 7.*)

7

23

There are 23 in the sperm and 23 in the egg. The fertilized egg has 46 chromosomes.

(*Turn the page to question 8.*)

8

The 23 chromosomes from the sperm and the 23 from the egg *pair up*. The baby grows from the fertilized egg. This means we have 23 pairs of chromosomes in our cells. All these pairs of chromosomes match in one sex. Which? *male* or *female?*

(Write your answer. Check across the page to see if you are right.)

9

In the cells of a male there is a pair of chromosomes that does not match. This is the XY pair. This is the pair of chromosomes in a male.

Suppose a fertilized egg has sex chromosomes that do *not* match. Will the child be *male* or *female?*

10

Suppose a fertilized egg has sex chromosomes that do not match. Then the fertilized egg will grow into a male. The sex chromosomes in the cells of the male that do *not* match are ̲ ̲.

11

What are the sex chromosomes in a girl? They are XX. They match.

XX is the pair of sex chromosomes in the cells of the ____________.

8

female

All 23 pairs of chromosomes match in the cells of a female.

(Go on to question 9.)

9

male

(Go on to question 10.)

10

XY

This is the pair of sex chromosomes in the cells of the male.

(Go on to question 11.)

11

female

(Turn the page to question 12.)

12

The female parent has an XX pair of sex chromosomes in her body cells. The egg cells she makes have *half* this pair of chromosomes. This means an egg cell has only one X chromosome.

Which parent has the sex chromosome Y? the *female* or the *male?*

13

The male parent has an XY pair of sex chromosomes in his body cells. His sperm cells have either an X or a Y. What does this mean? This means that half the sperm cells have the X chromosome. And the other half have the ____ chromosome.

14

Can we know the sex before the egg is fertilized? No. It's a matter of chance. The sperm cell can give either X or Y.

An egg has only an X sex chromosome.

What can a sperm cell have? Either the X or the ____ sex chromosome.

15

The male parent's sperm cells control the sex of the child. What sex will the child be? It is a 50-50 chance.

When a baby is born is there a chance for a boy? What chance is there for a boy?

12

male

Only the male can give the Y chromosome.

(Go on to question 13.)

13

Y chromosome

(Go on to question 14.)

14

Y

The sperm cell makes the difference.

(Go on to question 15.)

15

a 50-50 chance

The fertilized egg has a 50-50 chance of growing into a boy.

(Turn the page to question 16.)

16

The fertilized egg needs nutrients and oxygen to grow. It grows in the mother's uterus. The fertilized egg needs proteins, carbohydrates, fats and other things to grow. Where does the embryo grow?

It grows in the mother's u-----.

17

The embryo grows in the uterus. The embryo gets oxygen and nutrients through the placenta. It gets rid of wastes through the placenta.

The part that connects the mother and child is the p-------.

18

What keeps the embryo from getting hurt? The water bag. The embryo is growing in a water bag in the uterus. The water bag protects the baby.

The embryo grows in the uterus for about 9 months. As the baby grows each month, the uterus stretches and gets larger.

As the baby gets larger, the u----- stretches.

19

How do nutrients go through the placenta? The molecules of nutrients and oxygen can pass through the membranes of blood vessels. These blood vessels are in the placenta.

The baby gets its food through blood vessels in the ________________.

16

uterus

The uterus is the organ in the female where the fertilized egg grows.

(Go on to question 17.)

17

placenta

The baby gets nutrients and oxygen from its mother's blood stream.

(Go on to question 18.)

18

uterus

The baby grows in the uterus. The uterus stretches as the baby grows.

(Go on to question 19.)

19

placenta

(Turn the page to question 20.)

20

Suppose a pregnant woman takes a drug. The molecules of the drug can go through membranes in the placenta. The baby could become addicted to some drugs. The baby could become addicted while growing in the u-----.

21

Why would these baby addicts have withdrawal symptoms when they are born? Because their supply of drugs would be gone at birth.

When the baby is born, it would no longer get drugs from the mother's b---- stream.

22

When the baby is born, it is pushed out of the uterus. Then the baby gets its oxygen from the air it breathes. The baby uses its own lungs. It gets oxygen on its own.

How does it get its oxygen?

Through its ____________.

20

uterus

(Go on to question 21.)

21

blood

(Go on to question 22.)

22

lungs

The baby begins to use its lungs when it is born.

3. WHO WILL THE CHILD BE LIKE? FATHER? MOTHER? BOTH?

A baby girl may not look like her mother. She may have her father's eye color. Or, she may have a nose like her grandfather's.

All your 46 chromosomes make you what you are. Heredity passes from parents to child. You may also have some traits that were hidden in your parents. These traits may have been inherited from grandparents—or great-grandparents.

There is a shuffling of chromosomes every time an egg is fertilized. Every time a baby is born there will be a new mixture of chromosomes. The chromosomes in the eggs and sperms help make the child a member of the family.

Genes and Inherited Traits

You may remember, *genes* are special patterns of chemicals that make up chromosomes. Genes are lined up on each chromosome. There are thousands of genes lined up on each chromosome. And you have 46 chromosomes.

Suppose we look at only one pair of these genes. Let us see how this pair of genes is inherited.

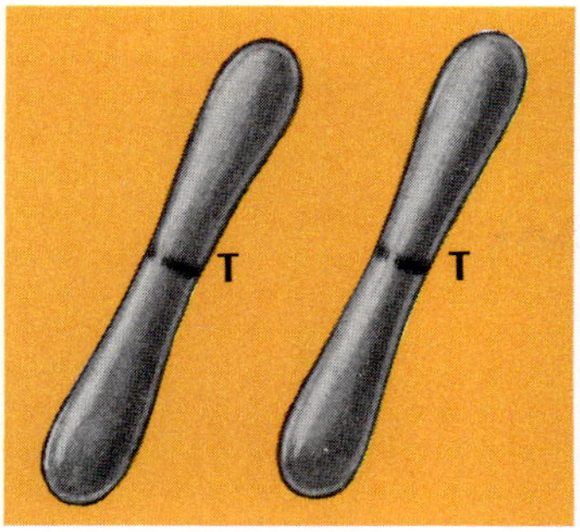

●

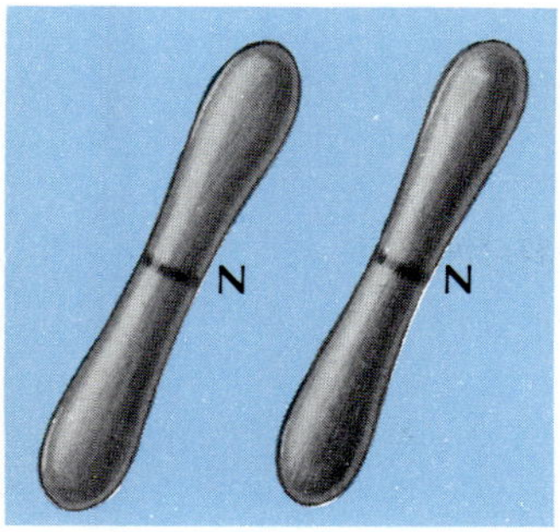

▲

Tasters and Non-tasters

Two scientists were once working with a certain chemical. The chemical is called PTC (PEE·TEE·CEE). It tasted bitter to one of the scientists. But the other scientist did not taste anything. Strange?

This chemical can be put on paper. When 100 people taste PTC paper, about 70 people taste the chemical as bitter. These people are called *tasters*. The rest of the people do not get a bitter taste. These people are called *non-tasters*.

A taster of PTC paper has genes for this trait. Let us say the taster's genes are *TT*. This means there is gene *T* on each chromosome of a pair. ● The pair of genes, *TT*, match. The taster is "pure" for this trait.

What genes would a non-taster have? He does not taste the PTC on the paper. He has a pair of genes *NN*. ▲ He is "pure" for this trait. He has no gene *T*.

Suppose a non-taster marries a non-taster. We can now show this. ◆ You can see that an egg gets one *N* gene. And one *N* gene is in a sperm cell. Egg and sperm cells have half the pair of genes. Each cell gets an *N* gene.

Now a sperm cell with *N* fertilizes an egg with *N*. ◆ The fertilized egg has a pair of genes *NN*. The child will be a non-taster. It makes no difference if it is a boy or a girl. The child will be a non-taster.

Let us say a woman and a man are "pure" for the trait taster. ★ Suppose *TT* marries *TT*. What will their children be? Tasters. Of course. The parents each have only a *T* gene to pass to a child. The child will be *TT*.

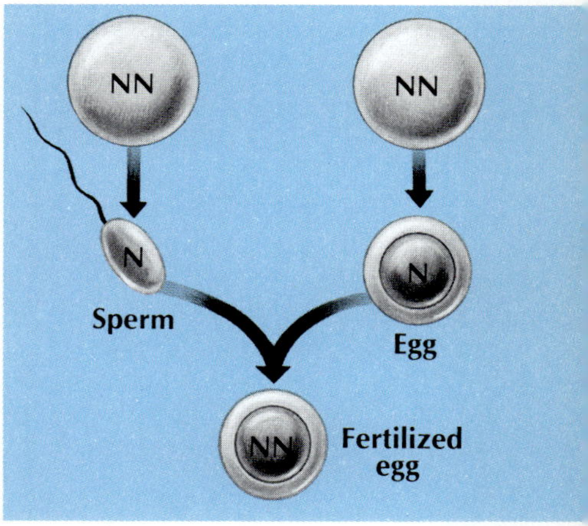

◆

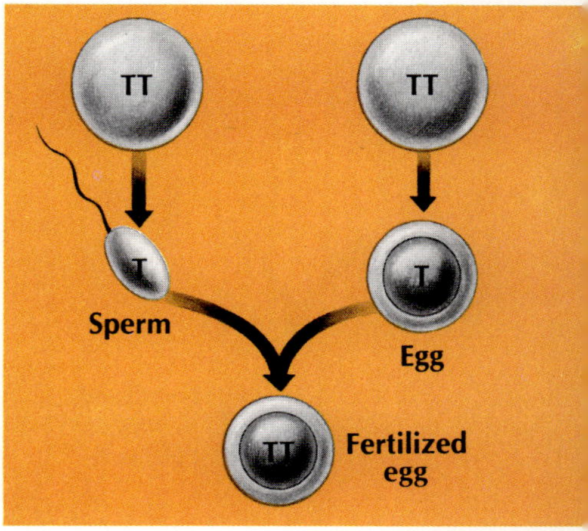

★

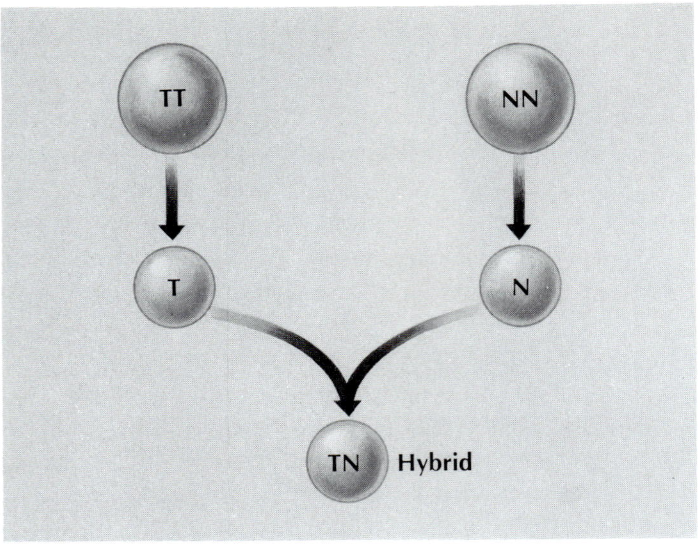

■

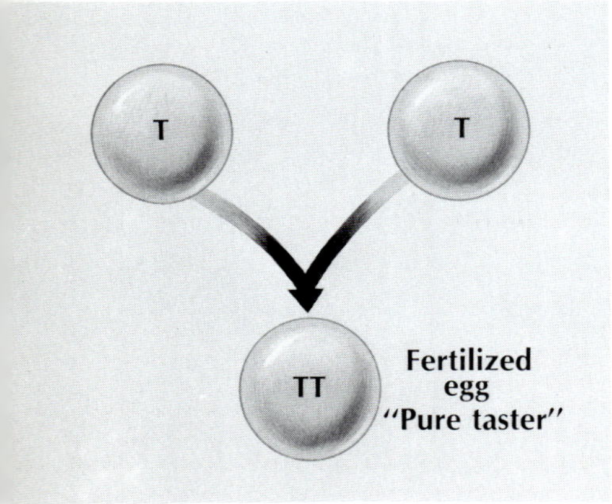

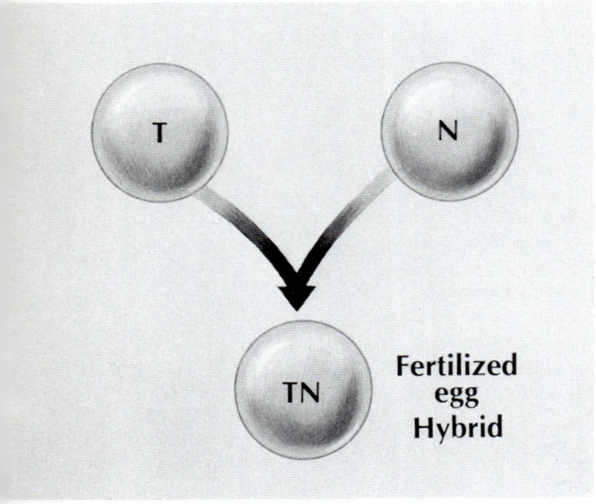

●

▲

A Taster Marries a Non-taster

Now suppose a taster, *TT,* marries a non-taster, *NN*. What genes will their children get? One parent can only *give a T*. The other parent can only give an *N* gene. ■

The child gets genes *T* and *N*. There is one gene for *T* and one gene for *N*. The genes, *TN*, are not the same this time. This child is a **hybrid** (HIGH·brid). ■ A hybrid has two different genes for a trait.

Now give this child a piece of PTC paper. What will the child be—a taster or a non-taster? Yes. The child is a taster.

Which gene shows up? In this case, the one that shows up is the *T* gene. This is called a **dominant** (DOM·in·ent) gene. The dominant gene is the one that shows up in the child.

The gene *N* in this child is hidden. This gene is hidden, or does not show up. It is called a **recessive** (ree·SES·iv) gene.

This means a taster can be *TT* or *TN*. A taster can be "pure" for the trait, or *TT*. ● A taster can be hybrid for the trait, or *TN*. ▲

The non-taster gene is the recessive gene. It stays hidden. In the hybrid it stays hidden.

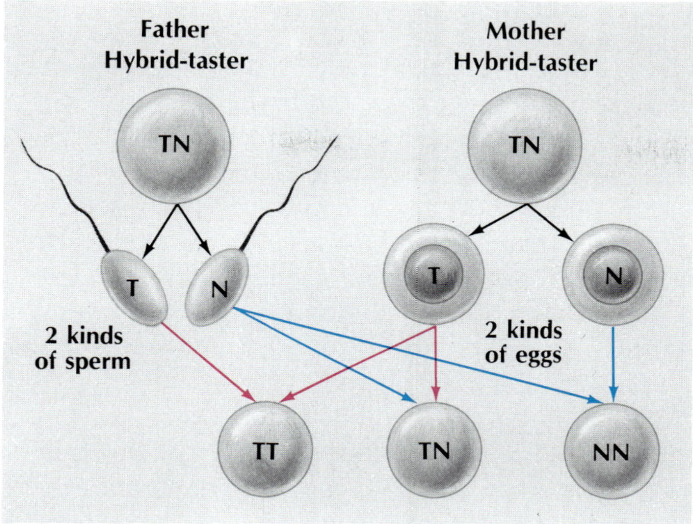

♦

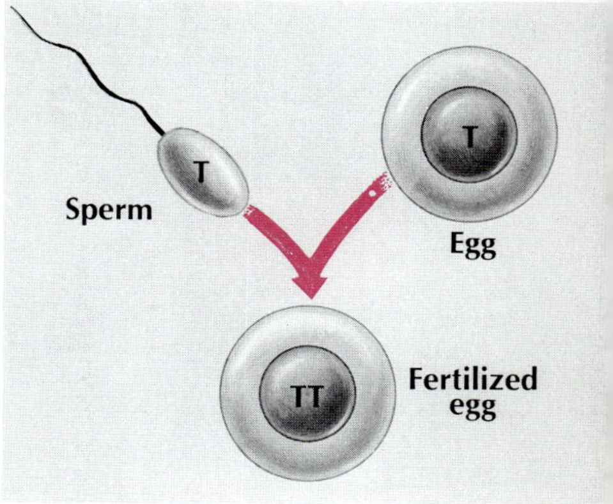

★

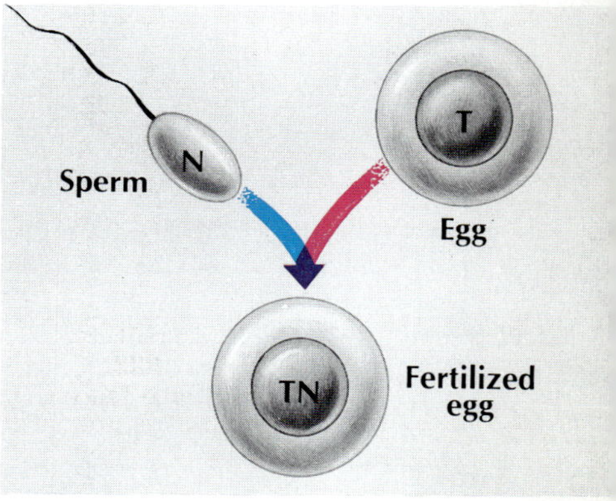

◇

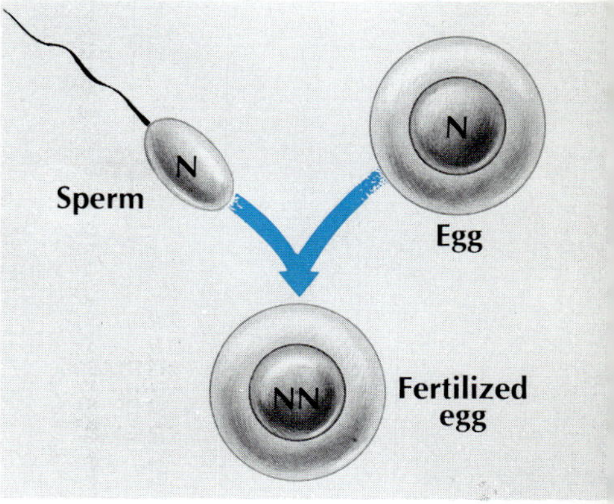

⊙

Can a Recessive Gene Show Up?

What does this mean: *TN* × *TN*? ♦ The letters stand for two parents. Each parent is hybrid for one pair of genes. In this case, they are both hybrid for the taster trait.

You can see that the mother has two kinds of genes. She can make two kinds of eggs. Some eggs will have the *T* gene. Some will have the *N* gene.

The same is true for the father. Some of his sperm cells have the *T* gene. And some have the *N* gene. Now it's up to chance.

A sperm with a *T* gene may fertilize an egg with a *T* gene. Then what will the fertilized egg be? As you can see, it is *TT*. ★

There is more to this. Half of the father's sperm cells may have an *N* gene. Suppose a sperm cell with an *N* gene fertilizes an egg with a *T* gene. The fertilized egg will be *TN*. ◇ And a sperm with an *N* gene may fertilize an egg with an *N* gene. What will this give? An *NN* fertilized egg. ⊙

You can see that chance plays a part here. It is a matter of chance which sperm will unite with an egg. The children can be *TT*, *TN*, or *NN*.

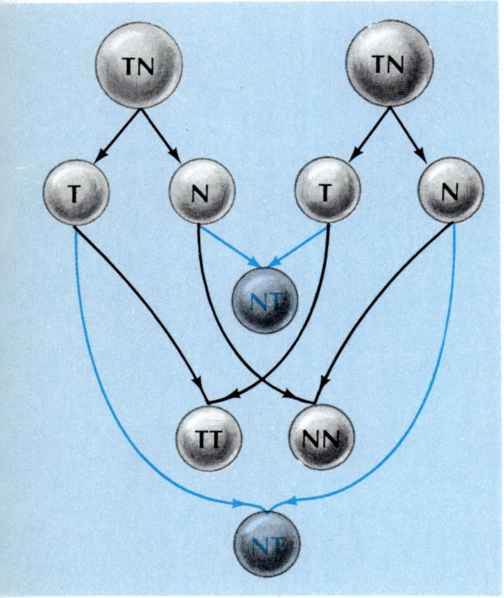

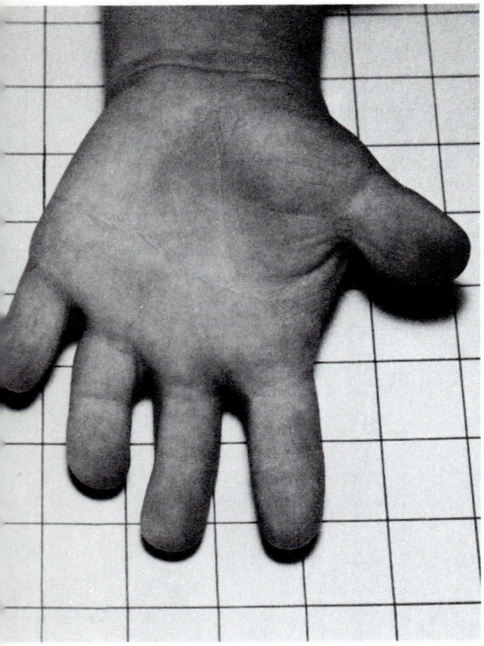

A Try at Explanation

Why is the taster gene called a dominant gene? You are right if you say it shows up in the child.

Which gene can be hidden, or recessive? the gene *T*, taster, or the gene *N*, non-taster? True, the gene *N*.

How do we show a hybrid taster? *TN*. Why will a *TN* child be a taster? It will be a taster because the gene *T* shows up. *T* is the dominant gene.

Now what kinds of children can there be when *TN* marries *TN*? ■ There can be *TT*, *TN*, and *NN* children.

What will the *TT* children be? "Pure" for taster. And the *TN* children? They will be hybrid tasters.

Which will be non-tasters? The *NN* children will be non-tasters. They will be "pure" for non-taster.

Out of 4 chances, how many tasters could there be? ■

■ Out of 4 children, how many *could* be tasters? Three out of every four children could be tasters. The children can be tasters *3 chances* out of every *4 chances*.

You may put this another way. When two hybrid tasters have a child, there are 3 out of 4 chances the child will be a taster.

What is the 1 chance out of 4? Do you see that there is one chance that the child will get the *NN* pair of genes? There is 1 chance out of 4 for the hidden trait to show up. There is 1 chance out of 4 for the recessive trait to show up.

Will a child with *NN* genes be a taster or a non-taster? He will be a non-taster. Of course, he has no gene for *T*.

Here is an example of a trait that did not show in the parents. The parents were hybrid tasters. Yet they may have a non-taster child. The child got this hidden gene, *N*, from each of its parents. The parents carried the recessive trait. These hidden genes were in the family.

The Case of Long Fingers or Short Fingers

Some people have very, very short fingers. This is because a bone is missing in each finger. ● Sometimes a bone is also missing in the toes. This is an inherited trait. It is a dominant trait. What does this mean?

We'll begin with a person who has short fingers. This person comes from a family in which everybody has *short fingers*. This person is "pure" for short fingers. He has the pair of genes, *SS*. He is "pure" for short fingers. ▲

Let us say this person marries someone with long fingers. This someone is from a family in which everyone has long fingers. Let's say this person has a pair of genes, *LL*. This person is "pure" for long fingers. ▲

What does each parent give to a child? A child will get one *S* gene from one parent. And one *L* gene from the other parent. The child will have the genes *SL*.

The child will be a hybrid. Why will the child be a hybrid? Because he has a pair of different genes, *SL*. The child will be hybrid for short fingers. Why short fingers? This is the dominant gene. Trace it yourself again. ▲

▲

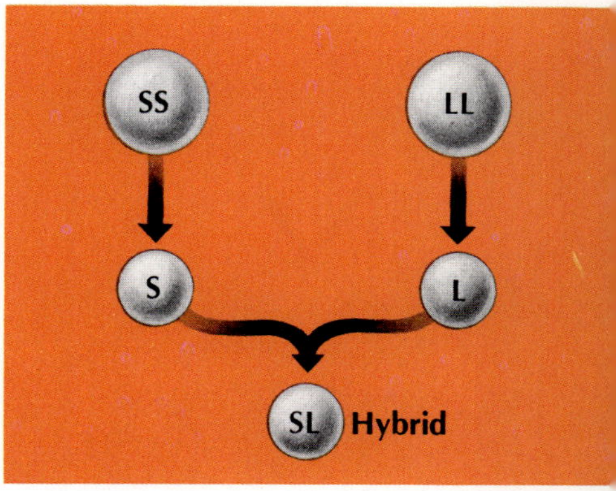

The Rh Factor in Blood

Let's suppose someone gets hurt. This person needs blood. He needs a certain type of blood. He must be given the *same* type of blood he now has.

People have different types of blood. People have different proteins on their red blood cells. People inherit blood types. That is why they can have different types.

One kind of protein on red blood cells is called the **Rh factor.** A person who has the Rh factor is called *Rh +*, or *Rh positive*. A person who does *not* have this protein is *Rh negative*, or *Rh −*. More people have Rh^{+} blood than Rh^{-} blood.

The Rh factor, like other blood types, is inherited. This means, of course, that the gene for Rh blood is on the chromosomes.

The *Rh +* gene is dominant. People who have an *Rh +* gene will have blood that has the Rh protein. Some parents with Rh + blood may be hybrid. ◆ They may have a hidden gene for *Rh −*. There is a chance, then, that some of their children will have Rh − blood. ◆

What difference does it make? No difference in daily living. But a doctor worries about a pregnant woman who is Rh^{-}. And he worries about her child.

◆

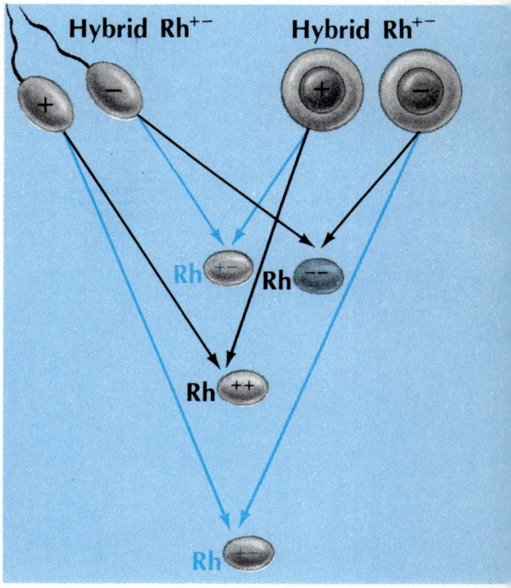

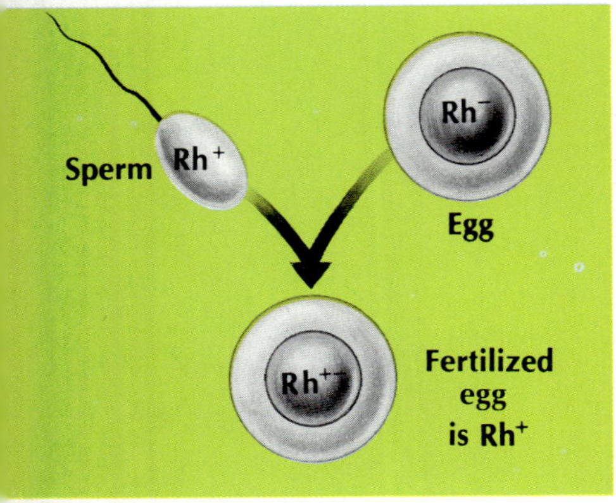

If a woman is Rh−, it means that she does not have the Rh factor. The Rh protein is not in her blood. But the growing baby might be making Rh+ red blood cells. It might have gotten this gene for Rh+ blood from its father.

Sometimes there are small breaks in the membranes of the placenta. Then Rh+ blood cells from the growing baby may get into the mother's blood.

These Rh + blood cells do not belong in the mother's blood. This means her blood will build up *antibodies against* the baby's Rh + red blood cells.

Probably there will not be enough antibodies built up to do any harm. There may not be enough antibodies to harm the red blood cells. (Red blood cells carry oxygen to all the cells in the growing baby.) The first baby is usually not hurt. But the antibodies stay in the mother's blood.

The second or third baby may be in trouble. By this time, the mother's blood may have a lot of antibodies. The mother's blood may have a lot of antibodies against the Rh protein. These may now get into the growing baby. How? Through the placenta.

What can happen to the growing baby's red blood cells? They can be destroyed by the mother's antibodies.

This means danger to the life of the baby. The baby's body cells will not get enough oxygen. This can also harm the baby's growing brain.

Doctors have a way to keep this from happening. They treat the mother. After she has her first baby, they treat her. They lower the number of antibodies in her blood. They lower the number of antibodies against the Rh protein. Then there are not many antibodies in her blood. Not enough to destroy the red blood cells of the next baby.

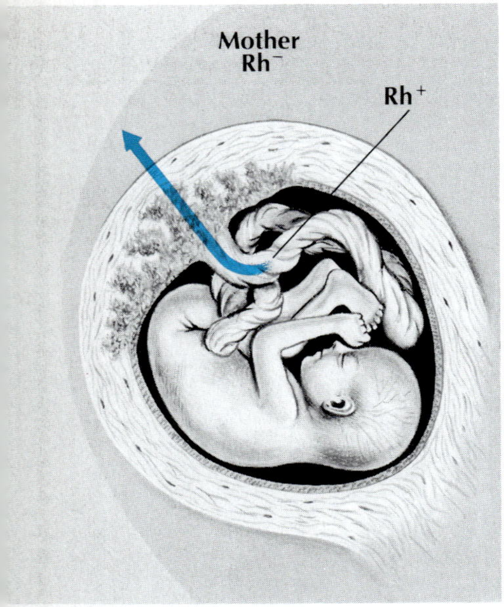

The Other Blood Types

Why do all men in the service wear dog tags? For several reasons. A sailor's or a soldier's dog tag tells his special blood type. Each dog tag tells the man's type. There are many kinds of blood.

This is a good thing to know if a person gets hurt. He may need blood. The kind of blood he gets should match his own.

There are genes for these groups of blood: group A, AB, B, and O. There are other blood types, too. You may not have exactly the same Rh factor or the same blood group as your parents.

Your red blood cells might have the Rh protein. And also the group A blood protein or the group B blood protein. There are many kinds of blood.

Genes That Add Up: Skin Color

Most people have genes for skin color. A person with little or no color in his skin is an *albino*. He has almost no pigment (PIG·ment) or color in his skin or hair. In fact, he has little pigment in his eyes. This lack of pigment is inherited. It is inherited as a recessive trait.

There are many shades of skin color. Shades go from black to brown to "white." Some people also have yellow or red pigments in their skin. ▲

Dark-skinned people have a lot of pigment in their skin. ◆ This pigment is called *melanin* (MEL·uh·nin). The many shades of skin color are the result of different amounts of melanin.

A dark-skinned person has more melanin than a brown-skinned person. ★

And a brown-skinned person has more melanin than a light-skinned person. A light-skinned person, or "white" person, has more melanin than an albino. ◇

The amount of melanin we have is due to genes. The color that shows up in our skin is due to many pairs of genes. We are going to use only two pairs of genes in our example. But there are more genes for skin color.

Let's call one pair of genes for melanin *AA*. And a second pair of genes for pigment *BB*. The two pairs of genes found in a dark-skinned person might be *AABB*. A light-skinned person would not have these four genes for color of skin. The light-skinned person might be *aabb*. The *a* and *b* genes stand for less melanin. A person, *aabb,* would be light-skinned, or "white."

▲

◆

★

◇

Suppose a person has only two of these genes for a lot of melanin. For example, what skin color would a person be with *AAbb?* Or with these two genes for color—*AaBb?*

Count how many *A* or *B* genes there are for pigment. You can see people with these combinations would be lighter than dark-skinned. They would be darker than "white." You can see they have only two genes for a lot of pigment.

In one of these examples, *AAbb,* there is a pair of genes, *AA*. In the other, *AaBb,* there are genes *A* and *B*. Both give a brown skin color.

It takes at least four genes—*AABB*—to get a dark skin color. The genes add up to give a dark skin color.

You know now that many genes help make us what we are. Many genes work together to make a trait. The story is much the same for height. Shades of flowers are also due to genes that add up. ■ And many genes act together to give fur color in animals.

It is the same story with plants and animals. The same as with human beings.

A bigger question is this. How do these many genes make the traits you have? How did they make your own special eyes, for example? or your dimples?

How do genes change the foods you eat into muscles, nerves, blood type, eye color, height, and all the other traits you have?

How does it happen? How do genes make the color of flowers? or the tallness of corn? or the rich milk of some cows, and the good steaks of some beef cattle?

You need more facts. You need to find out more about the genes. What chemicals make up the genes—the genes found on chromosomes? Let's see.

■

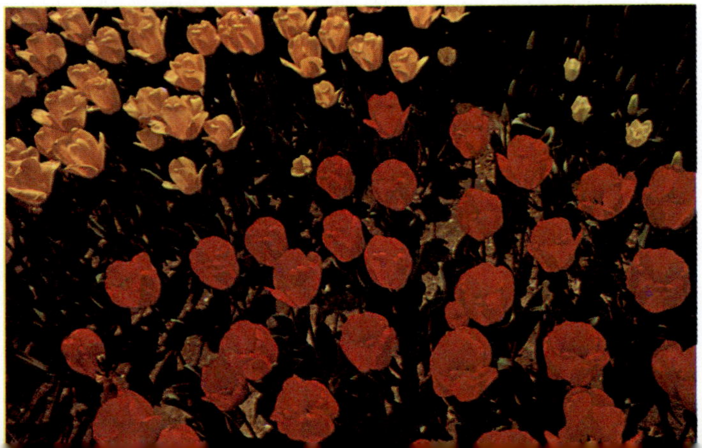

4. A CODE: A PATTERN OF HEREDITY

Have you heard of the Morse code? It is used to send messages. ● For example, three dots · · · and three dashes — — — and then three dots · · · (· · · — — — · · ·). This is the Morse code call for help. It means SOS. Many people know this code.

How does the Morse code work? Look at the chart. ● Different patterns of dots and dashes stand for the letters of the alphabet. You can translate (trans·LATE) the patterns into letters. And then words. If you know the alphabet letters in code, you can make words. And then whole messages.

The Morse code was invented by man. It is not a code found in nature. But there are codes that are found *in living* things. There is a special code in the nucleus of a cell. There is a code of chemicals in each chromosome in the nucleus. This special code helps to pass along the messages for intelligence, skin color, sex, and eye color. And all the other traits you have. It is a very important code.

Is this a code of dots and dashes? No, it's a code of chemicals. Where is it found? It's found in the nucleus of your cells. You can say that you inherit a special code of chemicals.

THE MORSE CODE

A	· —	N	— ·
B	— · · ·	O	— — —
C	— · — ·	P	· — — ·
D	— · ·	Q	— — · —
E	·	R	· — ·
F	· · — ·	S	· · ·
G	— — ·	T	—
H	· · · ·	U	· · —
I	· ·	V	· · · —
J	· — — —	W	· — —
K	— · —	X	— · · —
L	· — · ·	Y	— · — —
● M	— —	Z	— — · ·

A Chemical Code: A Pattern of Chemicals

You are special. You inherited your special chemical code. That is, the patterns of the chemicals are a little different in each of you.

All animals and plants inherit their own patterns of chemicals. All living things have their own code of chemicals. A code is in the egg nucleus. And in the sperm nucleus. The egg and sperm have a code in their chromosomes.

In each of us, some parts of the code are alike. Some of the patterns of chemicals are alike in all of us. We are all humans. Parts of our chromosomes are alike. We have some of the same patterns because we are all humans. Not lions, or potatoes.

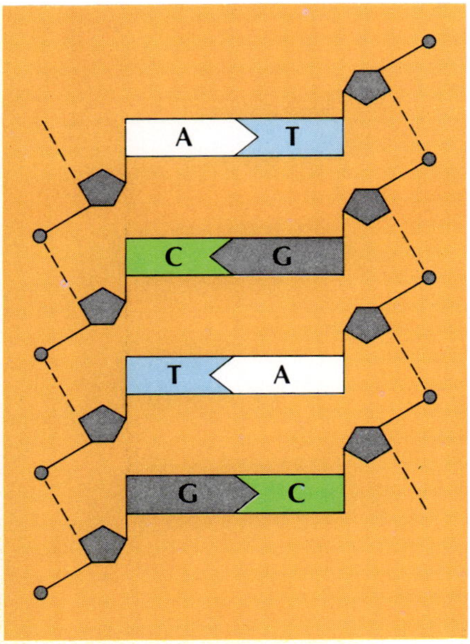

A Chemical Code — DNA

Think of it this way. In each chromosome there is a long string of chemicals. The chemicals are in patterns. The string of chemicals is called **DNA** (DEE·EN·AYE). It is called DNA for short. DNA is made of different patterns of chemicals.

Study the picture of part of a long string of DNA. ■ Here you see only a piece of one long string of DNA. ● This is only part of the DNA in a chromosome.

Can you see that there are four *different* kinds of chemicals that make up DNA? The letters *A, T, C,* and *G* stand for the names of the chemicals.

These four chemicals of DNA can be found in different patterns. They make up a code. So these chemicals can make up messages.

It's the same idea as with the Morse code. Suppose you know the patterns of dots and dashes for the alphabet letters. Suppose you know the alphabet letters in code. Then you can translate the patterns into words. And send messages in Morse code. In the chromosomes, four chemicals make the patterns. These patterns make up many messages. In fact, they make up messages for all kinds of life.

We are alike because we are humans. But we are all different, too. The messages give us different traits. The patterns of chemicals in your DNA code are different. They are different from any other person's. You are born with your own set of chromosomes. That's why you are YOU.

DNA, Genes, and Chromosomes

There are patterns of chemicals lined up on the chromosomes. What have we been calling these patterns of chemicals? That's right, *genes.*

Up to now we have called these chemical patterns *genes.* Now you know that genes are patterns of the chemicals of DNA. And that these patterns carry messages. You also know that genes are parts of chromosomes.

Changing a Code into Working Cells

Where are these patterns of chemicals in you? The DNA code is in the nucleus of your cells. How does the code work? How do letters like *C, T, A,* or *G* make up messages? How do chemical messages become different cells in YOU? And where does this happen?

First, let's say again that the patterns of chemicals make up messages. These patterns of chemicals are passed into the cytoplasm. It is in the cytoplasm that these chemical messages are translated (trans·LAT·ed). They are translated into proteins.

To learn how the code works, let's look at a game you may have played. How many other words can you make out of the four letters that spell meat? You can say *team* or *tame* or *mate*.

Suppose you use only *three* of the four letters. You can make *many* more words. Here are *only* a few: *ate, eat, tea, mat, tam,* and *met*. You can make many more patterns. You can make many more patterns of letters. And with words, you can make messages.

The chemicals in the DNA code work a little like the word game. Patterns of these chemicals make up messages. Not words, but *chemical* messages.

Let's work with the chemicals in the DNA code. Let's work with the letters that stand for the chemicals. Here are the rules. The chemicals *C, T, A,* and *G* must be used in groups of three letters. These might be *GAC* or *CAG* or *AGC*. And the letters must also be used in three sets of three letters.

Let's take three sets of three letters: *AAA* and *GCA* and *CCG*. This spells out as *AAAGCACCG*. Is this a word? or a sentence? No. Then what is it? This is a chemical message. It means *one* kind of protein. This message directs the cell to make one kind of protein.

Suppose the message for one protein is *AAAGCA-CCG*. A little change in this message would direct the cell to make a different protein. Suppose the C is missing and an *A* is in its place. The code would now be *AAAG*__A__*ACCG*. This would be the message for a *different* protein.

Different proteins are made in cells. How? From three chemicals put together in different sets of three. This idea is a lot like using the letters in the word meat.

In the cytoplasm the messages are translated into proteins. Proteins make new cells. Proteins will make up new nerve cells. And bone cells, blood cells, and skin cells. Some of these proteins make traits such as blood type. And your skin color, and all the rest of your heredity.

We don't know how all this works yet. But scientists are making new discoveries each year. They are learning how you become YOU.

A Try at Prediction

In some people the protein that gives the red coloring to red blood cells is a little different. Then the red blood cells cannot carry enough oxygen to cells. The round red cells change shape. They look like this. ■ This is called the *sickle* (SIK·kule) *cell* trait. This trait is inherited.

There is a gene for sickle cells. Suppose a person is *SS*. He is "pure" for the trait.

Most people have genes *ss*. This pair of genes makes the red-color protein most people have. Where? In the red blood cells. And their red cells look like this. ●

How did sickle cell protein happen? You are right if you say there was a change in a message.

How could the message be changed? Only a one-letter change makes the difference. It can make the difference in the proteins in the red blood cells. One letter makes the difference in the trait. The sickle cell trait or the trait for regular red-blood cell protein.

■

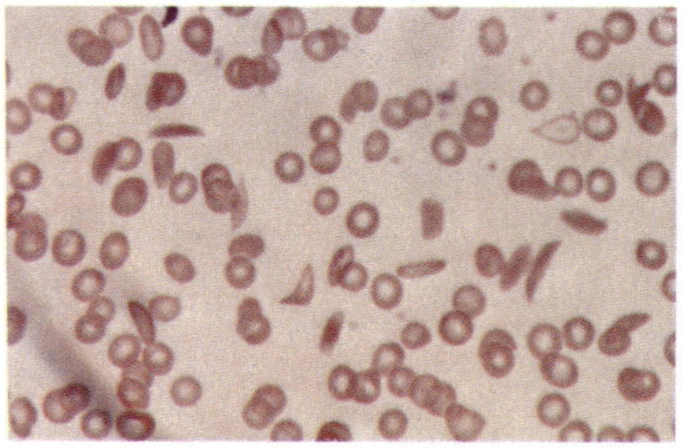

●

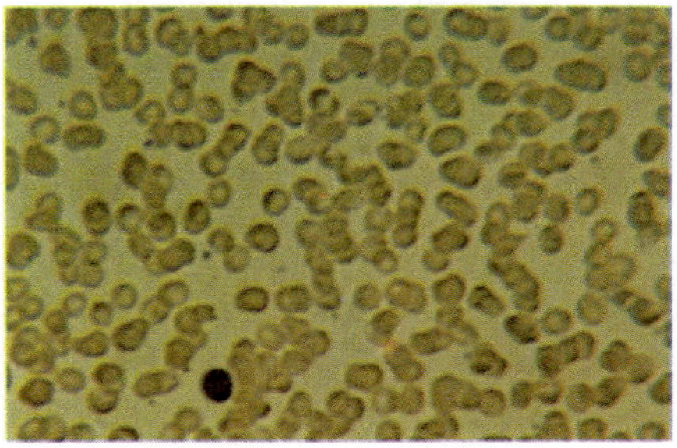

5. BETWEEN THE FERTILIZED EGG AND YOU

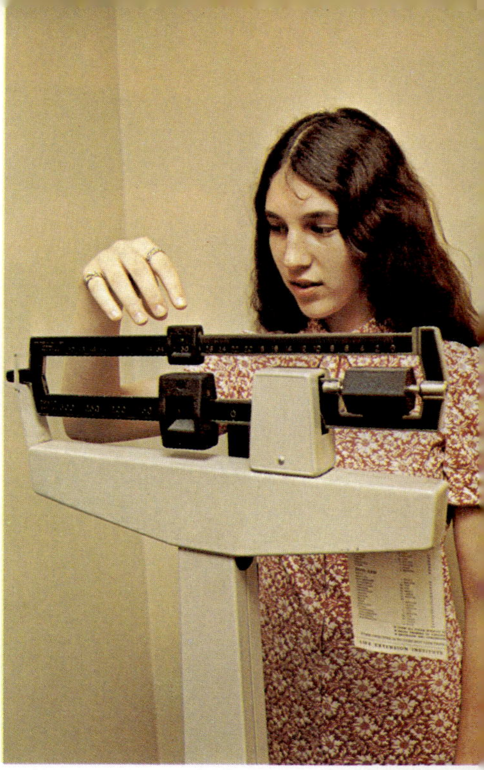

You cannot *see* proteins and enzymes working in your cells. But you see what happens. How much did you weigh when you were born? About 5 or 6 or 7 pounds? And now how much do you weigh?▲

All the millions of cells in you now were made from the fertilized egg. The fertilized egg divided to make more cells. The proteins you eat are used to make more cells. This is how you grow.

The DNA code is copied over and over. It is copied each time cells make more cells. Your cells have copies of the patterns in the DNA from the fertilized egg. The DNA is in the 46 chromosomes in the fertilized egg.

Sometimes there is a chemical missing in the cells. A gene or a chemical of the DNA code is missing. Then something goes wrong. Sometimes an important enzyme cannot be made. Here is an example.

One certain enzyme helps control how pigment is made. If this enzyme is missing, the skin has a little or no pigment at all. The person is an albino.

Other kinds of special enzymes may be missing. A *PKU* (PEE·KAY·YOU) baby may be born with some brain damage. Or it may become retarded if fed on a regular baby formula. A PKU baby is missing an enzyme. This enzyme is needed to use a certain nutrient. So a PKU baby needs special food.

A doctor can test a baby for the PKU trait. It is a very easy test. If the doctor finds the trait, he gives the baby a special diet. And then the baby will not be retarded.

We take for granted the way a baby grows. We take for granted that the patterns in the DNA code will be copied right. Each time a cell divides it makes a copy of the patterns in the DNA code. Each cell needs the messages.

When the code is not copied exactly, something could go wrong. Then the baby may need help. It may need special nutrients in its diet.

■

●

All in a Family

Here is a family of four children. ■ They all have about the same environment. They live in the same home. And they eat the same food. But they are different. Why? Each child has his own special heredity. Each has his own special inherited DNA code.

Are the children alike in any ways? Yes. They may get about the same training. They share many things every day. They share many of the good and bad things that happen each day.

The oldest child was an only child for a while. Now the oldest child helps take care of the younger children. ● She is no longer the only child in the family. And she has many jobs to do.

You can see children have different heredity in the same family. And they all have a little different environment. Even if they all live in the same family. Both heredity and environment have to do with how children grow up.

The Case of Identical Twins

What are *identical* (i·DEN·tih·kul) *twins?* These are two children that seem to be exactly alike. They come from the same fertilized egg.

How can two children come from one fertilized egg? As you know, the fertilized egg first divides into two cells. ▲ Then into four cells. And so on. Each time the cells divide, the DNA code is copied. Of course, all the cells then have the same code. Sometimes these cells separate into two equal parts. This happens early in the divisions of the fertilized egg.

The cells separate into two parts. Each part then develops. Each group of cells grows into a child. ▲ There are now two children from one egg. And each child has the same heredity. These children are called identical twins.

You cannot be in two places at one time. But suppose you had an identical twin. You and your twin would have the same heredity. What would happen if you and your twin were brought up in *different* homes or environments? ◆

Cases like this happen. Identical twins have been separated. These twins were brought up in different homes far from each other. Years later the twins found each other.

▲

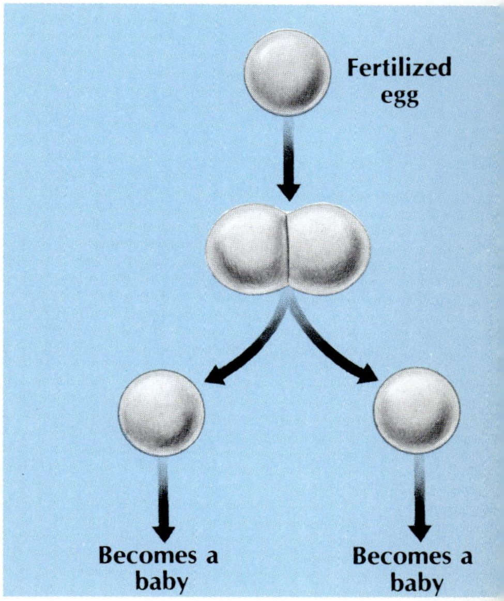

◆

■

●

After many years they still look very much alike. ■ You could guess this. You could guess this because these twins have the same heredity.

One twin may have had a chance to graduate from high school. The other did not have this chance. She had to go to work. So the twins had *many* different experiences.

Here is where the difference in environment comes in. Both twins were born with the ability to be educated. But how they lived made the difference. It made the difference in how well they learned. Their environment made the difference in what they learned, too.

One twin did *not* go to school or learn to read. But she learned to do other things. The other twin did go to school, and learned to read. These identical twins have the same heredity. But they have not learned the same things. Their environment helped to make some differences between these identical twins.

A Try at Explanation

Many twins are not identical. ● They may be boy and girl twins. Or they may be the same sex. They may have different eye color. They may be no more alike than other children in the family.

These are *fraternal* (frah·TUR·nal) twins. ● Why are they not alike? They come from two different eggs. Two eggs were fertilized by two sperm cells. They have different heredity. Why?

The Right Heredity and a Little Iodine

One child's genes may act so that he has a normal **thyroid** (THI·royd) **gland.** This is one of the glands that controls the way the body uses food.

Iodine from food is needed for the thyroid gland to work right. What would happen if the child does not get enough iodine in food? Then as he grows, his body will not use food in the right way. So he will not grow well. He will not learn very well. And he may not be able to go to school.

But a doctor can help this child. He can help if the child is *very* young.

A growing child needs the *right heredity and the right environment.* In this case, part of the environment was wrong. Iodine was missing from the child's food.

Lots of Sunshine

Here are two leg bones of a normal child. ▲ And here are two leg bones of a child with very soft bones. ◆ See how the bones are curved. This child has *rickets* (RIK·its). Rickets is a disease in young children. Bone cells cannot grow the way they should.

Is rickets due to heredity? Or is it caused by the environment? How can we find out?

We can treat a small child who has rickets. We can give him lots of sunshine. We can take him out in the sun to play.

Sunlight turns some fat-like matter in the skin into a new chemical. This new chemical is known to you as vitamin D. This chemical helps to put calcium into bone cells. Then the bone cells will be hard—the way they should be.

We could also give this child milk with this chemical. Or we could give him cod liver oil. Cod liver oil has a lot of this chemical called vitamin D.

Then the young child will get harder bones. He will have more calcium in his bones. And in his teeth, too. The right food in his environment can help this child. The child can grow up better.

▲

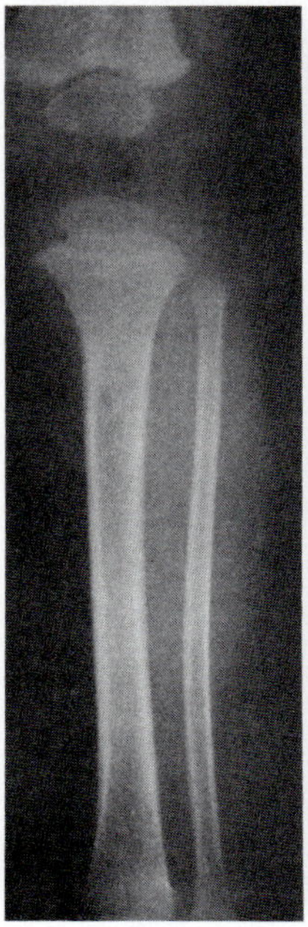

◆

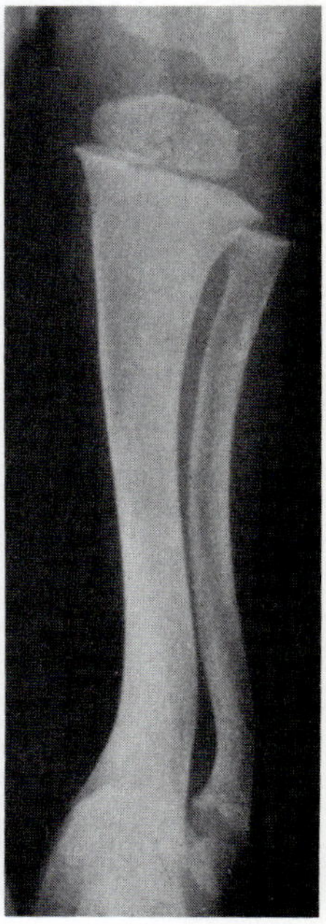

■

●

▲

You = Heredity + Environment

We are all what our heredity makes us. And we are all part of what our environment makes us. We need the right environment for our special heredity to show up.

The environment may make a difference in how we grow. It may make a difference in our height and weight. And in our ability to learn. It also may make a difference in many other ways. A poor environment may cause some diseases.

But for some things the environment makes no difference. The blood type you have has nothing to do with your environment. If you are color-blind, or have brown eyes—this cannot be changed by environment.

Yet many, many times the right environment is needed to shape the right heredity. You need both.

6. LIKENESSES AND DIFFERENCES

There are $3\frac{1}{2}$ billion human beings in the world. They have all inherited a code of DNA. These humans are all made of cells. And organs that all work the same way. The DNA code of all these people is a lot alike. That's what makes these $3\frac{1}{2}$ billion people all humans. And not horses or dogs or mice.

But they are not *exactly* alike. The DNA code is not exactly alike in each child in one family. There is always a reshuffling of heredity in every fertilized egg. This is how we all begin.

People

When people have many of the same traits, we sometimes say they belong to a race of people. Oriental (or·ee·EN·tel) people, for example, make up a race.

Oriental people have most of the same genes all humans have. As a group, they also have some genes that mostly only Orientals have. ■ They have many of the same kinds of genes in their group, or race.

Populations of black people share many special genes. ● Populations of white people also share many special genes. ▲

How Differences Arise

People have many of the same messages in the DNA code. People have many of the same genes. But changes in these genes happen once in a while. Humans have been on earth for a million, or maybe more, years. Over this time some genes have changed. New genes have appeared in people. New messages have appeared. Every once in a while this happens.

When a group of people lived in one place in the world, they shared many genes. When a new gene appeared in this population, it spread among the people. This was because the people married and lived in the same group.

This is the way a new blood group might have first appeared. The new genes spread among the people. People married and lived in the same group. People in this place now had the new blood group.

But in a different place on a land area the story might be different. A different gene or genes could show up in this group of people. This different gene or genes could spread out in this population. For example, the sickle-cell trait showed up more in one population.

What happens then? People on different continents may be different. (Turn the page.) Some different genes spread in one population. This population would be different from populations on other continents. This population would be different only in certain ways.

Today the world is changing so fast. People don't stay separated in different lands. Jet planes take people from one continent to another in a few hours. People are meeting together. They are trading together.

As a result, people from different continents are learning from each other. And they are learning to respect each other. ♦

In fact, it's getting to be a smaller world. The distances are shorter when people can travel fast.

And it's getting to be a more crowded world. There are few places today that have only one kind of population. People have to learn to live together with their likenesses and their differences.

♦ 

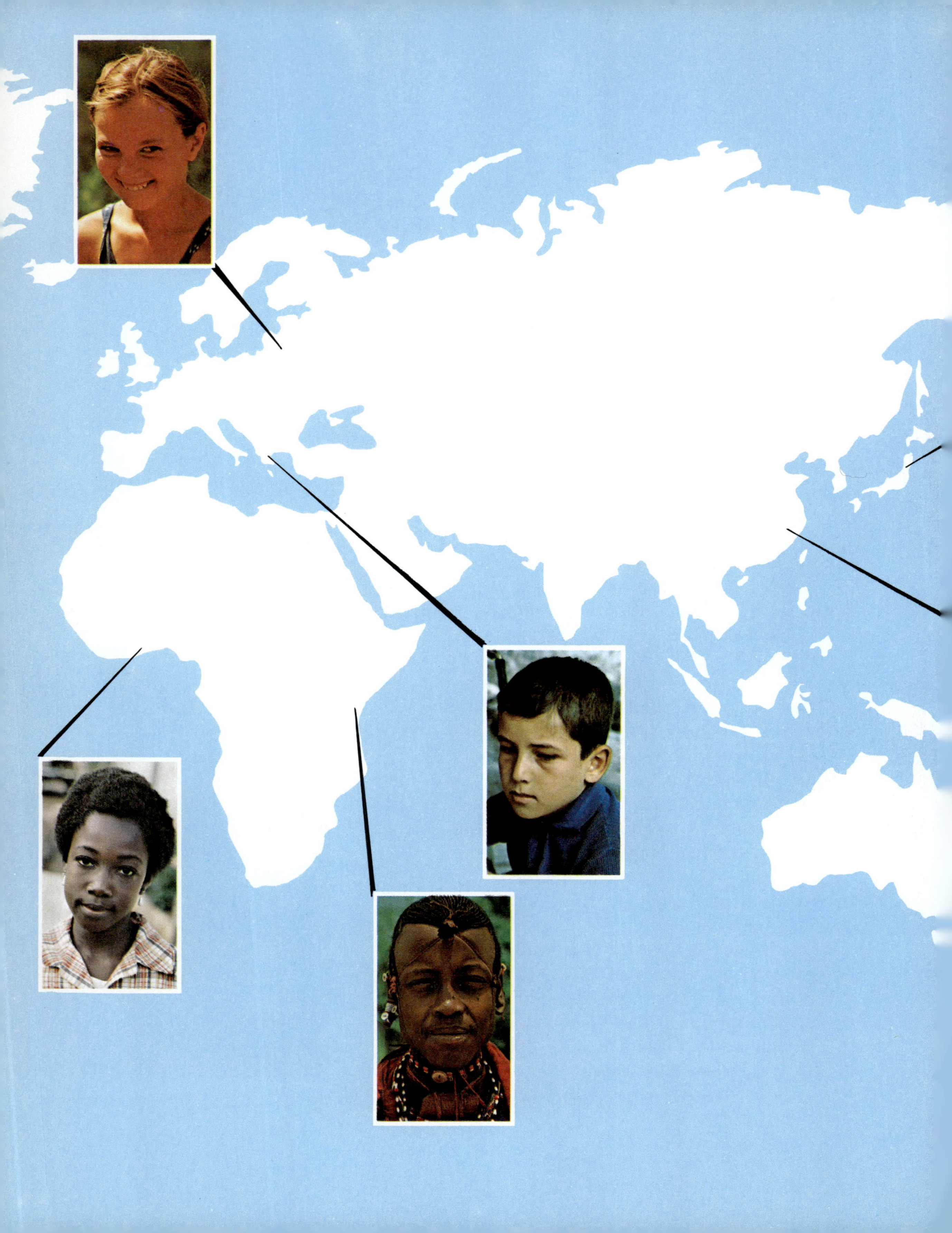

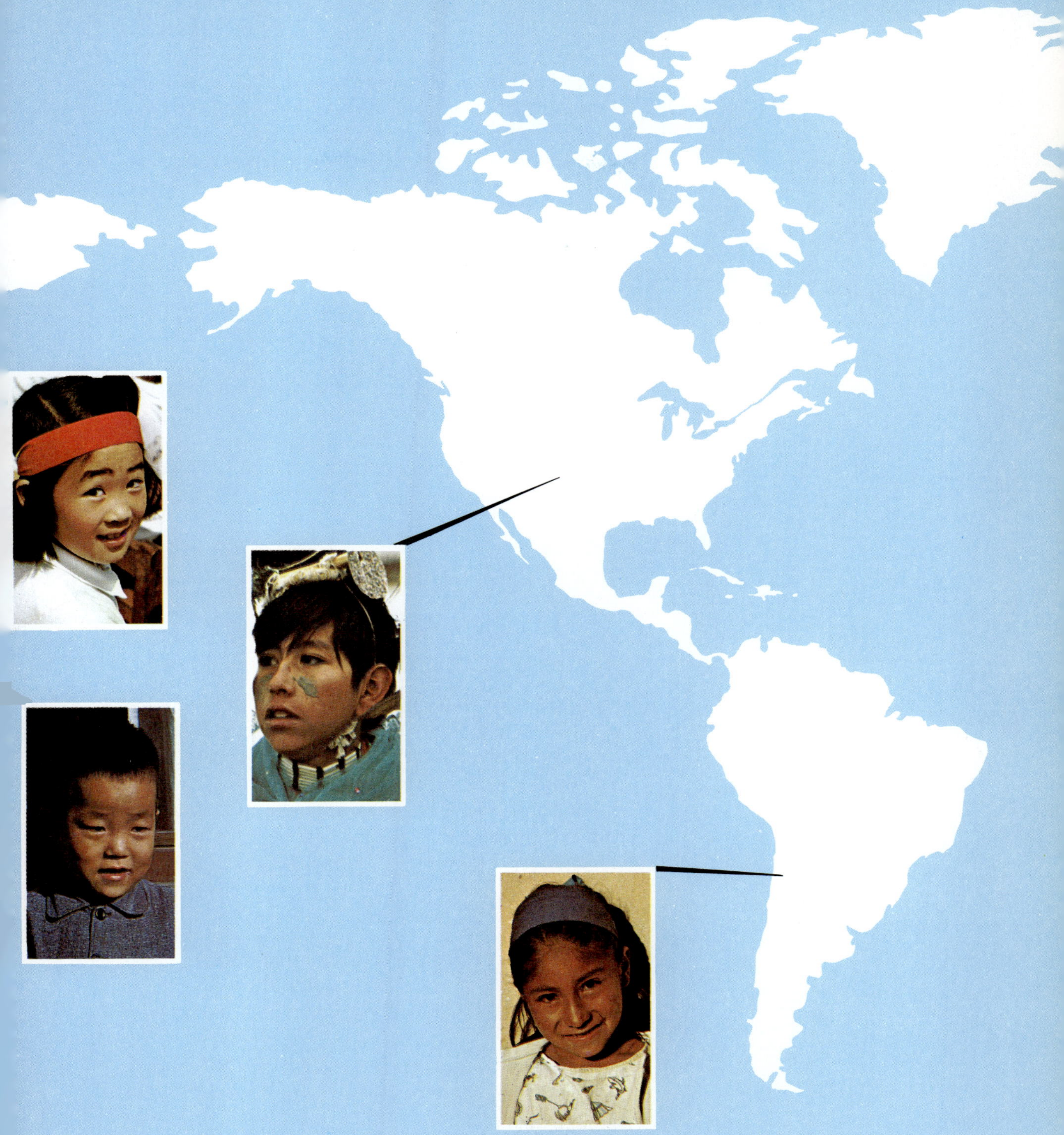

AN INDEPENDENT STUDY

You and Your Heredity

What traits do you inherit? How do you inherit traits? See how much you have learned. First read each question. Then write your answer on your answer sheet. To see if you are right, check the answer across the page.

1

The chromosomes of a fertilized egg hold the code of heredity. ■ The pattern of heredity you get is due to genes. Genes are on the chromosomes.

If you have brown eyes, we say you have g---- for brown eyes.

(Write your answer down. Check across the page to see if you are right.)

2

Suppose you have genes for long fingers from *both* parents. You would be "pure" for long fingers. If both parents had short fingers, you would be "pure" for short fingers. You would be *SS*.

What would you most likely have inherited from your short-fingered parents—genes for *long* or *short fingers?* Which?

3

Suppose one parent has normal pigment, *NN*. And the other parent is albino, *AA*. A child might get one gene, *N*, for normal pigment. And one gene, *A*, for albino.

This child would be a hybrid *NA*. The genes in this pair are different. Which is the hybrid, *NN* or *NA?* Which?

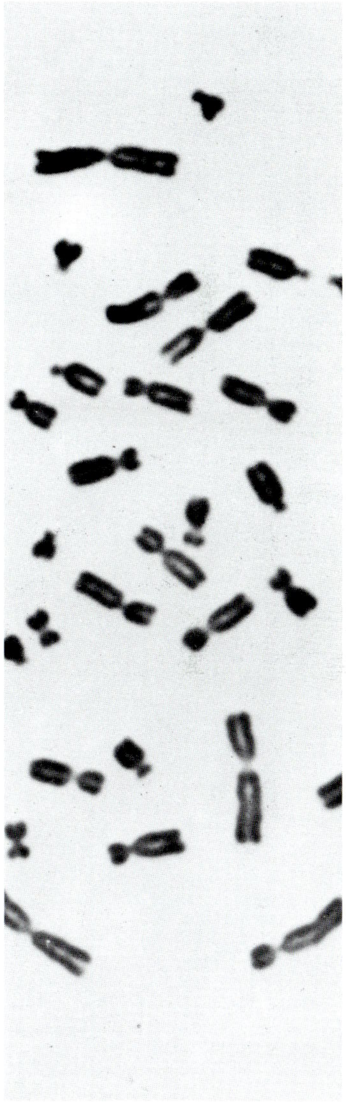

1

genes

Genes have the code for our heredity.

(Go on to question 2.)

2

short fingers

You would have inherited genes for short fingers, *SS*.

(Go on to question 3.)

3

NA

This is a hybrid pair of genes. The two genes are different.

(Turn the page to question 4.)

4

What is this *NA* hybrid going to look like? Genes for skin pigment are dominant over albino. A hybrid like *NA* would have pigment in his skin.

But there is a hidden gene there. This hidden, or recessive gene A, is in the person with *NA* genes. This person is h-----.

5

Suppose both parents are hybrid. That means each parent would have a hidden gene for albino. Like this: *NA* × *NA*.

Could some children get a gene *A* from each parent? Could they be *AA*? Yes. Some children could be albinos. They would have little or no pigment in their skin, hair, or eyes.

Suppose the children were albinos. Which gene did they get from each parent: the *N* or *A* gene? Which?

6

There is an Rh factor on red blood cells. This Rh factor is inherited as a dominant gene.

If you have this Rh protein, you are Rh^+ (positive).

If both parents are "pure" for Rh^+, the children have Rh^+ blood. Like this: $++ \times ++ = ++$ children.

If both parents do not have the Rh factor, they are Rh^- (negative). Like this: $-- \times -- =$?

If two parents have Rh^- blood, what will the children be? Will they have Rh^+ or Rh^- blood? Which?

7

Let's go back to the Rh + factor in the blood. Let's suppose two parents are hybrid: $+- \times +-$. The children can be $++$ or $+-$ or $--$.

Which pair of genes is hybrid? $++$ or $+-$ or $--$? Which?

4

hybrid

That's right. The gene for albino is recessive. It may be hidden in a hybrid.

(Go on to question 5.)

5

the A gene

They got the gene for albino from each parent. That's good thinking so far.

(Go on to question 6.)

6

Rh^-

Yes, the parents do not have genes for Rh^+ blood.

(Go on to question 7.)

7

$+-$

Yes. The hybrid pair has two different genes.

(Turn the page to question 8.)

8

Suppose you know that a person has Rh^- blood. Can you tell what kind of genes he has? Does he have −− genes or +− genes? What do you think? Which does he have?

9

Now see what you know about sex chromosomes. What do you know about the chromosomes that make a male? a female? Suppose a fertilized egg has the XY pair of sex chromosomes. What sex will the child be?

10

The fertilized egg with XY sex chromosomes will grow into a male. Suppose this same fertilized egg had two genes for albino. And suppose this fertilized egg mass separated into two parts. And each part grew into an embryo. There would be two embryos —identical twins.

Would they both be albino boys? *Yes* or *no?*

11

Why do both embryos have the same sex and other traits? They are *identical twins.* They have to be the same. Why?

They have the same DNA code. Both embryos have the same DNA code. Twins that have the same heredity are ________________ twins.

8

— —

He cannot have any Rh^+ genes. He has only Rh^- genes.

(Go on to question 9.)

9

male

(Go on to question 10.)

10

yes

The fertilized egg had the XY sex chromosomes for a male. And this same egg had *AA,* or two genes for albino. When the egg mass separated, the genes were the same in both embryos.

(Go on to question 11.)

11

identical

Identical twins have the same DNA code in their chromosomes.

(Turn the page to question 12.)

12

Not all twins are identical. What are twins called that look different? These twins are called fraternal twins. They are as different as any children in that family. They came from two fertilized eggs. Two eggs were fertilized by two sperms. Twins from two fertilized eggs may be girl-boy or both the same sex.

Fraternal twins are different from identical twins. Which twins come from *one* fertilized egg that separates—*fraternal* or *identical?* Which?

13

What is the DNA code? It is the code that is in *each* chromosome. It is the chemical code that carries the heredity. It is the code that carries the heredity for eye color and skin color. And for all of your traits. Identical twins would have the same D-- code.

14

And fraternal twins? Fraternal twins would each have their *own DNA code*. They come from two different fertilized eggs. Each fertilized egg has a *different* DNA code? or the *same* DNA code? Which?

12

identical

Identical twins come from one egg. Fraternal twins come from two eggs.

(Go on to question 13.)

13

DNA

Identical twins have the same code in their chromosomes.

(Go on to question 14.)

14

different

Each twin has its own DNA code.

9

On Earth: The Chain of Change

INTRODUCTION

Reading the Stories in the Rocks

These salamanders (SAL·uh·man·derz) you might find in the woods. ■ They live near water the way frogs do. You could guess this. They have thin skins like frogs have. In fact, they are related to frogs.

About 300 million years ago, a big salamander like this one was moving around. ● How can scientists say it lived 300 million years ago? They found only the jawbone of this salamander in rock. The rest of the animal was gone.

Here is something even more exciting to find out. The jawbone was found about 350 miles from the South Pole! There is ice and snow all around the South Pole now. There are only a few plants and animals found near the South Pole.

What does this jawbone tell us? What must it have been like at the South Pole about 300 million years ago? The South Pole must have had an *environment* that salamander-like animals could live in. What kind of environment would this be? It would have been hotter than it is today.

There are stories in the rocks. There is the story of the animals and plants that lived in the past! And the story of the climates and environments of the earth that have also changed. Scientists have learned to read these stories in the rocks.

You will learn a little of this story now.

1. GOING BACK INTO TIME

This is the way it was. The bones in this picture are real. ■ They were once part of animals. They belonged to animals that lived long ago.

Animals of this kind are not alive today. They died out long ago. But sometimes the animal bones did not turn to dust. Very old animal bones are found today. How is this so? You know that plants and animals *decay* when they die.

Sometimes animal bones do *not* decay. Sometimes animal bones are quickly covered over. Sometimes the animal bones are covered with mud, or sand, or tar. When this happens, no bacteria reach them. Then the bones may not decay.

Here is what can happen to bones. A bone may be covered with tar. Or a bone may be buried in the mud. First, minerals (MIN·er·alz) in the mud fill in the spaces in the bones. Then minerals harden the bones. Then the bones feel like rock.

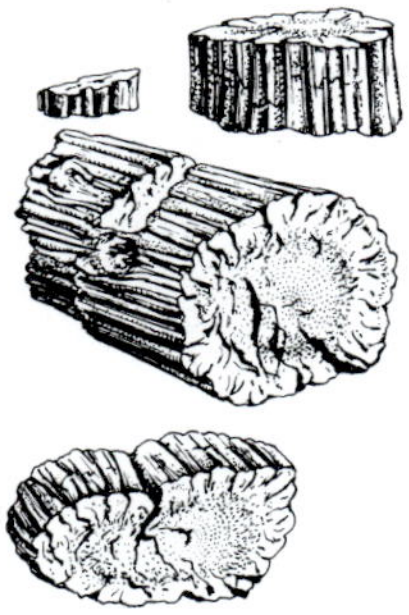

Petrified tree

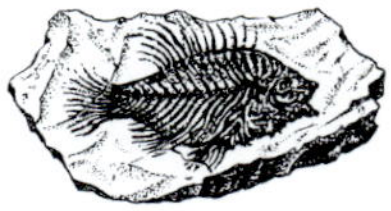

Fossil

This tree stump looks and feels like a rock. ● A hard and heavy rock. A tree that has changed in this way is called *petrified* (PET·rah·fied). This tree fell into a swamp about 200 million years ago. It was quickly covered over with mud. The tree is now like rock. You can find many trees like this in the Petrified Forest in Arizona. ▲

Both the bones and the trees are petrified. Each of the bones and trees is a **fossil** (FOS·ill). A fossil is the remains or trace of something that once lived.

The Story in Fossils

A petrified plant or animal part is one kind of fossil. Any hard part of a plant or animal can become a fossil. A tree stump or a bone can become a fossil. The hard part is left behind when the rest of the body of the plant or animal decays.

There are sometimes more than bones that are left by animals. Sometimes the teeth of an animal are left behind. Hard parts like teeth and bones have a better chance to become fossils. They have a better chance than soft parts, like skin or lungs.

Teeth and bones decay slowly. They have a better chance to be covered over by mud, sand, or tar. Before they decay, they can become fossils.

There are even fossils of animals without any bones. There are fossils of clams. ◆ And there are fossils of starfish. ★ What parts of a clam can become a fossil? A clam has no hard bones. But clams do have hard shells. Starfish have hard, little spines.

Suppose some clam shells or some starfish spines fell into sand. Then this sand was slowly, slowly, slowly pressed into rock called sandstone. This took thousands of years. Where would the shells be now? In the rock. Now the shells or starfish spines are fossils. They are fossils found in sandstone.

◆

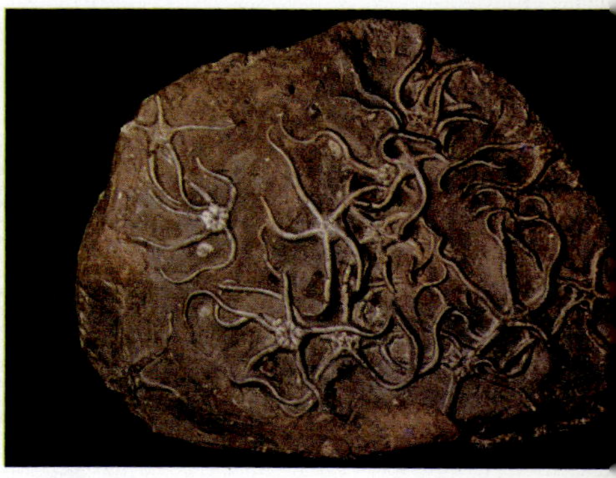

★

Mammoth

Here you can see another kind of fossil, a *mammoth* (MAM·eth). ◇ Big woolly mammoths have been found frozen solid in ice. These woolly animals were larger than elephants. They had long hair. Whole animals were preserved (pre·SERVED). This means they did not decay.

These animals were preserved for a few million years in frozen soil and ice. The whole animals were preserved. You can say they were in a deep freeze. Bacteria did not decay the frozen animals. Most bacteria can't live in freezing temperatures.

◇

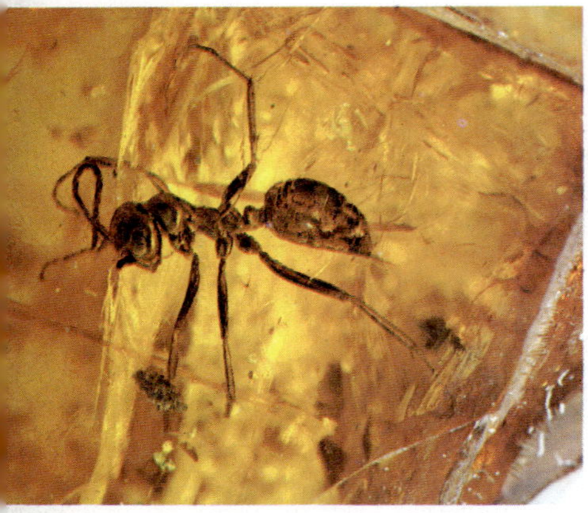

What can we tell from fossils preserved over many years? We can tell what plants and animals of long ago looked like.

Ant in amber

Insect fossils are also found in rock. Fossil insects are found in other places, too. Look at the ant in this piece of *amber* (AM·burr). This ant is millions of years old.

This ant was trapped in a drop of sticky sap from an evergreen tree. Can you think how an ant might get stuck in sap as it crawls along the bark of an evergreen tree? Over a long time, the sap hardened. The hard sap is called amber. Amber is a fossil, too.

The fossil insect did not decay. No bacteria got to it. The whole insect was preserved. This ant looks like ants today. Ants have not changed much over millions of years.

Sometimes we know about a plant or an animal even if it did not leave bones or teeth. Even if it was not frozen. Or even if it was not preserved in rock or amber. Sometimes we can tell about the soft parts of an animal or plant.

Leaf print

Have you ever put your hand in wet, soft cement (see·MENT) on a new sidewalk? Then you know what a handprint is. After the cement gets hard, the handprint is still there. This is about the same way that a leaf of a tree makes a *leaf print*.

Some leaves fall from trees into soft mud. Even if the leaves decay slowly, they make a print in the mud. Then the mud may get hard. The mud with the prints in it may be changed into thin slabs of rock.

These prints in the rock can tell us about the kinds of plants that lived millions of years ago. Fossil leaf prints like this show us how ferns looked. ▲ They show us how ferns looked long ago.

Even a soft animal like the jellyfish can leave a print. This is a fossil print of a jellyfish-like animal. ◆

Fossils can tell us about plants and animals that lived long ago. They also tell how long ago something lived. They tell us how long ago this jellyfish lived. Let's now look into this.

Layer on Top of Layer

Suppose that you stack four sponges like this. ★ Starting with the yellow one, stack one on top of the other. You now have four layers of sponges.

Now look at the picture. ★ Which sponge was put down first? The yellow sponge. The next layer is the green sponge. The blue is the third layer, and the last, or top sponge, is the pink one.

The other sponges were put down after the yellow one. Which sponge was put down last? The pink sponge.

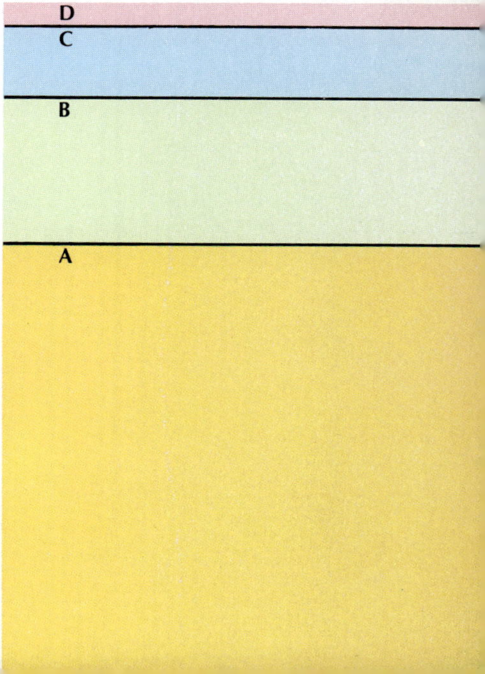

Now suppose that instead of sponges, there are layers of rock like this. ◊ You can see the different layers A, B, C, and D. It takes millions of years for rock layers like these to form.

Which is the youngest rock? The top layer? Yes, layer D. It was the last layer to be put down.

The oldest layer of rock would be on the bottom, layer A. Layer A is the first layer. Like the yellow sponge, it is on the bottom. It was put down first. That makes it the oldest layer. You can see how thick the bottom layer is. Why is it so thick? It must have taken a long time for this much rock to form.

Dinosaur

Suppose you look closely at layer B. You can find many fossil clam shells in rock layer B. But you find no fossil bones of a *dinosaur* (DIE·nah·sore). Dinosaurs *were* animals that belonged to the same group as lizards. They were like lizards except in size. Some dinosaurs were *very* large. And some were small.

There are also no fossil animals like birds, or like elephants, horses, or dogs in layer B. How do we explain this? Can you explain this?

Animals like dinosaurs, birds, and furred animals did not live that long ago. They did not live when layer B formed. Clams lived in the oceans long before birds came. Long before horses or dogs came.

Now look at a younger layer of rock. Here is layer C, the third layer in the picture of all the layers. It's under the newest layer, D. There are fossil bones of birds and dinosaurs in layer C. What does this mean? These animals must have lived together at one time. They died, and some left fossil remains in rock layer C. Then soil and mud covered them. The new layer of soil on top became layer D.

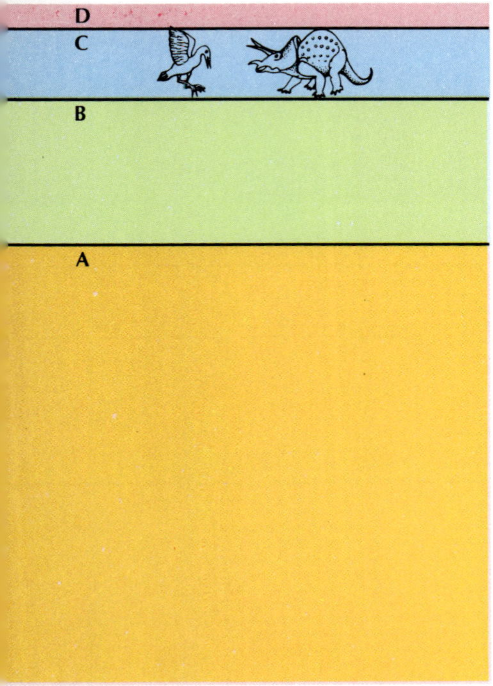

Suppose you know the age of the rock in layer C. What else, then, do you know about layer C? You know the age of the fossils in the rock. If you know how old the rock is, you know how old the fossils are.

It is hard to imagine that there was a time when the oceans had no fish. Or when there were no lobsters, no crabs. Were there plants? Yes, but only small plants. There were plants, such as algae of all kinds, in the oceans. And there were some kinds of animals, too. Sponges and jellyfish of many shapes and colors lived in the oceans. They lived in the oceans long before lobsters came. They lived there millions of years before fish came.

There was also a time when the land had no plants. There were no plants and no animals on the land. Slowly, small plants appeared on the land. But there were no trees for millions of years.

Look again at the layers of rock A-B-C-D. ▲ Now you are looking at some of the kinds of fossils in each rock layer. Which fossils are in the oldest layer—clams or birds?

Dinosaurs, birds, trees, and horses appeared on land millions of years *after* clams. The oceans were filled with living things millions of years before life came to the land. You would find the fossils of dinosaurs, birds, trees, and horses only in the upper layers of rocks. They have not lived on the earth as long as clams or fish have.

Man has not been on earth as long as birds or dinosaurs have. Man is the latest arrival on the earth. How do we know? Fossils of man are found only in the very top, newest layers of rock, like layer D. ♦

Fossils tell us a great deal about living things as they were. *Fossils tell us the order animals and plants appeared on earth.*

▲

The Age of Rocks and Fossils

How can we know the age of a fossil? How can we know the age of a clamshell fossil? or a jellyfish print? How can we say a fossil bone is from a horse that lived 58 million years ago?

We can tell the age of rock layers over the world. In the rock layers are fossils. If we know the *age* of the rock, we can say *when the fossil lived.*

Or we may find a new fossil and know when it lived. How? Scientists know the age of a few special kinds of fossils. Then they also know the age of the rock the fossils are in. So it is not hard to take a newly found fossil and tell the time it lived.

Suppose a new fossil is found *with* one we know. We know the age of one fossil. Then we say the new fossil is about the same age. We can then say the new fossil is as old as the one we already know. Can we also tell the age of the rock? We now know the age of the new fossil. This tells us how old the rock is.

It works both ways. If we know the age of the fossil, we know the age of the rock. And if we know how old the rock is, we know how old the fossil is.

♦

Telling Time

The sand in this hour glass moves from top to bottom in a certain time. ■ It may take a minute for the sand to move from one part to the other. This would be a minute clock. Or it might take an hour. Then it would be an hour clock.

You read clocks that tell time in minutes or hours. There are also clocks that tell time in *thousands of years*. And clocks that tell time in *millions of years*.

These clocks are radioactive (ray·de·oh·AK·tiv) clocks. These are special chemicals that give off rays we cannot ■ see. These chemicals break down into other matter. As they break down, they give off rays.

Scientists know how long it takes for a radioactive chemical to break down. They know how long it takes a radioactive chemical to become other matter. So they can use radioactive chemical clocks to tell time.

Some of these radioactive chemical clocks are used to date rocks and fossils. Others are used to tell the age of things made by early man.

A Radioactive Clock: Carbon-14

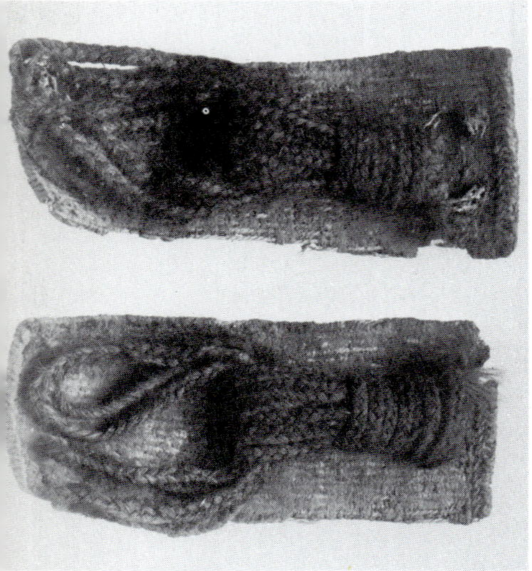

Carbon-14 is one of these radioactive chemicals. There is always a little carbon-14 in the air around you. This is how this radioactive clock works.

These sandals were made by man. ● They are more than a *thousand* years old. They were made from parts of plants. Some thousands of years ago these plants took in carbon dioxide. Some of the carbon in the carbon dioxide was carbon-14. A little of this radioactive carbon in the carbon dioxide became part of the plants. So the sandals have a little carbon-14 in them.

The carbon-14 breaks down slowly. ▲ It breaks down into other matter. It takes *about* 50,000 years for carbon-14 to break down completely. We can read the time by how little carbon-14 is left in the sandals.

Scientists know the age of the sandals by studying how little carbon-14 is left. They know how long it ● takes carbon-14 to break down.

The time comes when all the carbon-14 is broken down. All the carbon-14 is gone. Then we cannot read the time.

A carbon-14 clock can be used to date *some* things. It reads the age up to *about* 50,000 years. So it can date these sandals made by man. But it can't be used for *very* old things.

A Radioactive Clock: Uranium

To tell time in *millions* of years we need another kind of clock. One radioactive chemical that breaks down *more* slowly than carbon-14 is *uranium* (you·RAY·nee·um). There are other radioactive clocks, too.

Scientists know how long it takes for uranium to break down. It gives off deadly rays we do not see. And then what is left is a special kind of lead.

Remember the sand clock? We know how long it takes for the sand to run down into the bottom. We also know how long radioactive clocks take to run down. Scientists know how long it takes uranium to give off rays and turn into lead. If we know how much lead there is, we can tell how long it took to make this much lead.

Yes. We can tell the age of fossils. Or we can tell the age of the rock. The rocks and the fossils in them are about the same age.

A Trip in a Time Machine

Suppose you could go forward in time. Suppose you could see the future. What will the world be like in the year 2000? What is your guess? The year 2000 is not too far away. It will be your world. And it will be the world of your children. Will it be a better world?

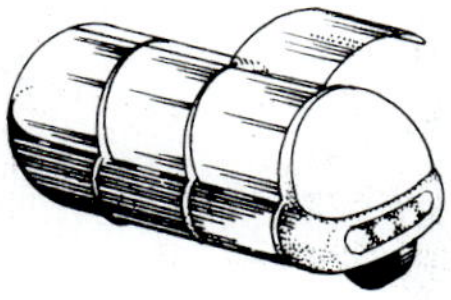

Maybe man will go to many other planets. Perhaps man will even dig for minerals on the moon. Maybe man will live there.

What will the cars be like? Will they run on atomic (uh·TOM·ik) fuel?

Maybe man will have found answers to many problems by then. Maybe man will be able to find and break up hurricanes (HURRY·canes). Maybe man will do this before they hit our towns and cities.

How about the year 3000? What will the world be like? What is your guess? What will man be like? Will man be taller? or heavier or thinner? Will it be easy to replace worn-out organs like a heart or a lung?

Will some animals of today have disappeared by the year 3000? Could there be new kinds of plants and animals that are not found today? Will there be another Ice Age?

You cannot know for sure what will happen. Or how possible these ideas may be. Living now, *we cannot predict the future*. We cannot tell the future. We can only guess. We cannot say for sure.

2. A TRIP INTO THE PAST

Suppose you could take a trip that would take you *backward* in time. You could be more sure of what the past was like. Suppose you went back only 500 years. To the 1400's. What was it like in America?

In those days there were many Indian tribes living in America. There were thick forests and grasslands. There were buffalo and deer. And birds of many kinds. The plants and animals were about the same as now. Only there were more of them.

But 500 years ago is not a long time. Not for the kind of trip you are going to take back into time.

Let us go further *back into the past*. You will take a trip back 3 billion years into the past. This is a long way to go. Your first stop on the long trip back into the past will be at 25,000 years ago.

What was it like 25,000 years ago? That's not really so long ago. That is, if you think about the age of the earth. Scientists believe the earth is about 5 billion years old. Life may have started on the earth about $2\frac{1}{2}$ billion years ago.◆ So, 25,000 years is not so long.

25,000 years ago people much like us lived on earth. These people were much like modern man. How did modern man begin? What do we know about this?

25,000 Years Ago: Cro-Magnon People

Cro-Magnon

Suppose you could really go back in time—25,000 years. What kind of life would you see? You would see men, women and children. They might be near their cave. This is what one artist thinks they were like.★ They were the **Cro-Magnon** (kroh·MAN·yon) people.

The Cro-Magnons lived 25,000 years ago. But maybe they first appeared 50,000 years ago.◇ They were one kind of people that lived at this time.

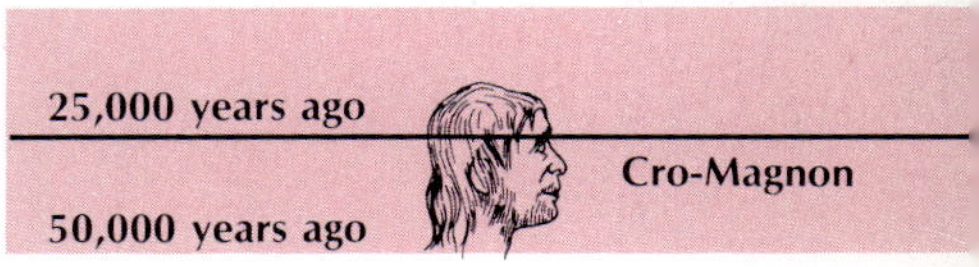

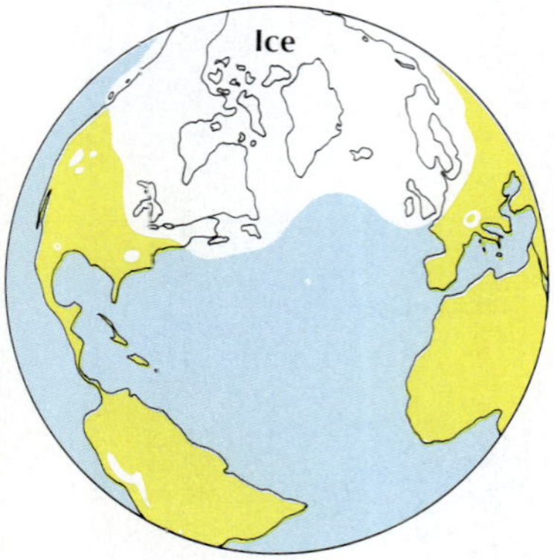

They lived in an Ice Age. There have been several Ice Ages on earth. At least four times, sheets of ice covered part of North America and Europe. ■

Men before these Cro-Magnons probably roamed over the world. They probably roamed in warmer times—between the cold weather of the Ice Ages. Perhaps the first Cro-Magnons came to Europe from Asia. Perhaps they came to Europe in one of these warm times.

■

Cro-Magnon Man

Do we really know what the Cro-Magnon people were like? Yes. Many fossils have been found. They seem to have been tall. The men were almost 6 feet tall. And the women were about 5 feet 5 inches. They had long legs. Maybe they were fast-moving people.

See the kind of forehead and the chin these people had. ● They left records of how they looked. You see, they buried their dead. The fossils of the buried have been found. ▲

They must have had many skills. Some were craftsmen. They chipped rocks to make sharp tools. ◆ They used stone axes. They carved in rock and in bone. They used bone needles.

They made clothing and sandals from plants, leather, and fur. We can even tell the age of the remains of sandals and leather clothes. How? By using radioactive carbon clocks.

●

◆

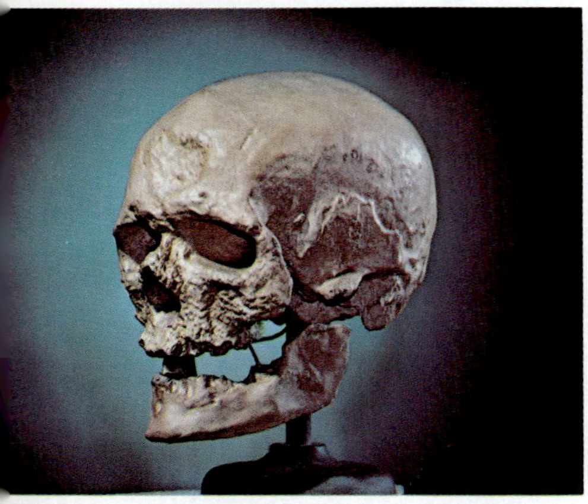

▲

★

There were artists among them, too. They made paintings in caves. Caves in France and Spain and Africa have paintings like this one.★ There is a story in this painting.

See the spears in the animals.★ Groups of hunters used these spears. They hunted on foot. They knew how to hunt well. What kinds of animals are they hunting? They look like buffalo.

Some of the Cro-Magnons took time to draw and to carve. This may have been their job. We are not sure. The Cro-Magnons cooked their food. We have found the remains of burned bones at their campsites.

Some Cro-Magnon people lived near water. They probably found they could use empty tree trunks to move on the water. People also moved from place to place on foot. They moved in small groups, or tribes.

The Cro-Magnons were like man today. In fact, if a Cro-Magnon man walked down the street in modern clothes, you might not notice. You might pass him on the street without a second look!

Scientists have found fossils of human types much older than Cro-Magnon man. To see this type of man, you will have to go back *twice* as far into the past.

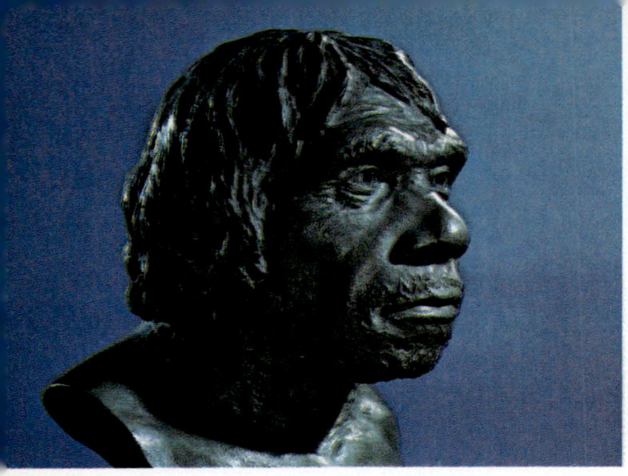

50,000 Years Ago: Neanderthal People

This is a picture of a man who lived in the valley of the Neander (nee·AN·der) River in Germany. He lived there 50,000 years ago—maybe as far back as 100,000 years ago.

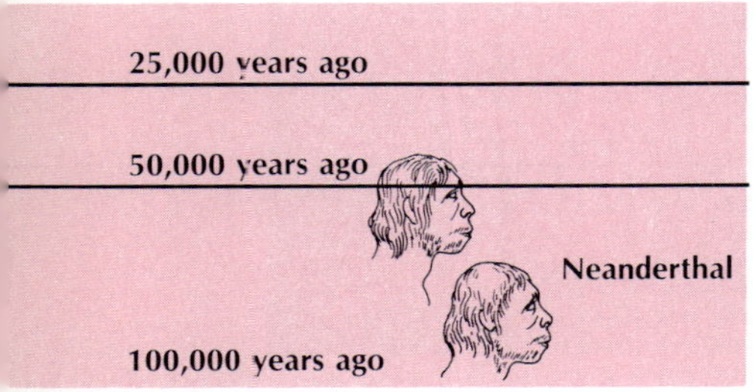

These **Neanderthal** (nee·AN·der·thahl) people, or people very much like them, must have roamed around over thousands of years. Many whole fossils of these people have been found over Europe. And in parts of Asia and North Africa. These fossils show us they buried their dead. In this way, their bones had a chance to become fossils.

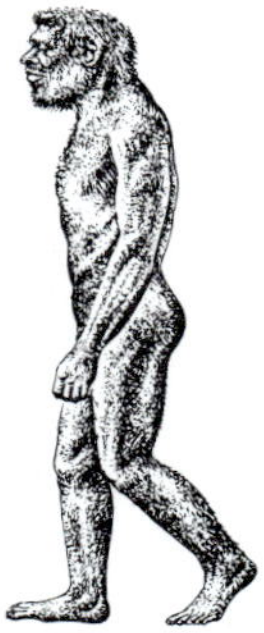
Neanderthal

These people were short and stocky with broad shoulders. They had very short legs. They were about 5 feet tall.

They were hunters that sat around their campsites. We have found partly burned bones and rocks around their campsites. They used some fine-chipped rocks as tools.

Suddenly the Neanderthals died out, or so it seems. No one knows why they died out. Did it have something to do with the Cro-Magnons? Did they battle? Or maybe they intermarried, producing new combinations of genes in populations?

Were There Even More Ancient People?

Java and Peking man

Half a million years ago, there were some small groups of people who lived in China and Java.★ Look at the picture of **Java** (JA·va) **man.**◆ The fossil of **Peking** (PEE·king) **man** from China looks very much the same.⊙ Do you see any differences?

But the story of man goes back at least 1 million years.★ It may even go back almost 2 million years!★ We are not sure. Nearly human-like fossils have been found in East Africa and Central Africa.⬟ We say nearly human. They walked upright. They were about 4 feet tall. They had the same kind of teeth humans have. But they had small brains. They were not like you.

We think *at least* 1 million years ago some early kinds of man had appeared. At that time, there were very few of these kinds of men on the earth. We know these men roamed around. We know this because their fossils are found in many places.

These early people moved in small groups, or tribes. They probably mated within their own group. There were always new patterns showing up in children. All kinds of combinations (kom·bih·NAY·shuns) of genes showed up in their children.

You know children usually look a little different from their parents. Over a long time, a really new pattern might appear in a group of people. This new pattern, or change in heredity, could be passed on from parents to their children.

It probably took a long time for the many small changes to make people in any one group look different. It probably took a long time for changes in heredity to show in the group of people.

In one group, or tribe, there were many likenesses among the people. But there were also some differences among the people.

Other groups of people, who lived in other places in the world, might look a little different. They might have some different traits. They had their own likenesses, too.

★
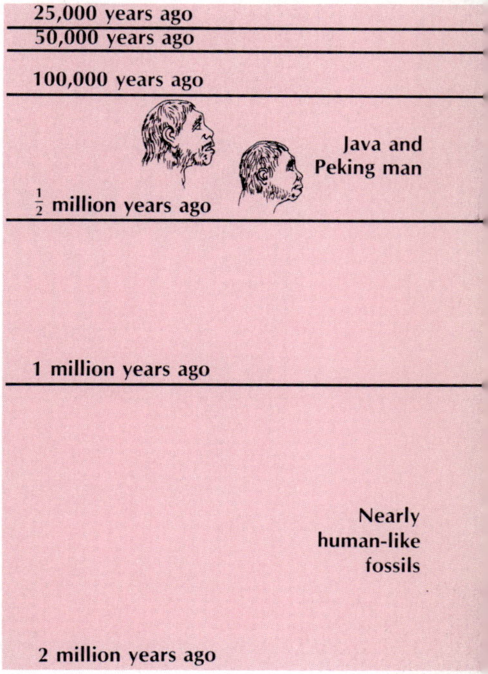

◆

⊙
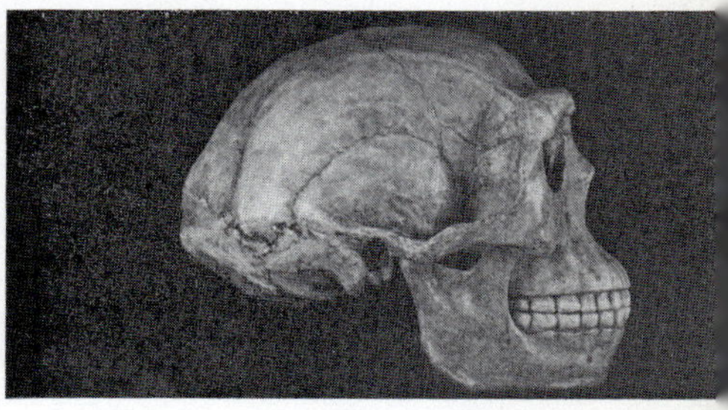

⬟
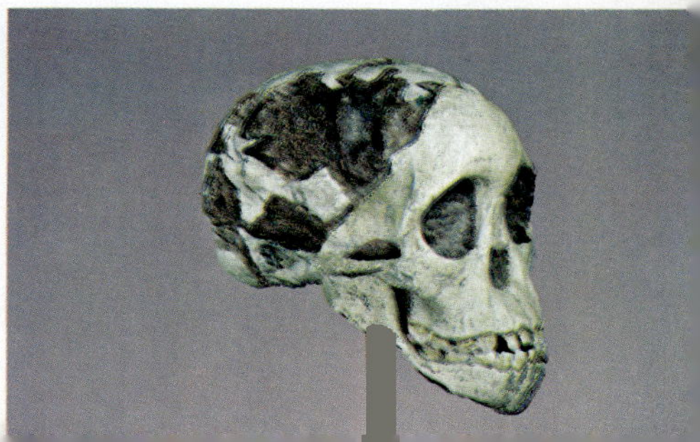

One group might have been taller, like the Cro-Magnons. Or shorter, like the Neanderthal people. Or they might have had different eye color. Of course, we can't tell the eye color from fossil remains.

It is the same today. People over the world today have their own likenesses. And their own differences. Even within the same groups there are many, many differences, too.

But man is still man. Humans today are all much alike. All humans, no matter what their shape of head, blood group, type of hair, or color, are one kind.

●

●

Before Man

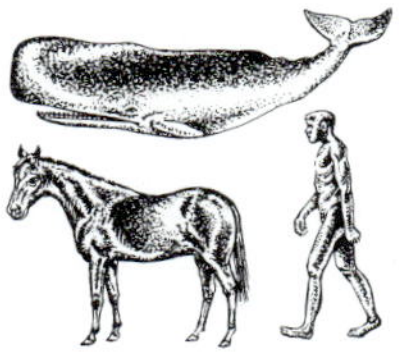

Mammals

All warm-blooded animals with fur or hair are called **mammals.** The young of mammals are fed on the milk of the mother. This means that horses, elephants, seals, whales, apes, and man are all mammals. ■

Of these animals, apes and man are the most alike. ● Apes and man are *more* alike than the other mammals. Apes and man are called **primates** (PRI·mates). From the pictures you can see that man has much in common with the other primates. You can also see that monkeys are primates. ▲ And so are the little animals called lemurs (LEE·murz). ◆

▲

▲

◆

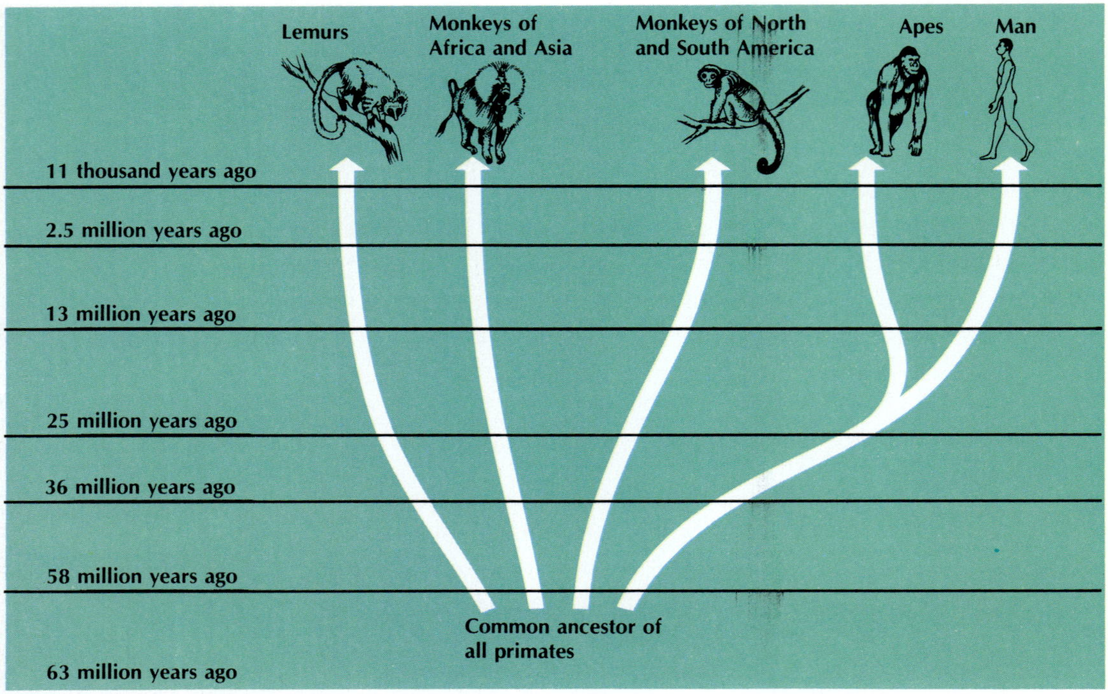

What were the ancestors (AN·ses·terz) of these primates like? What kind of animals might have been the ancestor of lemurs, monkeys, apes, and man? We can't trace the exact road. In fact, there were probably many roads. There may have been many roads from a primate ancestor to modern lemurs, monkeys, apes, and man.

One branch in the tree of ancestors led to the modern lemurs. ■ Other branches on the primate tree led to the monkeys in Africa and Asia. And to monkeys in North and South America. Still more branches led to the many kinds of apes. And another branch, or branches, led to early human types. ■

But what about man? Some people think that man came from apes. But this is not true. Man did not suddenly arise from apes. But man and man-like apes are related in many ways. Other animals are *also* related. For example, birds and lizards are related. But no one says a bird with feathers suddenly came from a lizard.

Reptiles

Look at this fossil. ● It is a bird. See the print of the feathers? But this bird looks a little like a lizard or an alligator. And these animals are **reptiles** (REP·tils). Reptiles are air-breathing animals with scales.

See the wings and the long tail on this fossil bird? ● It also had a long head like that of an alligator or lizard. This bird had an ancestor that was more reptile-like. But look at the feathers! We know it had feathers. The imprint of feathers was found in rock. We call it a bird. Why? It has something no reptile ever has—feathers.

So it was, we think, with man and his early primate ancestors. One road from a common ancestor led to the apes. Another road led to early, almost human types. ▲

You know we don't know the whole story about this. We have not found the whole fossil story. Did human-like animals appear only 2 million years ago? How far does the road to human types go back?

We are sure that human-like animals lived at least 1 million years ago. Possibly 2 million years ago. We have found fossils this old in Africa. Then later, some kinds of man came into Europe and parts of Africa. We know because we have found some of their fossils. These men are like Java man and Peking man. Later came Neanderthals. And finally, the humans most like us, the Cro-Magnons, appeared. ▲

●

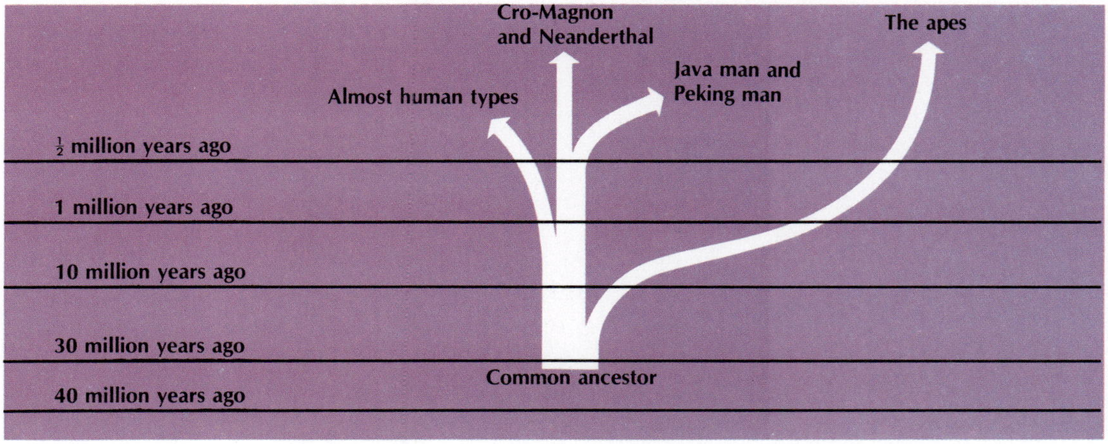

▲

3. LIFE ON THE EARTH BEFORE MAN

We have talked a lot about man. Of course. It's only natural that we should be interested in ourselves.

However, you have only gone back a million years or so. You have only seen the history of man. Was there a time when man was not on earth? Was there a time when there were no primates? Most certainly. There was a time when *other* mammals were the highest kind of life on the earth.

About 25 million years ago, there were many large grazing mammals that were like camels and rhinoceroses (ri·NOS·er·us·ez). And there were those that looked a lot like elephants. There were some middle-sized horses like these. They were also grazing animals. We know a lot about the background of horses.

60 Million Years Ago

Turn the time machine back to 60 million years ago. This was a time of great change. There were many changes going on. There were changes in the earth's weather. And there were changes in the earth's crust. No one knows why, but there were many changes going on.

The Rocky Mountains were forming.▲ They did not form all at once. The mountains took a long time to form—over a few million years. There had been rumblings and stirring in the crust of the earth. Then came the Rocky Mountains. About this time, some of North America was also covered by seas.▲ It is hard to imagine that these changes in the earth took a long time. That a mountain range took such a very long time to form.

This small horse was a leaf-eating forest animal.♦ This is eohippus (ee·oh·HIP·is). It lived *about* 60 million years ago. It is one of the oldest fossil horses we have found. It was a mammal. It stood about 1 foot high. It was *much* smaller than middle-sized horses.

These horses had 4 toes on their front feet and 3 on their back feet. Look closely.♦

Some of these little horses were successful in their environment. But the climate changed. It got cooler, and the forests disappeared. The environment then was a grassy plain.

The animals that had shorter toes or fewer toes may have gotten along better on the grassy plains. They may have been fit for running on the grassy plains. And they were able to eat grass instead of tree leaves. So they lived. Horses of several kinds have lived on the earth. The modern horse arose from some of these early kinds of horses.

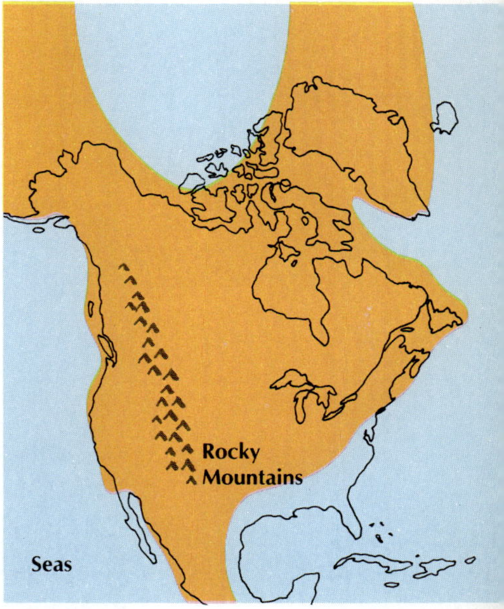

▲

♦

And Back Further Again

Things were changing. New kinds of plants and animals appeared. And some disappeared. Living things like flowering plants—the trees and weeds you know—were appearing on land. First in small numbers.

In the ocean were many kinds of fish.

Small mammals and birds also began to take over the land environment. The first mammals were small in size. They were about the size of a mouse.★ You would hardly notice them among the many big reptiles that were beginning to die out. These big reptiles are called dinosaurs.

★

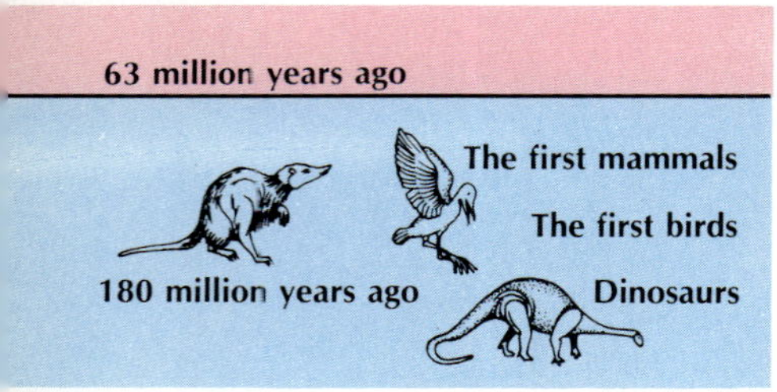

180 Million Years Ago

Let us go further back into time. Let's go back to a time when there were few mammals. The birds were just appearing. The dinosaurs were living then. ■

The dinosaurs came in all sizes and shapes. They were reptiles. One of the biggest was the plant-eating *Brontosaurus* (bron·toe·SORE·us). It was very heavy. It probably could not move well on land. So it probably spent time floating in lakes and swamps.

Brontosaurus

Other dinosaurs were almost bird-like. They ran on long hind legs, like the *Allosaurus* (al·oh·SORE·us). Look at the teeth. They were long and sharp. With those strong teeth, they could rip apart prey. These animals were meat-eaters.

Allosaurus

See this animal gliding through space? ● It looks like it is in free fall. Dinosaurs like these were gliding reptiles. They did not have wings like a bird. They had scales on their thick skin to protect them. They could survive. The scales and thick skin protected them from drying out. They laid eggs with heavy, leathery shells. Just as reptiles do today. The eggs could not dry out.

Some dinosaurs lived in the ocean rather than on land. You might mistake them for huge fish. ▲ They had the shape of fish. But these swimming dinosaurs were still reptiles. They had lungs. They were air-breathing animals.

Dinosaurs were special kinds of reptiles. They lived in the air, the ocean, and the land environments. ♦ They seemed to live successfully for a long, long time.

4. CHANGING – EVERYTHING SLOWLY CHANGING

Let's stop the time machine for a short while. Let's go over some of the ideas we have talked about.

Whether we talk about plants or animals or human beings, the story is the same. When new kinds of living things appeared on earth, there were only small numbers of each kind.

These small groups were found in a certain place. If ■ the place, the environment, was good, the animals and plants had a chance to live. They had a chance to grow into large populations.

Most changes in living things have been slow. When a change happened in a plant or animal, it might have been small. A change might have been due to a new combination of genes. Or a gene might have changed. Then the DNA passed to the children would be different. This inherited change may not have an effect on the everyday life of the plant or animal. *But if the environment changed,* this new trait could be a matter of life or death.

● Even the way animals reproduce has changed over millions of years. These changes made animals fit for new environments. Fish and frogs lay eggs in the water. ■ Eggs of reptiles and birds are protected by a shell. ● Shells protect growing embryos from drying out. This change allowed animals to lay eggs on dry land. And some eggs grow in the body of the mother. ▲ This change protected developing eggs from all changes in the environment.

These changes in the way animals reproduce were very slow. They took millions of years. There were changes in DNA. These new patterns of reproduction fit animals to live in new environments.

All these slow changes are called **evolution** (ev·oh·LOO·shun). Evolution is the slow change that has taken place in the earth itself and in animals and plants. Slow changes have gone on for millions and millions of years.

You can say the earth and the living things on earth ▲ have gone through many evolutions. Mountains have been made. And mountains have been worn down again. And there have been many parades of living things on the earth.

Some parades of living things have disappeared. For example, dinosaurs died out. Did their environment change too fast? Was there no food supply? We don't know.

Now let's go further back in time. Let's go back about 300 million years ago.

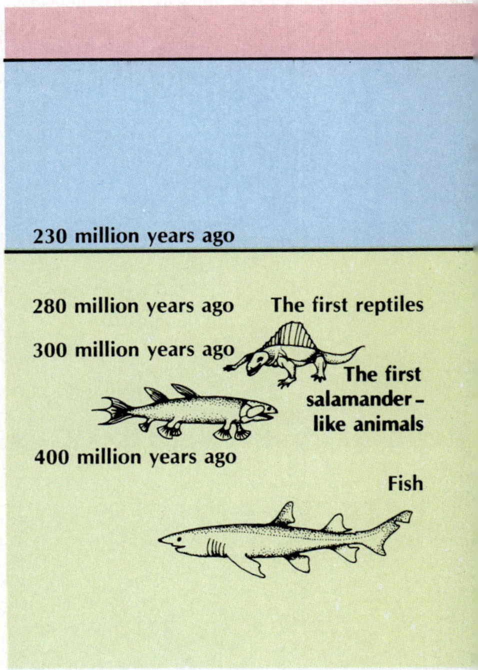

Back About 300 Million Years Ago

You stopped awhile to think about all kinds of plants and animals. About all living things that have slowly appeared on the earth. And about some that have disappeared.

You are now far back in the past. It is now about 300 million years ago.♦ What kinds of life were on earth so long ago?

Stop the time machine for a minute. There is a quiet stillness in the air. Only the sound of many kinds of insects.

The climate is wet and hot. Around you are many swamps. Giant dragonflies are flying over the swamps.★

There are forests of huge ferns. The ferns are as high as big trees. Giant cockroaches live in the forests. And spiders and snails.

Here you can see some animals that look like salamanders. They are slow and clumsy as they pull themselves over the ground. They are the ancestors of salamanders. And of frogs and toads that we see today.

Now you'll go back still another 100 million years. Life on earth is far different now from life you have seen so far.

The Time the Fish Looked Over the Land

There is almost no life on *land* 400 million years ago.♦ Animals have not come on land at this early time. In this early time of the earth and living things, there was no animal life on land.

But look in the oceans. The waters are full of life. There are large sharks and many kinds of bony fish. And you can see seaweeds, clams, coral, jellyfish, and starfish.

But how did animals with backbones come to live on land? We have some ideas. We know how some animals may have come to land. For instance, here are some strange fish that can take gulps of air. These fish live today. They are called African lungfish. They have large air bags, or sacs, in their bodies. These air sacs are like small lungs. These fish can breathe air just as you do.

A few kinds of air-breathing fish live today. We may guess that air-breathing fish have lived in the past. One of the air-breathing fish of long ago may have been the ancestor of land animals.

We do not know why, but some fish long ago came out on land. Perhaps the climate was getting drier. Maybe the rivers and lakes dried up. The climate was changing, we know. We can tell this from rock layers.

Maybe these strange fish pulled themselves along the land looking for another lake or river. They could take gulps of air and live. Some may even have been able to stay on land for awhile.

Gradually, over thousands of years, some may have been able to stay on land all the time. They were fit for life on land. They had better air sacs. They could breathe air. They could *stay on land.* There was food on land. By this time there were insects and worms. There were also many green plants that could be food for insects and worms. Now animals with backbones were on land. They were fitted for life on land. They had the right genes. And they were in the right environment. So they lived and reproduced their kind.

Living Fossil Fish

You may say this is all a good story. But is it really true? Could some kind of a fish have come on land? From the fossil remains we know there were fish that had very thick fins. Maybe they could pull themselves along on the land.

Scientists thought such fish had died out about 70 million years ago. But only about 30 years ago, a strange fish was caught in a fisherman's net.

Coelacanth

Out of the deep water around South Africa, a *coelacanth* (SEE·luh·kanth) was caught. ▲ Several more have been found since that first catch. This is why scientists call this queer fish a living fossil. It looks like the kind of fish they thought had died out. It has heavy scales and thick fins.

Coelacanths may be living proof. They may be living proof that fish with leg-like fins once lived. Coelacanths today look like fossils in very old rock. ◆

Animals like the coelacanths may have given rise to the first salamander-like animals. Salamanders, frogs, and toads still stay close to the water even today. They have small lungs. They can use them to breathe air. And they lay eggs in the water like fish.

▲

◆

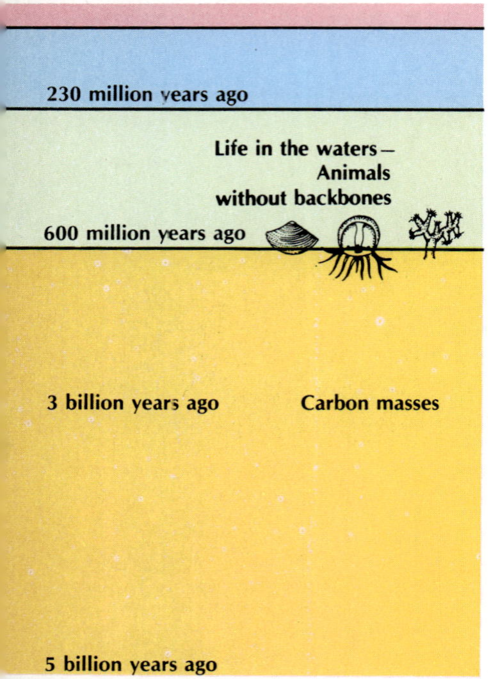

600 Million Years Ago

Now you are back about 600 million years ago. ■ Is there life? Yes. In the oceans and in the many lakes and rivers. Many animals without backbones lived in the oceans, lakes, and rivers.

In North America, much of the land is flooded by seas at this time. The rest of the land is bare.

There were long periods when land was flooded. Rivers ran into the oceans. Minerals from the land were washed into the oceans. These minerals made oceans saltier.

When whole mountain ranges rose out of the ground, flooded land rose and became mountains. Can you explain how fossils of water-living animals can be found on the tops of mountains? ●

3 Billion Years Ago

■ The time machine stops at about 3 billion years *ago*. ■ Now there are no fossil remains anymore. We really do not know when life started. For example, we don't know when one-celled animals and plants first appeared.

●

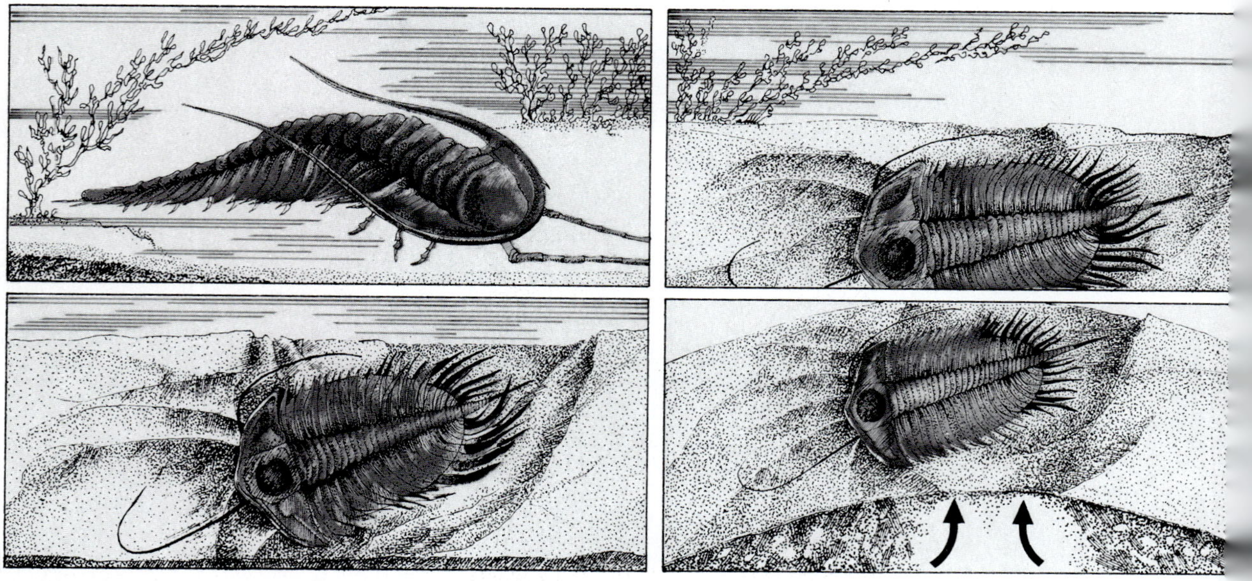

But scientists have found masses of carbon deposits in some very old rocks—rocks that are almost 3 billion years old. ▲ Living things are mostly made up of different kinds of molecules that have carbon in them. These carbon masses may be the remains of microscopic kinds of life. They might be the crushed remains of the earliest kinds of cells. This part of the past we do not know.

This is the end of the story of fossils. You have gone from modern living things to the first kinds of life. These first kinds of life may have been one-celled. They were at the beginning.

▲ 

A Try at Explanation

1. Let's say someone in class says cavemen hunted dinosaurs. What evidence could you give to show him he is wrong?

2. Remember the strange fish in the fisherman's net, the fish called a coelacanth? Why are coelacanths called "living" fossils?

3. Suppose you find a fossil. It's possible. In a big lump of coal you may find a leaf print. Can you explain how it got in the coal?

4. Write a story about how some giant mammal or reptile died out. Be sure to use the words *right genes, fit for the environment,* and *unfit for the environment* in your story.

A Try at Prediction

1. A friend asks you to put these animals in the order they came on earth: dinosaurs, fish, mammals, clams, and one-celled animals.

In what order would you put them? Why?

2. Are fossils being formed today? Why do you think so? or why don't you think so?

5. STARTING AT THE BEGINNING

Nothing is sudden in the slow evolution of the earth. Nothing is fast in the parade of living things. Time is measured in millions and hundreds of millions of years.

Some successful plants and animals have continued since they first began. ■ Many animals without backbones have changed very little over time. Especially if they live in the water. In the oceans, the environment does not change very much. Many fish have not changed either. ●

Even some insects, such as ants, have not changed much over millions of years. Scorpions have not changed much over millions of years either. ▲

But many things *have* changed. The environment and many animals and plants have changed over the history of the earth.

You have gone back in time almost 3 billion years. You have gone back to the beginning of living things. Now you have to come *back again to the present time.* ♦

Turn the pages that follow slowly, very slowly. Study each picture. The pictures are like the scenery you pass on a trip from the beginning of living things to now.

Can you see that plants and animals have gotten spread around? From oceans to land. You will see the mountains that appear. And the changes in the climate of the earth.

You will watch the slow changes, the changes of living things. You will see the changes on earth over the millions of years up to now.

■

●

▲

TIMETABLE OF ARRIVALS

2 million years ago	Nearly human-like fossils
70 million years ago	Spreading out of mammals
135 million years ago	Flowering plants
180 million years ago	Mammals Birds Dinosaurs
230 million years ago	First dinosaurs
345 million years ago	Reptiles
405 million years ago	Salamander-like animals
425 million years ago	Insects
	Land plants
500 million years ago	Backboned fish
600 million years ago	Animals without backbones
2 billion years ago	Algae
3 billion years ago	Carbon masses
3.3 billion years ago	Oldest rock found

You are here now. You have gone back in time nearly 3 billion years. You now have to return to the present.

The Age of Animals without Backbones

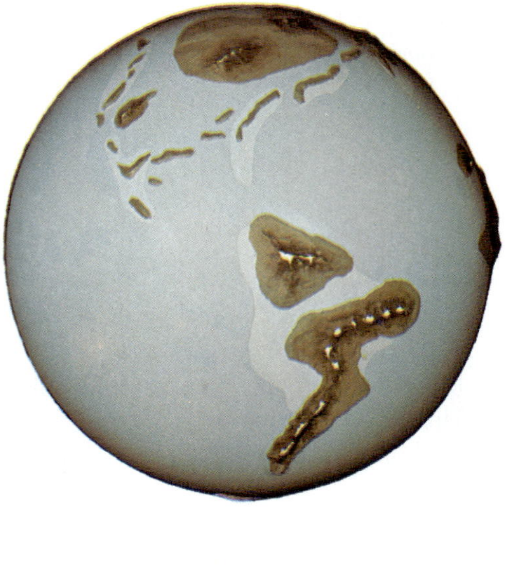

Living things first appeared in the rivers and oceans. There was no life on the land. There were many microscopic living things in the oceans. There also were sponges and jellyfish. There were worms and relatives of starfish. The rivers and oceans were full of many kinds of life.

What did all these living things have in common? They had no backbones. But many had hard parts like shells, spines, and tubes.

How do we know they lived? They left their remains in the rocks. Their hard parts became fossils. And the soft bodies of animals like jellyfish became prints. So we know that these animals lived.

This was a long time ago. And these were the first kinds of life. It took millions and millions of years for all these living things to appear. There were green plants too. All kinds of algae were in the rivers and oceans. They were the producers of the time.

The Step To the Land

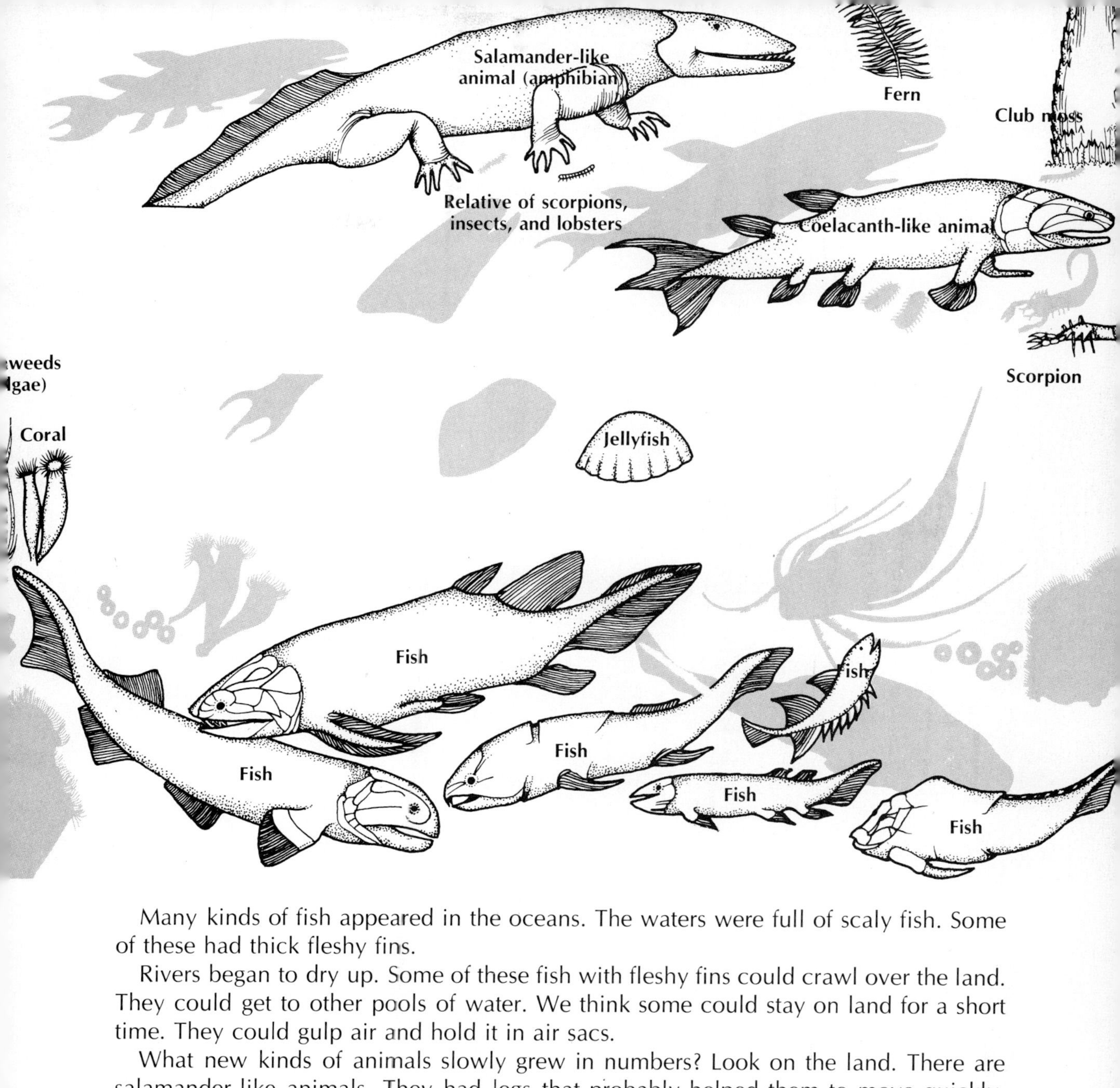

Many kinds of fish appeared in the oceans. The waters were full of scaly fish. Some of these had thick fleshy fins.

Rivers began to dry up. Some of these fish with fleshy fins could crawl over the land. They could get to other pools of water. We think some could stay on land for a short time. They could gulp air and hold it in air sacs.

What new kinds of animals slowly grew in numbers? Look on the land. There are salamander-like animals. They had legs that probably helped them to move quickly. And they breathed air. They lived near water.

Even today the salamanders haven't gotten very far from water. It is the same for their relatives, the frogs and toads. They need to keep their skin wet. And they lay their eggs in water. They all return to the water to lay their eggs.

We also see scorpions on land now . What are the producers? Some of these animals fed on the many kinds of green plants on land. Now there is a food chain on land.

The Victory Over the Land

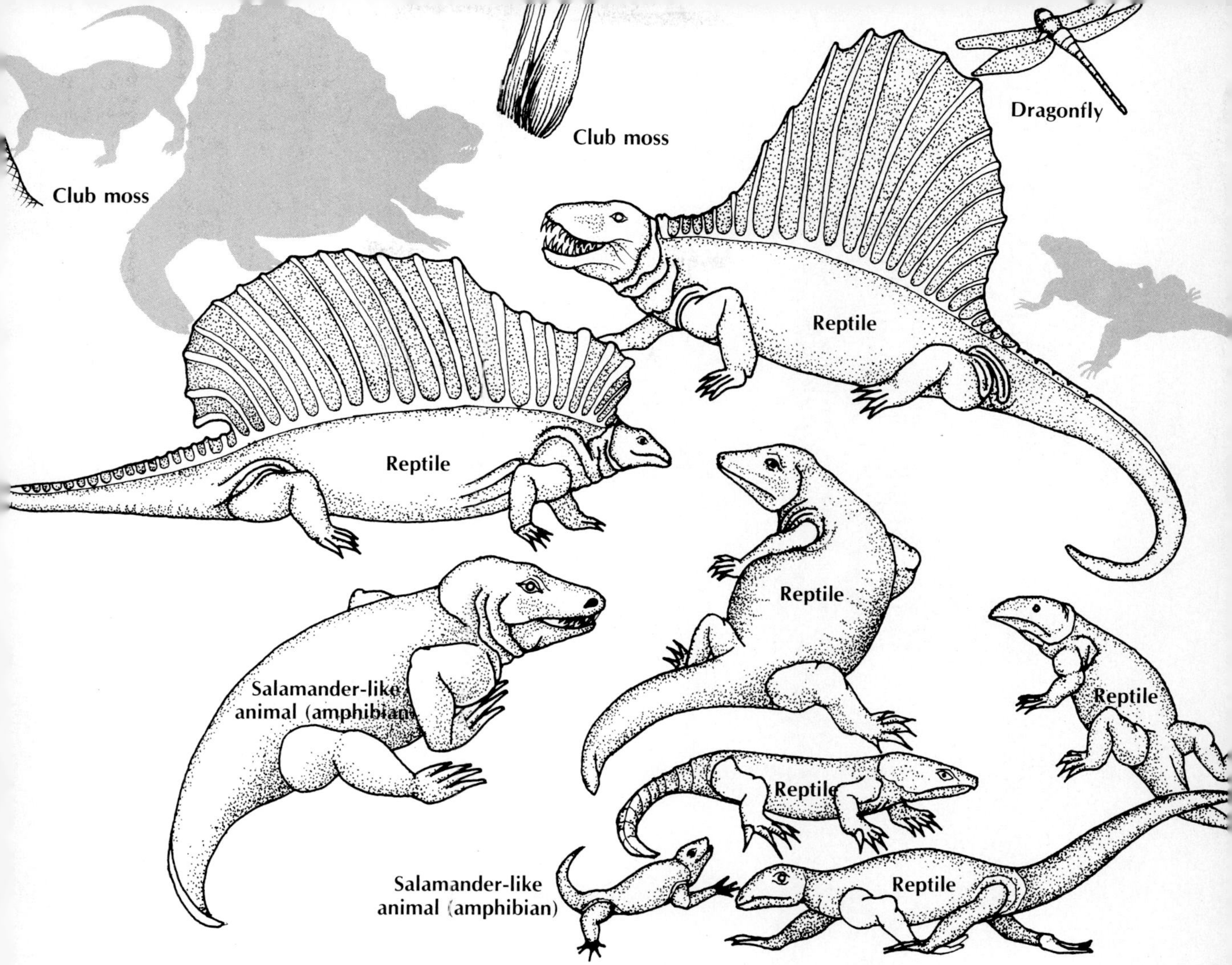

The scene was changing. There were many more plants now. Some ferns were tall as trees. But there were no trees as you know them. There was plenty of food. New kinds of animals slowly appeared. They still had the shape of salamanders. But they had thick, scaly skin that did not dry out on land. They also had good lungs and strong legs. Many had claws. They were reptiles. They did not have to go back to the water. Their eggs had thick shells. Their eggs did not dry out on land.

Some of the first reptiles were small, like lizards. And some kinds were quite large. Some ate plants. Others ate meat.

Reptiles began to be very important animals on earth. Can you name some other living things that lived on land, too? Yes. There were still the salamander-like animals.

How do we know these living things lived together about the same time? We found their fossil remains in the same layers of rocks.

Up to Man

Climates have changed. The surface of the land has changed too. Plants and animals that were fit for the environment lived. Others were not fit and died.

The dinosaurs died out. But some of the smaller reptiles lived on. And birds and mammals spread out over the earth. Some of these ate plants. Others ate the flesh of animals.

Mammals were fit for the new environments. Most kinds of mammals kept their fertilized eggs well protected. Then the young were cared for by the parents.

What kind of mammals do you see on the land? There are camels, elephant-like animals, horses, and bison. And there are men and other kinds of mammals. But some of these mammals did die out. And some of these only died out in North America. How many of these animals live in North America today? They all once did. Did changes in the environment make some of these animals unfit? Did man cause some of these mammals to die out?

AN INDEPENDENT STUDY

Hunting Fossils

Museums (myoo·ZEE·ums) have most of the fossils that have been found. They are always looking for more fossils.

Suppose you went on a fossil hunt. Where would you look for fossils? What might you find?

Do these next pages on your own to find out what you know about fossils.

Read each question and put your answer on your answer sheet. Then check across the page for the answer. Be sure to check the answer each time before you go on to the next question.

1

One of the early mammals was the woolly mammoth. ■

Where might the fossil remains of a woolly mammoth be found? *In rocks in Africa*? or *In frozen soil and ice in Siberia*?
Where?

(Write your answer and check across the page.)

2

Why didn't these mammoths decay? Because they were covered over with snow and frozen before they got a chance to decay. The air and bacteria could not reach them.

How is it that these mammoths have been found today?

Because the mammoths did not ____________.

3

Another cold place is Greenland. There is snow and frozen soil in Greenland. But *plant* fossils have been found in Greenland. Men have found fossils of plants that lived in *warm* climates there. They were found deep in the rocks in Greenland.

What do these fossils tell us? What was the climate of Greenland in the *past*?

Was it *warm* or *cold*? Which?

1

In frozen soil and ice in Siberia

(Go on to question 2.)

2

decay

These mammoths were covered with snow. They were preserved.

(Go on to question 3.)

3

warm

The climate was once warm in Greenland. Fossil plants tell the story.

(Turn the page to question 4.)

4

Climates have changed over many years. There were Ice Ages for long periods of time. Then this changed over a long time.

Once it was warm in Greenland. It was as hot as near the equator. We know icy Greenland had a warm climate way back in the past. We know plants from hot climates once grew in Greenland.

How do we know? The fossils are there in the rocks.

These remains of plants are called f------.

5

We know the age of many rock layers over the earth. It takes a long time—millions of years—for these rock layers to form.

Plants or animals are trapped in layers of soil. This soil then is turned into rock. Parts of plants and animals may be trapped in this rock. Shells or bones may be in the rock.

We can tell the age of *rock* layers. We can also tell the age of the plant and animal ____________ in the rocks.

6

Let's say you found some rock that was 500 million years old. Suppose you see fossils in this rock.

If the rock is 500 million years old, *about how old* are the fossils in the rock?

7

Suppose a fossil of some early man is found. This fossil is found in deep rock. Suppose we know the age of this kind of rock on all continents.

How do we know the age of this human fossil? We can tell from the age of the ____________ it was found in.

4

fossils

Good. You have the idea that fossils tell us that the climate has changed in some places.

(Go on to question 5.)

5

fossils

The age of the fossils is about the same as the age of the rock layer.

(Go on to question 6.)

6

about the same – 500 million years old

The age of the fossil is close to 500 million years. Not 5 million. Not 500 *billion* years.

(Go on to question 7.)

7

rock

This way we know about when this early man lived.

(Turn the page to question 8.)

8

Look at the fossils in these layers of rock. These layers were put down at different times.

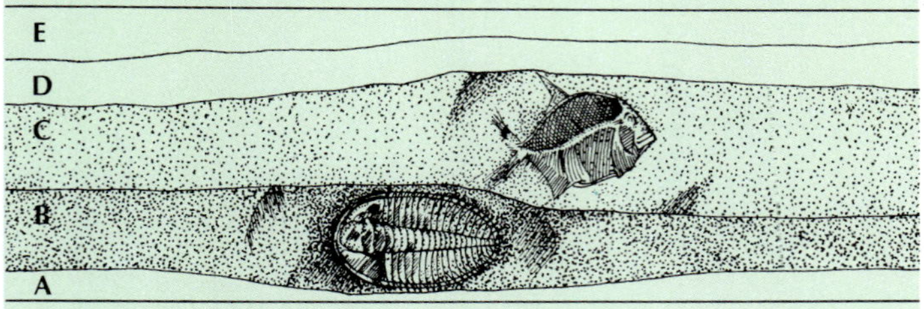

Which is the *oldest* layer of rock: *Layer A* or *E?* Which?

9

Of course. Layer A is on the bottom. The other layers of rock formed on top of layer A. They were made later.

Suppose trilobite (TRY·low·bite) fossils were found in layer B. Trilobites were sea animals without a backbone. No trilobites were found in *lower* layer A. None were found in *higher* layer C. Does this mean that trilobites *lived and died out* in layer B? *Yes* or *no?*

10

At one time there were many kinds of trilobites on earth. Maybe the climate changed. We're not sure why they died out, but they did.

How do we know they died out? Because they are not found in the newer layers of rock.

Which is the newer layer? *Layer B* or *C?*

11

Look again at layer C. Fossils of fish are found in layer C. But there are no fossils of frogs or birds.

This means that fish were on earth before frogs or _____.

8

Layer A

Layer A is the oldest. It was put down first.

(Go on to question 9.)

9

yes

Trilobites first appeared in layer B. They lived a long time—see how thick layer B is. Then they died out.

(Go on to question 10.)

10

Layer C

Layer C came after layer B. There were no trilobites in layer C.

(Go on to question 11.)

11

birds

(Turn the page to question 12.)

12

Scientists can tell the age of plants that have flowers and seeds. They say these plants spread out on earth about 135 million years ago.

How do they know? They know from the age of the _____ the fossil plants are in.

13

Scientists have found fossils of plants and animals that lived in water. They have found these fossils on *mountain tops* in the United States.

How can this be? This means this place was once, long ago, under _____.

14

How did this land get to be mountains? Land has been pushed up in many places over the earth. The crust of the earth has changed many times. In some other places land has been washed away. Or land has been flooded over.

Fossils of clams have been found in rock that is now on top of mountains. These fossils tell us that this land was once under _____.

15

Some land is now under water. If you could dig down some places under the oceans or lakes you might find fossils. These are fossils of animals and plants that once lived on land.

Fossils also tell us which plants and animals *lived together at the same time*. For example, there were crawling salamanders on land. Suppose they ate insects. The insects must have had food.

What does this tell us? What do insects eat? There must have been producers on land. These producers were g____ p_____.

12

rocks

(Go on to question 13.)

13

water

Before the land rose, the land was under water.

(Go on to question 14.)

14

water

This land that was under water was pushed up to become a mountain.

Clam fossils were already in these layers of land.

(Go on to question 15.)

15

green plants

That's good thinking.

(Turn the page to question 16.)

16

Of course, there had to be green plants to make food. Animals could not live if there were no plants.

Some dinosaurs ate meat. They ate other kinds of dinosaurs. What did the grazing dinosaurs eat? They ate producers.

What were the producers? The producers were _____ ______.

17

Many plants and animals lived in the seas long ago. The water was not as salty as now. Later on, plants and animals appeared on land.

Green plants first. Then insects. Then salamanders and frogs. Then reptiles, and later dinosaurs.

Human fossils were not found with dinosaur fossils. What does this mean?

Did man come on earth *before* or *after* dinosaurs? What do you think?

18

How do we know?

Fossils of humans go back at least *1 million* years. Fossils of dinosaurs are in rocks that are *many millions* of years old. In fact, they died out about 135 million years ago.

It is the *hard* bones or shells that become fossils. Which parts decay slowly and may become fossils?

Hard parts or *soft parts*? Which?

19

Ants trapped in sticky plant sap did not decay. No bacteria or air got to them. They were preserved for millions of years.

These ants were covered over before they could start to ____________.

16

green plants

Of course, there had to be food chains. Energy from light was trapped by green plants.

(Go on to question 17.)

17

after

Man appeared on earth after dinosaurs.

(Go on to question 18.)

18

hard parts

Soft parts decay fast.

(Go on to question 19.)

19

decay

These ants are fossils. And so is the sticky sap. The hardened, sticky sap is called amber.

(Turn the page to question 20.)

20

Look across the page at this picture. ■ These are different animals that came to a water hole to drink millions of years ago. This picture is of tar pits at La Brea (la BRAY·ah) in California.

In these tar pits hundreds of different kinds of fossils of animals have been found. All their bones were found in the tar. Bones and teeth of saber-toothed cats were also found here.

What do these fossils tell us? These animals got trapped in the sticky tar when they came to drink water. What animals with long teeth might have been waiting for them?

21

Where are other places to find fossils? Deep down in rock. Even deep layers of rock may come to the surface today.

Sometimes a crack appears in the crust of the earth. A whole slice of the earth may be pushed over. Look below to see this. Now we can see the older layers of rock. Man can easily dig for fossils in layers A and B now. These were the deepest, oldest layers. Now man can dig in these old layers.

20

saber-toothed cats

Some of them also got trapped in the tar. Their fossils were also found there.

(Go on to question 21.)

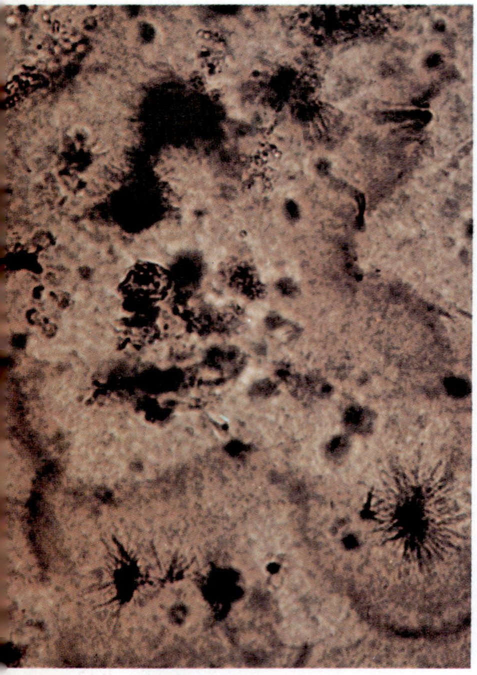

■

●

6. FIT FOR THE ENVIRONMENT

▲

What is the point of the story of the past? Let's see if we have the big ideas. There has been a parade of plants and animals. One-celled animals and plants probably started the parade of life. ■ These one-celled living things had genes that could be inherited. Some of these genes changed. Slowly, more and different kinds of plants and animals appeared.

There have been changes in genes. You know that the early oceans and streams were teeming with life. There were jellyfish, worms, sea weed and coral. ● And there were giant trilobites at one time. ▲ Much later, there were fish of all kinds. New patterns of heredity appeared.

On land there were many different environments. Dry air. Cold air. Warm air. Dry land. Swampy land. Snow-covered land. The environments kept changing. Plants and animals lived if they had the *genes* that fit them to live in the environment.

If they didn't have the genes that fit them, they died out. Or if they couldn't move, or travel to the environment they needed, they died out.

All these slow changes have gone on for almost 3 billion years of life. It has taken millions of years for some genes to show up. You can say successful plants and animals have genes that took millions of years to show up. There were new combinations of genes. There were also changes in genes—once in a while. Sometimes new combinations or changes in genes made an organism better fit. Or sometimes the changes caused the death of the plant or animal. Of course, plants or animals unfit for a place could not live.

The new combinations or the new genes could make plants and animals better and better fitted. For what? Fitted for life in a new environment. For life in the deserts. Or fitted for life in warm lakes. Or fitted for life in salty oceans. Or fitted for life in a swampy place. Or in a dry place.

Look at how well-fitted some animals and plants are for their special environments. Can you say why it is good for a moth to look like the bark of a tree? ◆ Then birds that feed on these moths can't see them.

What is this bird's environment? ★ Can you see him in his environment? How does looking like the surroundings help an animal?

Where is the fish? ◇ Can you find it? What helps the fish blend with his environment? How does this protect the fish?

◆

★

◇

Which bird's beak would be best fitted to crack hard seeds? Which beak or bill would be good to pick out insects?

Yes. It's taken millions of years for the genes for these traits to show up. These genes make plants and animals fit for their special environments. Plants and animals pass on these fit genes to their children.

A Try at Prediction

Suppose some bear cubs have very light fur. And a few have white fur. Suppose we now say that white fur is a new good trait. It makes the cubs more fit for their environment. Can you say where these cubs live? What is their environment? Is it the Arctic or a green forest?

Now try this problem. Suppose we say that white fur is a new trait that is *not* good. The genes for white fur do not help the animal to be fit. Let's say bear cubs with white fur are quickly spotted by enemies. Can you say what this environment is like? Is it the Arctic or a green forest?

Prize Genes

Slow changes over time have heaped up to produce good traits for plants and animals. That is how plants and animals became fit for their environments.

Man tries to do the same thing. Man tries to produce good traits in living things. Man picks out prize genes in plants and animals. Remember how man picked out the healthiest cattle to get better beef? Man has prize cows for milk. There are prize chickens that lay *many* eggs.

Man waits for new genes to show up. Or for new gene combinations to show up in cattle, for example. Or in hogs, or in wheat or corn. Sometimes man makes genes change. Seeds or plants may be put near special radioactive chemicals. Rays you cannot see may change the DNA code in the seeds or plants. That is how the peanut on the right was made. ● The plant was grown near radioactive chemicals.

Farmers keep watching the new crops. Sometimes they find new combinations or new genes showing up. There may be good genes that make vegetables larger. And genes that make fruits big. Or good genes that make wheat seeds richer in proteins. Or good genes that make plants grow quickly. It may take several years for these good genes to show up. Farmers keep watching the crops. This is part of the story of prize genes in plants and animals.

Man can pick out these good genes. He can do more. He can pick out the genes that make wheat grow better in a special environment. Some kinds of wheat have genes that make them fit to grow in parts of Italy. Or in places in the United States. Or in India and Pakistan.

There is "miracle" rice that grows well in places in the Philippines. And in West Africa, too. ▲ This rice grows *well* where it is hard to grow anything at all. It grows fast. It produces big crops. It has the right color and taste. It can produce food for many, many people. So it is a "miracle" rice.

Man can pick out the good genes. He can also put the plants and animals in special environments. He can put plants and animals with fit genes into fit environments. Man has done these wonderful things. He knows a lot about heredity.

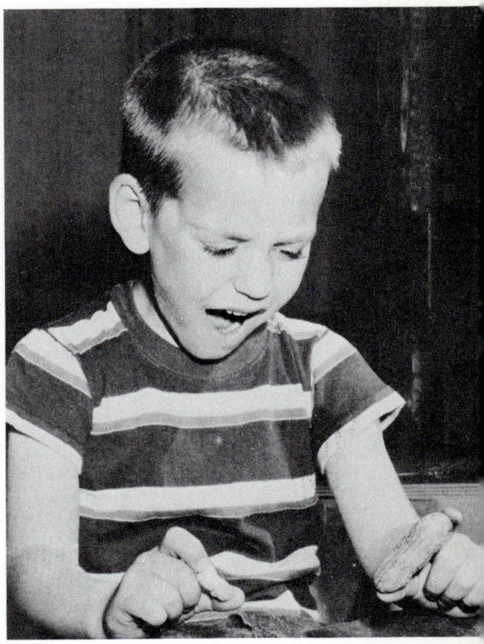

●

▲

■

●

▲

◆

Unfitting the Environment

You know all this. At the same time man does some good, he is doing great harm. Man is poisoning the environment. *Man is "unfitting" the environment.*

It has taken millions of years for plants and animals to become fit. Now man is destroying these fit animals and plants. He is killing off the fit genes. Man is changing the environment so fast. ■ He is changing the environment so fast that plants and animals are no longer fit. Their genes are not fit. And the environment is no longer fit for many plants and animals. ●

How is man "unfitting" the environment? How is man poisoning the environment? How is he fouling the air? the lakes, rivers, and oceans? the land? You know.

You now know about the poisonous gases in our air. You have learned about poisonous chemicals that get into animals. And into the air and oceans. You know how man spoils the environment.

Man the Spoiler

Man is destroying many plants and animals now. Man killed off the dodo birds by clubbing them. ▲ The passenger pigeon is gone. ◆ And the rhinoceros is on the way out.

Look at *some* of the endangered animals. These animals are in danger of dying off. Man is harming them.

The grizzly bear is on the way out. ★ The eagle has been poisoned by man. ◊ Some ranchers think that they kill too many other animals. The crocodile has been killed to make shoes and bags. ⊙ There are not many of them left. The Florida manatee is dying out. ⬟

Man is the spoiler. He fouls the outside environment. You know what is happening to the air, water, and the soil. You know even more than this.

Man is also fouling his own inner environment. Man takes into his body food additives, chemicals, and many, many drugs. We do not know the harm these may cause us.

★ 

◊ 

⊙ 

⬟ 

You can begin to ask: *Who is becoming endangered?* Is it perhaps MAN?

Should we add man to the list of the endangered animals?

What do you think?

Acknowledgments

Key: *T*—top; *C*—center; *B*—bottom; *L*—left; *R*—right.
AMNH—American Museum of Natural History; NAS—National Audubon Society; PR—Photo Researchers.

ART

Harbrace: pp. **34-49, 57, 58, 76, 93, 95***(B)*, **113, 114, 118, 146***(T,B)*, **147***(BL)*, **150***(T,C)*, **152, 195, 196***(T)*, **199, 205, 206, 207***(T)*, **219***(T)*, **252, 254, 255, 264***(CL)*, **316, 341, 345***(TR)*, **347***(T)*, **348, 355, 357***(T)*.
Harbrace-Berelson: pp. **267***(C)*, **342, 343, 347***(C)*, **350**(second from *T*), **351***(T)*, **354, 356***(B)*, **358***(T)*, **361, 364***(T)*, **367.**
Howard Berelson: pp. **33, 61, 62***(T)*, **64***(B)*, **65, 79, 89, 90***(T)*, **96***(C)*, **102-103, 139***(T,BL)*, **183, 190, 191, 210***(T)*, **225, 226***(T,B)*, **227, 277, 339, 345***(B)*, **346, 356***(T)*, **357***(C,B)*, **358***(B)*, **362***(T)*, **363***(L)*, **364***(B)*, **368-375, 380, 386.**
Howard Friedman: all dictionary illustrations and pp. **27, 64***(T)*, **66, 172***(B)*, **173-181, 236***(C)*, **266, 267***(B)*, **271, 274, 278, 280, 284-312, 321.**
Graphic Arts International: pp. **201, 209, 211-218, 219***(B)*.
David Lindroth: p. **50.**
Jean Zallinger: pp. **347***(B)*, **350***(B)*, **359.**

PHOTOGRAPHS

Harbrace: pp. **1***(T,B)* Courtesy of Sony Corporation of America, **2***(R)*, **3***(B)*, **5, 23-26, 27***(R)*, **28***(B)*, **30***(L)*, **31, 32***(T)*, **52-55, 58-60, 63**(second from BR, *BR*), **70***(T)*, **71***(B)*, **86-94, 95***(T,C)*, **98-105, 106***(BL,BR)*, **108-125, 137***(BL,BC)*, **140***(TL)*, **147***(T)*, **149***(T)*, **150***(B)*, **162***(T,C)*, **163***(B)*, **164-171, 172***(T)*, **178, 179**(top three), **194***(T)*, **203, 204***(T)*, **210***(B)*, **222, 223***(C,B)*, **224, 228, 231***(T)*, **232, 234***(T)* Courtesy of Hubbard Scientific Co., **238, 241***(C)*, **250***(BR)*, **251, 263***(T)*, **269***(T)*, **282***(BL)*, **284***(C)*, **291***(L,R)*, **295, 306, 313***(T)* by Jangoux, **313** (second from *T,B)*, **319, 321***(R)*, **322***(L,R)*, **324***(B)*, **326***(T,BL)* by Jangoux, *(C)*, **327***(TL)* by Jangoux, **341***(C)*, **353***(C)*, **394** (black and white).

UNIT 1

pp. **ii** Raymond a Mendez; facing **1** NASA; **2***(L)* Rogers, Monkmeyer; **3***(T)* Fields, PR; **18***(T)* Thomas, PR, *(C)* Willis Peterson, *(B)* Rue III, NAS; **19***(T)* Boulton, NAS, *(B)* Miller, NAS; **20***(T)* Grant Heilman, *(C)* Pollock, NAS, *(B)* Kinne, PR; **21***(L)* Francesconi, NAS, *(R)* Davidson, NAS; **22***(T)* AMNH, *(BL)* Rue III, NAS, *(BR)* Allan Roberts; **27***(L)* Grant Heilman; **28***(T)* Parker, NAS; **29**, PR; **30***(R)* Rue III, NAS; **32***(BL)* White, NAS, *(BR)* Rue III, NAS.

UNIT 2

pp. **50** Grant Heilman; **57** UPI; **62***(B)* United Nations; **63***(TL)* Richard Silver, M.D., N.Y. Hospital-Cornell Med. Center, *(CL, BL)* Hoppock Ass., *(TR,* second from *TR)* Eric Gravé; **67** Eric Gravé.

UNIT 3

pp. **68***(T)* Dr. E. R. Degginger, *(C)* Baglin, PR, *(B)* Erika, PR; **70***(B)* USDA; **71***(T, C)* USDA; **72***(T)* Eric Gravé, *(C)* Craig, NAS, *(B)* Spencer, NAS; **77***(L)* Runk/Schoenberger, Grant Heilman, *(R)* Jerome Wexler of Visual Teaching; **96***(T)* General Biological Supply House, Inc., Chicago, *(BL)* Thornton, NAS, *(BR)* Miller, NAS; **97***(L)* Gerard, NAS, *(R)* Staffan, NAS.

UNIT 4

pp. **106***(T)* Grant Heilman; **134***(T)* Grant Heilman, *(B)* Mercieca, NAS; **135** *(L)* Pitcairn, Grant Heilman, *(R)* Dr. E. R. Degginger; **136** Grant Heilman; **137***(T)* Hoppock Ass., *(BR)* Hugh Spencer; **139***(BR)* Kinne, PR; **140***(TR)* Grant Heilman, *(B)* Kinne, PR.

UNIT 5

pp. **142** all Runk/Schoenberger; **144** Harvard College Observatory; **145** Lick Observatory; **147***(BR)* NASA; **149***(B)* Beebe, PR; **151***(T)* Josef Muench, *(B)* Grant Heilman; **153***(T)* Shostal Ass., *(BL)* Steve & Dolores McCutcheon, *(BR)* Josef Muench; **154***(T, CL, CR)* Josef Muench, *(B)* Pitcairn, Grant Heilman; **155***(TL)* Judith Litvitch, *(TR)* Australian News and Information Bureau, *(BL)* Allan Roberts, *(CR)* Michael Smith, NAS, *(BR)* Milton Heiberg; **156-157** Shostal Ass.; **158***(T)* Slater, Monkmeyer, *(B)* David Muench; **159***(T)* Josef Muench, *(B)* David Muench; **160***(T)* Kinne, PR, *(B)* Douglas Faulkner; **161** DPI; **162***(B)* Douglas Faulkner; **163***(T)* General Biological Supply House, Inc., Chicago; **174** Willis Peterson; **177***(T)* Shostal Ass.; **179***(B)* Milton Heiberg; **180***(B)* Milton Heiberg; **196***(C)* Rue III, NAS, *(B)* Grant Heilman; **197***(T)* Grant Heilman, *(BL)* Dr. E. R. Degginger, *(BR)* Grant Heilman.

UNIT 6

pp. **198***(T)* Thornhill, NAS, *(B)* Ingeborg Lippman; **200** USDA, Forest Service; **202***(T)* International Silk Ass., *(B)* Triarch, Inc.; **204***(B)* Parke-Davis & Co., **207***(B)***-208** March of Dimes; **211***(T)* L. M. Drusin, M.D.; **212** Pfizer; **213***(T)* UPI, *(B)* Tuberculosis & Respiratory Disease Ass., Wagner Int. Photo Inc.; **216** UPI; **217***(T)* Allan Roberts; **218***(L)* Susan Johns; **220***(T, B)* Grant Heilman; **221** Armed Forces Institute of Pathology; **223***(T)* USDA; **226***(C)* Syndication International, Gilloon Agency; **229***(TL)* Latta, PR; *(TR)* Susan Johns, *(BL)* Breuer, Shostal Ass., *(BR)* Susan Johns; **230***(T)* Jordan, DPI, *(B)* DPI; **231***(BL)* Pictorial Parade, *(BR)* Wisconsin Dept. of Nat'l Researchers; **233** Porterfield-Chickering, PR; **234***(B)* Don Monaco; **235-236** Grant Heilman; **237** Shapiro, DPI; **239** Winter, PR; **240** Everett, N.Y. Botanical Gardens; **243***(L,R)* Eric Gravé.

UNIT 7

pp. **250***(TL)* European Art Color Slide Company, *(C)* The Pierpont Morgan Library, *(BL)* Koner, PR; **252** United Nations; **253***(T)* AMNH, *(B)* Judith Litvitch; **254** The Metropolitan Museum of Art, Rogers Fund, 1920; **255** USDA; **256** J.C. Allen & Son; **257***(TL,TR)* Walter Chandoha, *(BL)* USDA, *(BR)* Grant Heilman; **258***(T)* USDA, DPI, *(BL,BR)* UPI; **259***(TL)* Soil Conservation Service, *(TR,B)* Grant Heilman; **260***(T,CL,CR)* Chace, NAS, *(BL,BR)* Grant Heilman; **261** USDA; **262***(T)* Grant Heilman, *(C)* Thompson, NAS, *(B)* Hay, NAS; **263***(C)* J.C. Allen & Son, *(BL,BR)* USDA; **264***(TL)* Museum of Modern Art, Joseph, *(TR)* Levitt & Sons, Platnick Photos, *(CR)* Levitt & Sons, *(B)* Blower, NAS; **265***(L)* J.C. Allen & Son, *(TC,BC,BR)* Heinrichs, Grant Heilman, *(TR)* Grant Heilman, *(B)* Walter Chandoha; **267** Triarch Inc.; **268***(C)* Lanks, Monkmeyer, *(B)* White, NAS; **269***(B)* USDA; **270***(T,B)* J.C. Allen & Son; **271***(T)* USDA.

UNIT 8

pp. **282**(*TL,TR*) Susan Johns, (*CL*) General Biological Supply House, Inc., Chicago, (*BR*) Walter Chandoha; **283** Annan Photo Features; **284**(*T*) Chace, NAS; **290**(*T*) Marjorie Jacobson, Camera M.D. Studios, Inc., (*C*) Donald Yeager, Camera M.D. Studios, Inc., (*B*) Maternity Center Ass.; **292**(*T,C*) Dr. Maimon Cohen; **310**(*B*) Richard Hoefnagel, M.D., Dartmouth Medical School; **313**(second from *B*) Curtis Roseman; **314** Charles Rotmil; **316**(*T*) AMNH; **318**(*L*) Clay Adams, Division of Becton Dickinson & Co., (*R*) Hoppock Ass.; **320**(*L,R*) Susan Johns; **321**(*L*) Grant Heilman; **323**(*L,R*) Lester Bergman & Ass.; **324**(*T*) Peter Vadnai, (*C*) Susan Johns; **325** Riboud, Magnum; **326**(*BR*) Dick Phillips; **327**(*TR*) Jacques Jangoux, (*BL*) Brake, Rapho-Guillumette; (*BR*) Jacques Jangoux; **329** Triarch, Inc.

UNIT 9

pp. **336**(*TL*) Staffan, NAS, (*TR*) Harrison, Grant Heilman, (*B*) AMNH; **338**(*T*) Harry Robinson, Dept. of Interior Nat'l Pk. Service, (*C,B*) Josef Muench; **339**(*T*) Grant Heilman, North Museum, Franklin & Marshall College, (*C*) Donald Baird; **340**(*T*) Dr. F.M. Carpenter, (*B*) Josef Muench; **341**(*TL*) Donald Baird, (*TR*) Dr. Martin Glaessner; **344**(*T*) Gendreau, The Bettman Archive, (*B*) Museum of the American Indian; **346**(*T*) Ewing Galloway, (*B*) AMNH; **348** all AMNH; **349** Yan, Rapho-Guillumette; **350**(*T*) AMNH,(second from *T*), Dr. Ralph Solecki; **351** all AMNH; **352**(*TL,TR*) Rue III, NAS, (*CL*) Allen Cruickshank, NAS, (*CR*) Mohr, NAS, (*BL*) Dr. E.R. Degginger, (*BC*) Tanaka, Animals Animals, (*BR*) Evelyne Appel; **353**(*T*) Stevenson, Animals Animals, (*BL*) Rue III, NAS, (*BC*) Hermes, NAS, (*BR*) Mercieca, NAS; **355-358** AMNH; **360**(*T*) Burton, Bruce Coleman Ltd., (*C*) Allan Roberts, (*B*) Harbutt, Magnum; **361** Dr. Frank Carpenter; **362-363** AMNH; **365** E.S. Barghoorn; **366**(*T*) Dr. E.R. Degginger, (*BL*) Field Museum of Natural History, Chicago; (*BR*), Donald Baird; **368-374**(*B*) from an exhibit at the New Jersey State Museum, photo: Harbrace; **377-387** AMNH; **388**(*TL*) E.S. Barghoorn, (*TR*) AMNH, (*B*) Gerard, NAS; **389**(*T*) Howard, NAS, (*C*) Fearn, NAS, (*B*) Dr. Cooke; **390**(*L*) Kinne, PR, (*R*) Cruickshank, NAS; **391**(*T*) U.S. Chamber of Commerce, (*B*) The Rockefeller Foundation; **392**(*TL*) Kinne, PR, (*TR*) Elihu Blotnick, B.B.M. Associates, (*C*) AMNH, (*B*) NAS; **393**(*TL*) Rue III, NAS, (*TR*) Caulfield, Animals Animals, (*BL*) Judith Litvitch, (*BR*) Jim Annan; **394** Francisco Hidalgo (*color*).

Index

A number in **heavy type** means there is a picture on that page. For example, look for the word apples below. There is a picture of apples on page **263.**

C

D

G

H

I

U

V

W

Y

Z

A 2
B 3
C 4
D 5
E 6
F 7
G 8
H 9
I 0
J 1